ADVANCED FRAMING

TECHNIQUES, TROUBLESHOOTING & STRUCTURAL DESIGN

Sarah Knock

A *Journal of Light Construction* Book

First printing: February 1992

Editorial Director: Steven Bliss
Book Editor: John D. Wagner
Article Editors: Sal Alfano, Clayton DeKorne, David Dobbs, Don Jackson, Marylee MacDonald, Gary Mayk, Paul Spring, Wendy Talarico
Editorial Assistance: Susan Bliss, Kim Du Mond, Josie Masterson-Glen, Patti Poulin, Mary Tarrier

Production Director: Jim Romanoff
Designers: Barbara Ackerman, Theresa Emerson,
Lyn Hoffelt, Meredith Schuft, Susan Spinelli
Illustrators: Bruce Conklin, Tim Healey
Cover Photo: Paul Spring

Founding Editor: Michael Reitz

International Standard Book Number: 0-9632268-0-0
Library of Congress Catalog Number: 92-090097
Printed in the United States of America

A *Journal of Light Construction* Book

The Journal of Light Construction is a tradename of Builderburg Partners, Ltd.

The Journal of Light Construction
RR 2 Box 146
Richmond, VT 05477

Contents

Section 1: FIRST STEPS

9 Wood Basics
15 Getting Power to the Site
20 Tools of the Framing Trade
27 How Two Guys, In Two Weeks, Build a Custom Shell

Section 2: FLOOR SYSTEMS

33 Taking the *Bounce* Out of Floors and Beams
37 Out on a Limb with Cantilevers
41 Underlayments for Resilient Flooring

Section 3: WALL FRAMING

50 Fast & Accurate Wall Framing
56 Time-Saving Framing Tips
57 Bracing Walls Against Racking
61 When Walls Go Round
67 Seismic Bracing

Section 4: CUTTING ROOFS

74 Production Roof Cutting
83 Cutting Jack Rafters
87 Framing a Gable Dormer
90 Framing Tower Roofs

Section 5: REMODELERS' SPECIALTIES

97 Demolition with Care
102 Sill Repair & Jacking
109 Raising the Roof
112 Everything You Always Wanted To Know About Removing Collar Ties
118 Adding a Bay Window
122 When You Have to Cut a Truss

Section 6: TROUBLESHOOTING

131 Field Guide to Common Framing Errors
136 When the Framing Shrinks
140 The Cure for Rising Trusses

Section 7: ROOF TRUSSES

145 Fast Framing with Trusses
150 Stacking a Roof: The Production Approach
153 Shopping for Custom Trusses
159 Truss Collapse

Section 8: ENGINEERED LUMBER & HARDWARE

165 Engineered Beams & Headers
172 Roof Framing with Wood I-Beams
179 New Members, New Connectors

Section 9: PANELIZED SYSTEMS

189 Building with Structural Foam Panels
198 To Stick-Build or Panelize?
201 Building Walls in the Shop
205 Buying Walls from the Factory

Section 10: POLE & TIMBER

213 Hybrid Timber Frames
220 Wrapping Frames with Foam Panels
227 A Pole Barn with Style

Section 11: BUILDING WITH STEEL

234 Steelwork in Wood Frames
245 Metal Building Basics
251 Steel-Stud Partitions

Section 12: ENERGY: DESIGN & PRACTICE

259 High Performance Wall Systems
265 Strapped Wall Detail
267 Cathedral Ceiling Solutions
274 Tight Construction Simplified

277 Index

Introduction

Advanced Framing contains the best articles on framing and structure published by *The Journal of Light Construction* since its start in 1982. Our technical articles are written by tradesmen, engineers, and other construction professionals, based on their years of field experience. Our goal has been to capture their job-site wisdom and get it down on paper in plain English and with clear illustrations. In this way, the magazine serves as a forum for building professionals to discuss their fast-changing field with one another, and as a training tool for newcomers.

Rather than gloss over the tough spots, *The Journal* has always tried to address them squarely, with the hope that our readers can learn from the experiences—and mistakes—of their peers, rather than learn everything on their own the hard way. So whether you're an accomplished builder, remodeler, or designer looking for pointers—or a newcomer looking for a close-up view of current building techniques—we think you'll find much of value in these pages.

Steven Bliss
Editorial Director

Section 1. **First Steps**

Steven Bliss

Wood Basics

From forest to finished house:
How wood works

Historically, some woods have filled many purposes, while others served only one or two.

The tough, strong, and durable white oak, for example, was highly prized for shipbuilding, bridges, cooperage, barn timbers, fence posts, and flooring.

On the other hand, woods such as black walnut and cherry became primarily cabinet woods. Hickory was made into tough, resilient tool handles. Black locust was prized for barn timbers and tree nails.

Early builders learned by trial and error which species to use, and that wood from trees grown in certain locations under certain conditions was stronger, more easily worked—or finer grained—than wood from other locations. Modern research has substantiated that location and growth conditions significantly affect wood properties, and has given us the tools to understand and predict wood performance.

Fast-grown plantation trees originally developed for pulp sometimes turn up as framing lumber—with twisted results.

Sapwood and Heartwood

Sapwood is located next to the bark. It contains few living cells, and primarily stores food and transports sap.

Sapwood commonly ranges from 1½ to 2 inches in thickness. However, the maples, hickories, ashes, some of the southern pines, and ponderosa pine can have sapwood 3 to 6 inches—or more—thick, especially in second-growth trees. As a rule, the more vigorously growing trees have wider sapwood layers. Many second-growth trees of saleable size consist mostly of sapwood.

Heartwood, the inner part of the tree, consists of inactive cells that are formed by a gradual change in the sapwood. Heartwood may contain deposits of various materials that frequently give it a much darker color. These materials usually make the heartwood more durable when used in exposed situations. Unless treated, all sapwood is susceptible to decay.

In some woods, including redwood, western red cedar, and black locust, material deposited in the heartwood makes it heavier and more resistant to crushing.

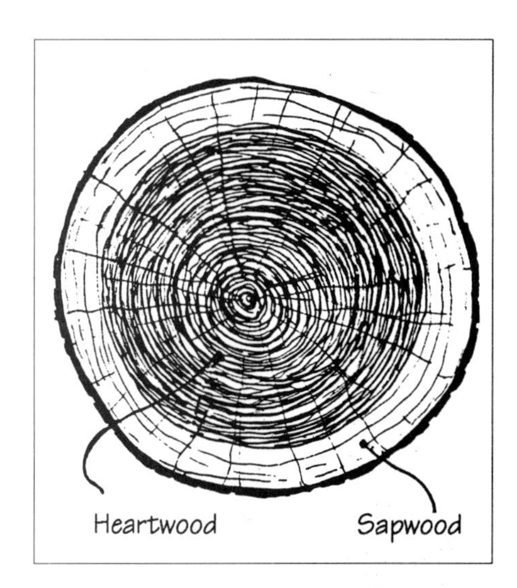

Physical Properties of Wood

A spectrum of characteristics is available among the many species of wood. Often, more than one property is important. For example, when selecting a species for a particular use, texture, grain pattern, or color may be evaluated against machinability, stability, or decay resistance.

Plain-sawed and Quarter-sawed Lumber

Lumber can be cut from a log in two distinct ways: tangent to the annual rings—called "plainsawed" lumber in hardwoods and "flat-grained" or "slash-grained" in softwoods—and radially to the rings—called "quartersawed" lumber in hardwoods and "edge-grained" or "vertical-grained" in softwoods.

Quartersawed lumber is usually not cut strictly parallel with the rays, and plainsawed boards are often far from being tangent to the rings. In commercial practice, lumber with rings at angles of 45 to 90 degrees with the wide surface is called quartersawed, and lumber with rings at angles of 0 to 45 degrees with the wide surface is called plain-sawed.

Moisture Content

The moisture content of wood is defined as the weight of water in wood expressed as a percentage of the weight of oven-dried wood. Weight, shrinkage, strength, and other properties depend upon the moisture content of wood.

In living trees, the moisture content may range from about 30 to more than 200 percent of the weight of the wood substance. The moisture content of sapwood is usually high. The moisture content of heartwood may be much lower than that of sapwood in some species, but greater in others.

Moisture can exist in wood as a liquid or vapor within cell *cavities*, or as water bound chemically within cell walls. The moisture content at which cell walls are saturated (all "bound" water), but no water exists in cell cavities, is called the "fiber-saturation point." This averages about 30 percent moisture content in wood.

The moisture content of wood below the fiber-saturation point is a function of both the relative humidity and the temperature of the surrounding air. The relationship

Quartersawed lumber, cut radially to the rings, is sometimes preferred for its straightness, but plainsawed lumber has more interesting grain patterns. The advantages of each are:

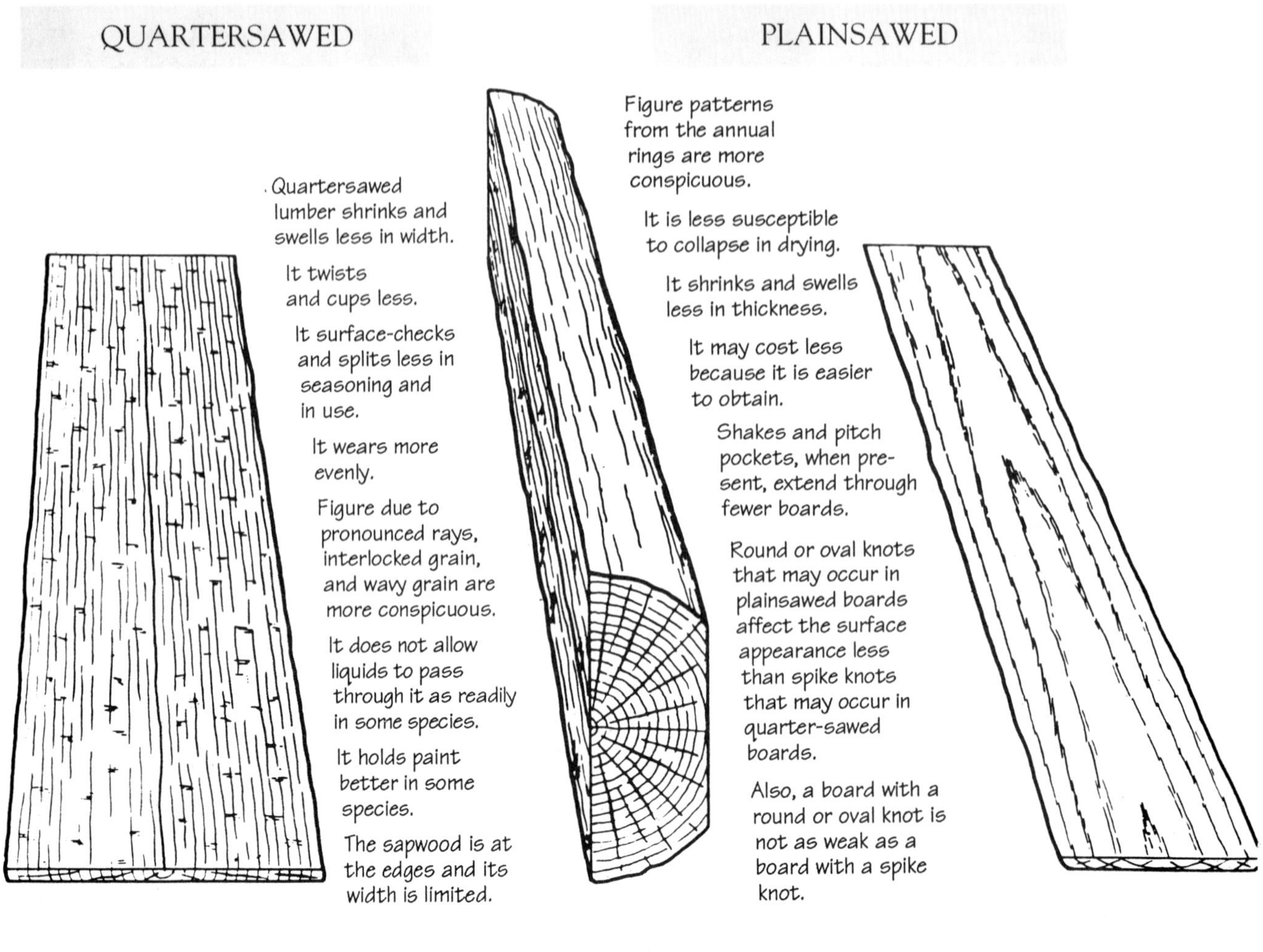

QUARTERSAWED

Quartersawed lumber shrinks and swells less in width.

It twists and cups less.

It surface-checks and splits less in seasoning and in use.

It wears more evenly.

Figure due to pronounced rays, interlocked grain, and wavy grain are more conspicuous.

It does not allow liquids to pass through it as readily in some species.

It holds paint better in some species.

The sapwood is at the edges and its width is limited.

PLAINSAWED

Figure patterns from the annual rings are more conspicuous.

It is less susceptible to collapse in drying.

It shrinks and swells less in thickness.

It may cost less because it is easier to obtain.

Shakes and pitch pockets, when present, extend through fewer boards.

Round or oval knots that may occur in plainsawed boards affect the surface appearance less than spike knots that may occur in quarter-sawed boards.

Also, a board with a round or oval knot is not as weak as a board with a spike knot.

Problems With Juvenile Wood

by John F. Senft

The housing industry has documented over 700 cases of a new and unpredictable problem—the rising truss. Researchers have linked the problem to juvenile wood.

Fast tree growth and shorter harvesting rotations are increasing the percentage of juvenile wood in timber harvested from plantations in the South and West. The impetus for plantations came from the pulp-and-paper industry's desire to maximize the output of timberland. The result was a successful effort to select, genetically screen, and manage tree species to produce more wood per acre and shorten the harvest rotation time.

Although fast-grown, short-rotation conifers may be fine for pulp, they lack the quality of wood fiber needed for structural lumber. Inevitably, though, wood cultured and managed for pulp is being processed into lumber, which returns a higher profit. As more commercial plantations reach harvestable size, juvenile wood—and a related problem, reaction wood—are becoming increasingly common.

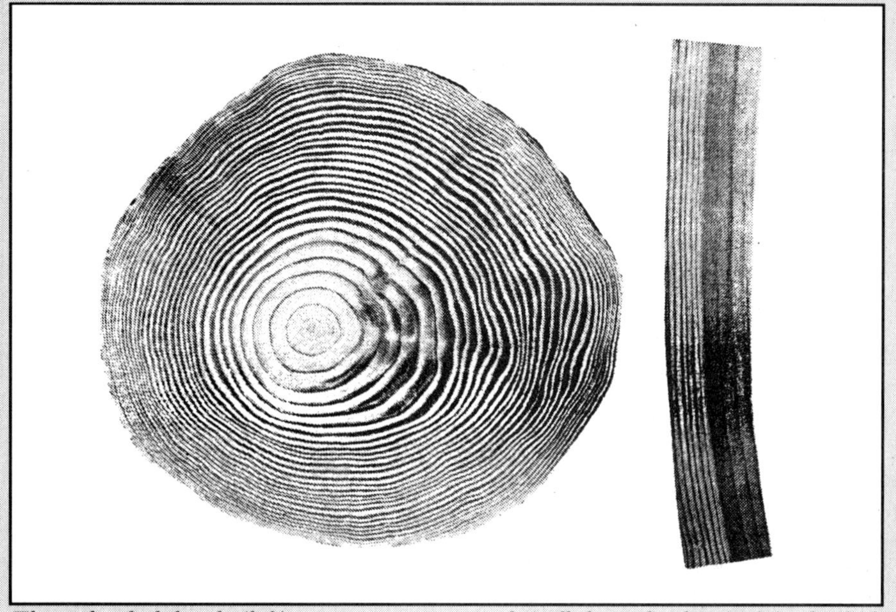

The wide, dark bands (left) are compression wood. Milled into lumber (right), the dark strip of compression wood shrank lengthwise, causing a crook.

Juvenile Wood

Juvenile wood encompasses the first 5 to 20 annual growth rings of a tree. Juvenile wood exists in the upper portions of the tree, even though the wood near the ground may have been forming mature wood for years. The transition from juvenile wood to mature wood is usually gradual.

Regardless of tree height, the growth rings nearest the center contain lower-than-average strength and density, compared to mature wood. The strength and stiffness are 30 to 50 percent lower. The cells are shorter than average. Shrinkage along the grain is much greater, and this longitudinal shrinkage tends to cause excessive warp.

Juvenile wood has a lower "modulus of elasticity," which can result in bouncy floors. Compression parallel to the grain is also seriously reduced, impairing the load-bearing capacity of columns. If juvenile wood is present, the result may be uneven floors, buckled walls and paneling, weakened joints, and poorly fitting doors and windows. Also, finger joints are inferior when significant amounts of juvenile wood are present.

Pieces that contain nonuniformly distributed rings of juvenile wood will warp readily during kiln drying. Severely warped pieces are usually suitable only for chipping.

With a practical rotation age nowadays of 20 to 35 years, a high proportion of harvest lumber will soon be juvenile wood. As rotation ages decrease, the proportion of juvenile wood appearing in products will increase.

Reaction Wood

Reaction wood often appears during the first few years of growth, and is frequently associated with juvenile wood. Reaction wood grows in response to growth hormones in stems that have become bent. The abnormal wood acts to restore the tree's uprightness. The wood's appearance and mechanical properties are affected.

Reaction wood occurs in softwoods as compression wood, and in hardwoods as tension wood. Compression wood has higher-than-normal density, less cellulose, and up to ten times more longitudinal shrinkage than normal. The wood is weaker and more permeable, and cracks may appear across the grain in drying.

Tension wood, scattered throughout the tree, is difficult to recognize. Like compression wood, it shrinks a great deal lengthwise—up to one percent.

Problems arise when a designer specifies dimension lumber based upon the allowable stresses for grade. Allowable design stresses are derived from the average strength values of a species. This process has traditionally produced lumber that has performed well—and safely.

Juvenile wood, however, often has strength that is well below average. A weaker-than-normal piece of lumber will usually shift at least part of its load to its stronger neighbors. But lumber from plantation thinnings—such as from large southern-pine plantations—may contain few stronger neighbors. ■

John F. Senft, B. Alan Bendtsen, and William L. Galligan. Adapted from "Weak Wood," in the Journal of Forestry.

Table 1. Weather vs. Wood Moisture Content

Temperature (°F)	Relative humidity (%) 10	20	30	40	50	60	70	80	90
30	2.6	4.6	6.3	7.9	9.5	11.3	13.5	16.5	21.0
40	2.6	4.6	6.3	7.9	9.5	11.3	13.5	16.5	21.0
50	2.6	4.6	6.3	7.9	9.5	11.2	13.4	16.4	20.9
60	2.5	4.6	6.2	7.8	9.4	11.1	13.3	16.2	20.7
70	2.5	4.5	6.2	7.7	9.2	11.0	13.1	16.0	20.5
80	2.4	4.4	6.1	7.6	9.1	10.8	12.9	15.7	20.2
90	2.3	4.3	5.9	7.4	8.9	10.5	12.6	15.4	19.8
100	2.3	4.2	5.8	7.2	8.7	10.3	12.3	15.1	19.5

The moisture content of wood changes as a function of the temperature and relative humidity. For example, if a space is 30° F and has 80% relative humidity (say, an attic in winter), the wood there will slowly rise to 16.5% moisture content.

between equilibrium moisture content, relative humidity, and temperature is shown in Table 1.

Wood is almost always undergoing at least slight changes in moisture content. These changes usually are gradual, and short-term fluctuations tend to affect only the wood surface. Moisture-content changes may be retarded by protective coatings such as varnish, lacquer, or paint.

The general goal in seasoning and storing wood is to bring the wood to the moisture that it will typically have in service.

Shrinkage

Wood is dimensionally stable above the fiber-saturation point. But below that point, it shrinks when losing moisture and swells when gaining moisture. This shrinking and swelling may result in warping, checking, splitting, or other performance problems.

Wood shrinks most in the direction of the annual growth rings (tangentially), about one-half as much across the rings (radially), and only slightly along the grain (longitudinally). The combined effects of radial and tangential shrinkage can distort the shape of wood pieces.

Wood shrinkage is affected by a number of variables. In general, greater shrinkage is associated with greater density. The size and shape of a piece of wood may also affect shrinkage. So may the temperature and rate of drying for some species. Radial and tangential shrinkage for a few common species are shown in Table 2.

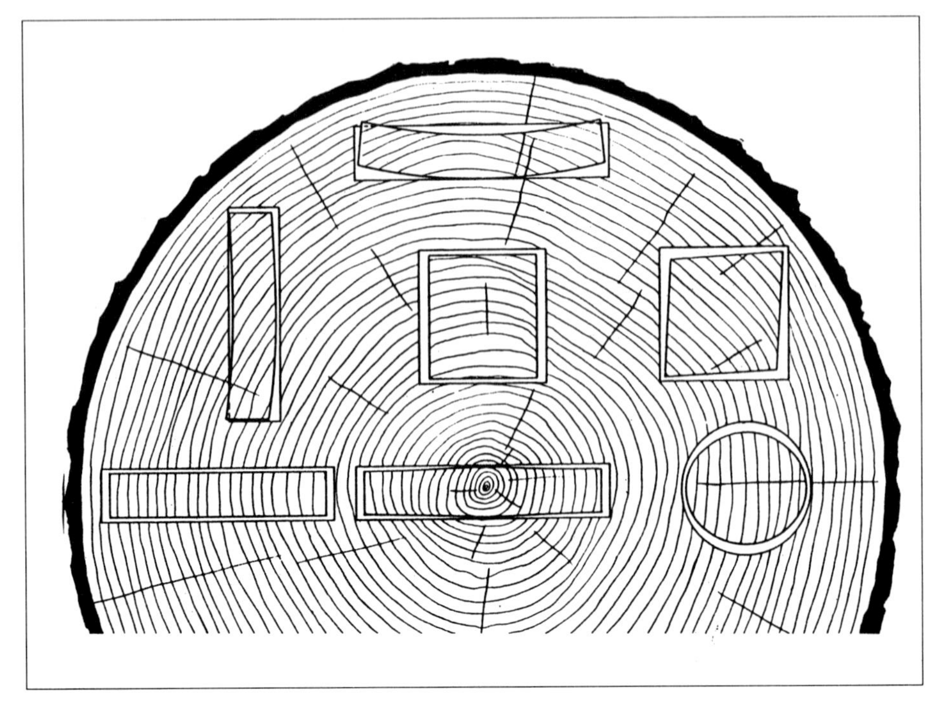

Characteristic shrinkage and distortion of flats, squares, and rounds. *Shrinkage along the rings is about twice as great as shrinkage perpendicular to the rings.*

Longitudinal Shrinkage

Longitudinal shrinkage (along the grain) is generally quite small—between .1 and .2% for most species of wood.

Certain abnormal types of wood, however, exhibit excessive longitudinal shrinkage. "Reaction wood," whether it is compression wood in softwoods or tension wood in hardwoods, tends to shrink excessively along the grain. Wood from near the center of trees (juvenile wood) of some species also shrinks excessively lengthwise. (See "Problems With Juvenile Wood," previous page).

Working Qualities

The ease of working wood with hand tools generally varies directly with the specific gravity (density) of the wood. The lower the density, the easier it is to cut the wood.

A species that is easy to cut, however, does not necessarily develop a smooth surface when it is machined. Consequently, many U.S. hardwoods have been tested for machining properties. The results for a few species are given in Table 3.

Three major factors other than density may affect the production of smooth surfaces during machining: interlocked and variable grain; hard mineral deposits; and reaction wood, particularly tension wood in hardwoods.

Interlocked grain is characteristic of tropical species. It can cause difficulty in planing quartered surfaces unless attention is paid to feed rate, cutting angles, and sharpness of

Table 2. Rate of Shrinkage

Shrinkage from green to oven-dry moisture content

	Radial %	*Tangential* %
HARDWOODS		
Cherry, black	3.7	7.1
Maple, black	4.8	9.3
Oak, northern red	4.0	8.6
Oak, white	5.6	10.5
SOFTWOODS		
Cedar, northern white	2.2	4.9
Douglas fir, coast	4.8	7.6
Hemlock, western	4.2	7.8
Pine, eastern white	2.1	6.1

Shrinkage values of selected woods along the rings (tangential)and across the rings (radial).

Table 3. Machining Properties

	PLANING % perfect pieces	SHAPING % good to excellent pieces	TURNING % fair to excellent pieces	SANDING % good to excellent pieces	NAILING % pieces free from complete splits
Ash	75	55	79	75	65
Basswood	64	10	68	17	79
Birch	63	57	80	34	32
Cherry, black	80	80	88	–	–
Chessnut	74	28	87	64	66
Maple, hard	54	72	82	38	27
Oak, red	91	28	84	81	66
Oak, white	87	35	85	83	69
Pecan	88	40	89	–	47
Walnut, black	62	34	91	–	50

Machining properties of selected hardwoods: the percentage of pieces with desirable characteristics in a typical load.

knives.

Hard deposits, such as calcium carbonate and silica, can dull all cutting edges. This is worse when the wood is dried before milling. Tension wood can cause fibrous and fuzzy surfaces and can pinch saws due to stress relief. The pinching may result in burning and dulling of the sawteeth.

Weathering

The color of wood is soon affected when exposed to weather. With continued exposure, all woods turn gray. This thin, gray layer is composed chiefly of partially degraded cellulose fibers and microorganisms. Further weathering causes fibers to be lost from the surface, but the process is so slow that only about 1/4 inch is lost in a century.

The chemical degradation of wood is affected greatly by the wavelength of light. The most severe effects are produced by ultraviolet light. As wetting and drying take place, most woods develop physical changes, such as checks or cracks. Low-density woods acquire fewer checks than do high-density woods. Vertical-grain boards check less than flat-grain boards.

Boards tend to warp (particularly cup) and pull out their fastenings. The greater the density and the greater the width in proportion to the thickness, the greater is the tendency to cup. Warping also is more pronounced in flat-grain boards than in vertical-grain boards. For best cup resistance, the width of a board should not exceed eight times its thickness.

Biological attack of a wood surface also contributes to color changes. When weathered wood has a dark gray and blotchy appearance, it is due to dark-colored fungal spores and mycelium on the wood surface. The silvery gray sheen often sought on weathered wood occurs most frequently where microorganism growth is inhibited by a hot, arid climate, or salt air.

The contact of fasteners and other metallic products with the weathering wood surface is another source of often undesirable color. ■

This article is excerpted and adapted from The Wood Handbook, *published by the U.S. Department of Agriculture Forest Products Laboratory.* The Wood Handbook *is available from the Superintendent of Documents, U.S. Government Printing Office, Washington, DC 20402. Stock Number 0100-03200.*

Getting Power to the Site

Whether you build, rent, or borrow it, plan your temporary electrical service with safety and convenience in mind

by Steve Lancaster

Although job-site power poles are temporary, they take a lot of weather and abuse. The one above, which taps into underground service, still requires rain-tight boxes and connections, and ground-fault interrupter protection.

Even in this day of battery-powered tools, you need power on site to build a house, if only to plug in your charger. There are four ways of getting electricity: buy it, make it, borrow it, or steal it.

Stealing it is frowned upon by local utility companies and district attorneys, and shouldn't be considered as an option. Borrowing it is an excellent idea (see "Borrowing Power," page 18), but it requires a cooperative neighbor within 100 yards—not always the situation. This leaves making it or buying it.

If you build in rural areas where utility poles are out by the highway, you're probably used to generating your own power. But for most of us building in cities or subdivisions, the best option is to use a temporary power pole and buy our electricity from the utility company.

A temporary power pole is nothing more than a metered, service drop that includes a conduit mast, a meter base, a disconnect switch, GFCI breakers, power outlets, and a ground. You can rent one or have your electrician make one up for you. Either way, it pays to use a pole that's set up right so that it delivers the power you need safely and reliably in all kinds of conditions. Construction sites are dangerous enough without adding the risk of frying somebody because of a jerry-rigged power source.

Rental Option

Renting is an economical alterna-

Figure 1. *Combining a job-shack or portable toilet with temporary power is common, particularly in rental units. Tying down the structure with steel foundation stakes or 2xs will keep it steady and, for smaller shacks, increase the chances it will be there in the morning.*

Figure 2. *This temporary pole makes up in bracing what it lacks in depth in rocky soil. In this case, the service drop is just a few feet away, making height minimums less important.*

tive if you use temporary power only occasionally. In my area, a 100-amp pole with one 220-volt and two 115-volt outlets runs about $50 a month, with a three-month minimum and a $45 set-up fee.

In a typical installation, the rental company will bore a hole on your site with a power auger mounted on the back of their truck (they charge extra if they hit rock), tip the pole into place, fill and tamp the soil, brace it, and drive the ground stake. The only thing that remains for you to do is to have it inspected by the local building department, and ask the utility company to set the meter and make the connection.

However, if your local utility doesn't rent temporary poles, you may have to make a few calls before you find someplace that does. Most tool rental yards in my area don't offer them. Strangely enough, the two companies that do are listed in the Yellow Pages under portable toilets (and one is an overhead door installer).

This brings up an interesting rental alternative: a combination toilet and temporary service. Here this runs about $120 a month including weekly pumping, but it kills two birds with one stone if you need a toilet on site. Another combination I have seen—both rented and homemade—is a job shack that includes a mast, service panel, etc. (see Figure 1).

Putting a Pole Together

The utility company will typically supply the meter, service wire, and insulator, but everything below that is on you. Your electrician and utility company should get together on the details and any local requirements. Here are the main elements.

The pole. My utility allows both metal and wood poles; the latter can be round (Class 6 as designated by the American Wood Preservers Association) or square (a solid 6x6 timber). Although wood poles don't have to be pressure-treated or made of a decay-resistant species if they are in use for less than a year, this makes good sense if you're going to all the trouble of wiring it. Your utility should be able to help with names of suppliers of treated poles.

In my area, poles have to be at least 20 feet in length (with 4 feet in the ground), but they are more typically 25 feet (with at least 5 feet in the ground) in order to meet minimum clearances from the service drop to the center of the street (18 feet) or to the curb (16 feet). Pole placement is also mandated by my utility: no less than 10 feet and no more than 100 feet from a permanent pole.

The pole also has to be braced against the pull of the conductors either with a guy and ground anchor, or a long 2x4 that is bolted to the pole and to a 3-foot stake driven in the ground. In practice, however, bracing is often less elegant than this (see Figure 2).

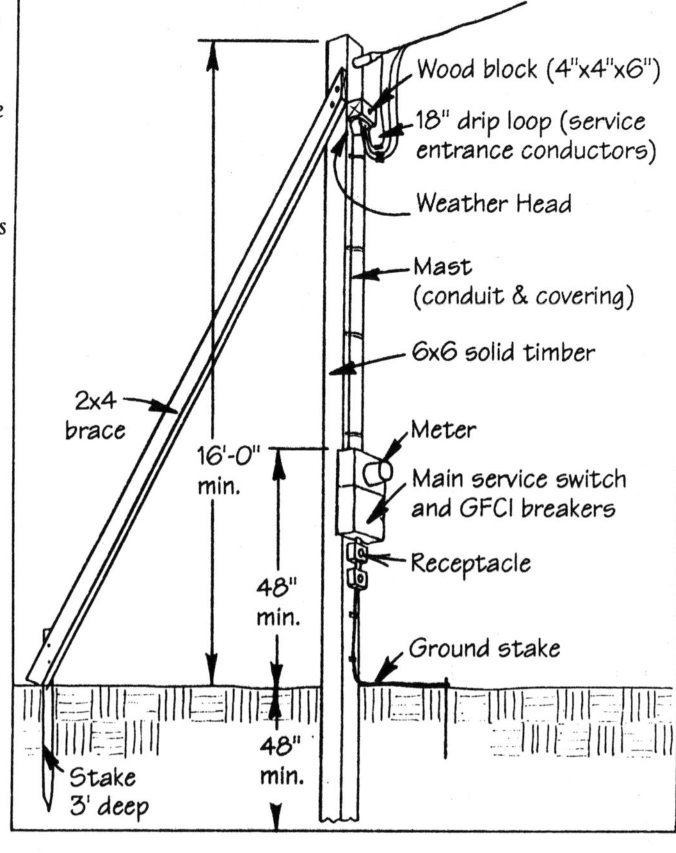

Figure 3. *Temporary power pole specs are often enforced by the utility that supplies the power. The required double mast protection—conduit and wood covering—seems like overkill until the first time a backhoe slams into the pole.*

Service entrance conductors. For a normal 240-volt system you'll need three conductors. The service entrance conductor should be large enough and of a type to satisfy Section 310 of the National Electric Code. For example, you'll need #4-gauge copper or #2-gauge aluminum wire for a 100-amp service. I like copper. It not only conducts electricity better, but it has none of the corrosion problems that require you to treat the aluminum conductors with a de-oxidizing agent when making connections.

Mast. The service entrance conductor is protected from construction site wear and tear by the service mast (see Figure 3). This is a metal or plastic conduit mounted to the pole. The mast should be further protected by another covering made of wood, fiber conduit, or PVC. This redundancy pays for itself the first time a backhoe boom strikes the mast.

The mast should be mounted to the pole with 2-hole pipe straps at each end, and every 3 feet in between. The covering should be attached in the same way. Place the weatherhead at the top of the mast, giving the service an 18-inch drip loop. A 4x4x6-inch wooden block is required immediately above the weatherhead. Rain-tight fittings should be used with all conduit connections.

Meter box. The meter box is mounted at the bottom of the mast, putting the meter at least 4 feet off the ground; eye level is ideal. The meter box should be exterior rated, and the main service switch should accommodate a padlock.

I use a combination unit that contains the meter mount, a main service switch, and space for several circuit breakers, but you can build it out of component parts if you choose. When it comes to breakers, I like GFCIs (ground-fault circuit interrupters) for the protection they afford everyone on the crew.

Receptacles. The outlets are mounted below the service box. These also should be mounted in waterproof boxes with wiring protected by conduit. You'll need at least two 115-volt outlets and one 220-volt receptacle in either universal or twist-lock versions.

Ground. The entire system needs to be grounded with an 8-foot stake driven into the ground at the base of the pole. The stake should be attached to the meter box with at least a #6-gauge ground wire.

Codes and Permits

Where I work, an electrical permit from the local building department is required before you can set a temporary power pole. Besides requiring you to fill out an application and pay a fee, most building departments (or electrical departments in some localities) also want to inspect the finished temporary service before approving it and issuing a "permit to connect."

Once this process is complete, the building department will notify the local utility company. When they get the word, you can make arrangements to have the power connected. This involves another fee, and still another inspection. Then, and only then, will a field crew come out and perhaps agree to run a line out to your pole.

Utility companies often make a utility service manual available to their customers. Its purpose is to lay out exactly what they require before they will hook you up. It is up to you to make them happy. (In my area, it's the utility company—not the building department—that really nails you on the details with temporary power poles.) The manual contains specs and drawings needed to engineer the

"Borrowing Power"

by John Flanders

It's a fact of life these days for urban builders that the vacant lot they're building on is bordered by existing houses, not other lots. So the sooner you can meet and establish good relations with your neighbors the better.

This meeting can also be an opportunity for you to discuss "borrowing" their power. The deal is a sound one for both parties: You pay their electric bill for the duration of the job in exchange for the use of their service drop.

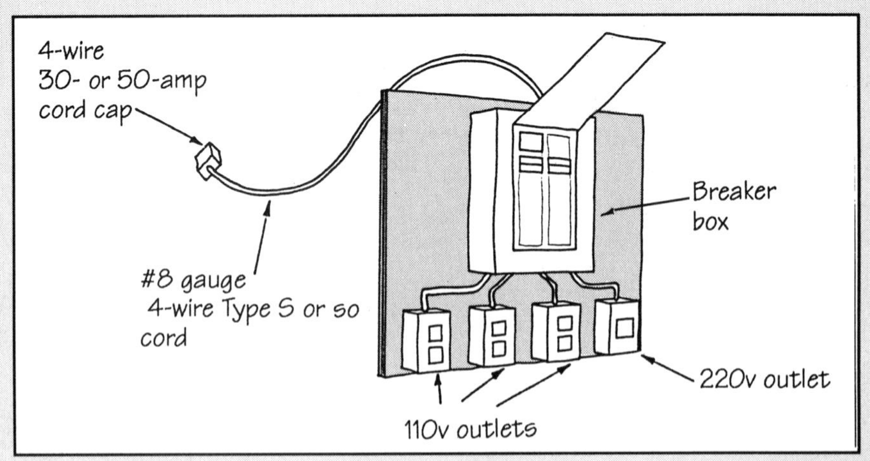

Kilowatts and Dollars

A little arithmetic shows how you come out ahead in this deal. In my area, a rented power pole with installation and removal costs around $200 for a four-month project. Add in the power company's hookup fee of $250 and minimal monthly charges of $15 and you'll find yourself spending $125 a month just to get the juice to the job.

The actual cost of the power you consume will be vastly less. Power tools tend to run only a small percentage of the time. As an example, if each member of a four-man crew operates a 12-amp tool five percent of the time, and shares a 10-amp, 220-volt compressor on the same schedule, they will consume a grand total of 64 kilowatts if they're working eight hours a day, 20 days a month. That's less than $100 at my utility's rates.

However, before you knock on the neighbors' door, you might discretely make sure that their service is up to the task. A 60-amp, 220-volt main switch is enough if you're not doing any heavy welding. If they have a three-wire service (which you can see at the weatherhead), or an underground service of fairly recent vintage, then you're on your way.

Of course, you need to exercise some caution. If you're going to offer to pay a neighbor's entire electric bill for some months, make sure they don't heat their house electrically, or have a hot tub in the backyard. Air conditioning is another load to look for. If you suspect that their bill is astronomical, offer to pay only a portion of it.

Temporary Hookup

Once you've made a deal to use the neighbor's service, your electrician can figure the least obtrusive way to tap into the panel and install a 30- or 50-amp four-wire outlet in a weatherproof box where you can get at it. Remind him that holes in finished surfaces aren't part of the program. But don't overlook the obvious. Plenty of houses have unused stove or dryer circuits just waiting to be put to good use.

On your end, you'll want a convenient way to break your subfeed into circuits. A Tempower outlet box from Harvey Hubbell, Inc., (P.O. Box 3999, Bridgeport, CT 06605; 203/333-1181), which contains a number of 110-volt ground-fault-protected outlets along with a 220-volt outlet for heavier draws, is the premium solution. All you need to add is an extension cord—usually #6- or #8-gauge four-wire cable, type S or SO—with ends to match your receptacle.

As an alternative, you can build up a portable panel from off-the-shelf components which does more or less the same thing (see illustration). Start with a rain-tight panel that has space for at least eight breakers. Mount it to a piece of plywood with enough extra space for several weatherproof outlet boxes. Bring your heavy-gauge extension cord into the panel through the appropriately sized hub being sure to use a waterproof cord connector.

From then on, you or your electrician will proceed as if it were a regular panel, wiring up ground-fault protected outlets to suit your needs. If you don't use twist-lock outlets, you can forego ground-fault circuit breakers, which are comparatively expensive, and use GFCI outlets instead. These have dropped in price dramatically in the last few years. And I've seen less "nuisance tripping" with the outlets than with the comparable breakers.

With this kind of setup, you can unplug the panelboard and store it with the rest of your tools at the end of the day. When the house is complete, the panel goes home with you to be used on another project. Then the only thing that remains is to pay the neighbor for that last power bill—lest you leave your clients with a snarling neighbor before they even have time to throw a house-warming party.

John Flanders is an electrician and general contractor in Berkeley, Calif.

temporary power system.

Both local building departments and utility companies use the National Electric Code (NEC) in setting their guidelines. The particular section of the NEC that applies to temporary power poles is Article 305, "Temporary Wiring."

Getting Power Around the Site

Once you've gotten power to the job site, practical distribution is the next step. Extension cords are the obvious answer, but they're not always handled as well as they should be.

First, they should be rated for outdoor use (this is designated by the letters W-A at the end of the cord designation) and checked frequently for breaks in the insulation and broken grounding prongs.

Second, unless your pole is within a few feet of the house, cords have to be sized to handle the *voltage drop* that comes from running current a long distance. This loss in power is due to the resistance that all wire offers electricity as it travels along. In order to minimize voltage drop over a long distance, you have to decrease the resistance by using heavier gauge wire.

Some tradesmen use #14-gauge or even #16-gauge extension cords because they're cheaper, light, and flexible. But over longer distances, higher amperage tools just aren't going to get the voltage their motors were designed to run on. For any trade that works off a power pole, a #12-gauge wire cord is a better choice and will take you up to 100 feet. This may sound cautious, but the charts that come with most power tools are even more conservative in their requirements (see Table below). They are based on limiting the line voltage drop to 5 volts at 150% of the tool's rated amperes for 115-volt service.

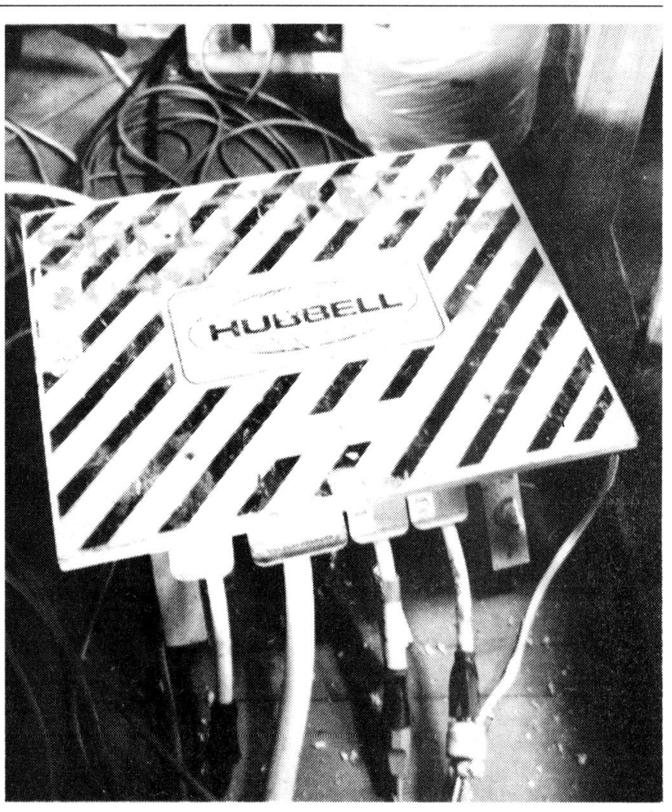

Figure 4. *A fixture on large commercial jobs, 220-volt spider boxes bring 110-volt and 220-volt receptacles right to the center of the job. This allows shorter runs of smaller-gauge drop cords, and keeps connections at the power pole to a minimum.*

For distances over 100 feet, consider a *spider box* (see Figure 4). It plugs into a 240-volt outlet on the pole and extends the pole's capability right down to the center of the job. It's basically a giant box with 240-volt and 115-volt outlets from which you can run all sorts of extension cords (the image this creates gave the spider box its name). This is a very versatile system because it can be moved from place to place, or even floor to floor, during the construction of a building, and shortens the run of 115-volt cords. It also helps prevent the incredible tangle of cords, "pigtails," and "Y"s that hang off a pole when a lot of trades are on the job at the same time.

Spider boxes aren't cheap—$200 to $700 depending on the number of outlets, and whether they include GFCI breakers, or twist-lock outlets, etc.—but they make a lot of sense. Not all electrical supply houses stock them, but they are widely manufactured for commercial construction and can easily be ordered. ■

Recommended Extension Cord Wire Gauge Sizes for use with portable electric tools

	Length of Cord								
115V:	25 Ft.	50 Ft.	100 Ft.	150 Ft.	200 Ft.	250 Ft.	300 Ft.	400 Ft.	500 Ft.
230V:	50 Ft.	100 Ft.	200 Ft.	300 Ft.	400 Ft.	500 Ft.	600 Ft.	800 Ft.	1000 Ft.
Tool Amp Rating									
0-2	18	18	18	16	16	14	14	12	12
2-3	18	18	16	14	14	12	12	10	10
3-4	18	18	16	14	12	12	10	10	8
4-5	18	18	14	12	12	10	10	8	8
5-6	18	16	14	12	10	10	8	8	6
6-8	18	16	12	10	10	8	6	6	6
8-10	18	14	12	10	8	8	6	6	4
10-12	16	14	10	8	8	6	6	4	4
12-14	16	12	10	8	6	6	6	4	2
14-16	16	12	10	8	6	6	4	4	2
16-18	14	12	8	8	6	4	4	2	2
18-20	14	12	8	6	6	4	4	2	2

Porter Cable Corp.

Steve Lancaster works as an electrician in the San Francisco Bay Area.

Tools of the Framing Trade

What one framing contractor uses daily on site and why

by Don Dunkley

To frame a typical custom home during the mid-1970s, I would roll up on the job site in my El Camino, pull down the tailgate, and unfurl plans that I had last seen when I bid them. To put the frame together, all I needed was my Skilsaw, extension cord, nail bags, and some seldom-used hand tools that resided in the bottom of a wooden tool box. Homes were typically built on a slab and came in around 2,000 square feet. They had 8-foot walls, flat ceilings, and 4/12 roofs with a couple of hips.

Times have changed, and so have the custom homes around here. They are no longer built on flat lots (now an endangered species), and they have grown to at least 3,200 square feet (see Figure 1). They include every conceivable ceiling treatment and window placement known to the design world, a first-floor plate height of 9 feet, a two-story entry, a winding staircase, and literally dozens of roof planes. To be a successful framing contractor, these days, you not only need a "Professional Attitude," but a truck full of tools.

In the next couple of pages I'll go into some detail about the tools I use and why I've chosen them. This isn't intended as a formula for everyone. I started as a framer doing piecework in California's tracts. So even though I now build custom homes, I use production methods, and that affects my choice of tools. So does personal taste, for which there's no accounting. But no matter how or where you work, it never hurts to hear another contractor's recommendations.

The Basics

For remote power, I use a Kubota

5,000 generator. It has an electric starter and automatic idle which keep down the pain in my shoulder and the noise level. It also has oversized wheels that help on uneven ground. I paid $1,400 for it a couple of years ago.

I carry four, 100-foot, 12-gauge extension cords with twist-lock ends (about $50 each) to get the power to the center of the site, and I use lots of lighter cords from there. This also means carrying Y's and adaptors (universal to twist-lock and vice versa) by the dozen.

My truck usually has three Model 77 Skilsaws in it (see Figure 2). Although I require my employees to have their own circular saws, it makes sense to carry spares since they are less than $150 apiece. Two saws can go down in a day with cut cords or faulty triggers.

For anyone who doesn't use worm-drive saws, I won't try to convince you of their superiority. As far as brand goes, I've tried others that are arguably lighter or more durable, but I've stuck with Skil. They're familiar, they do the job, and, where I build, the price stays consistently low. (They are such a staple that most tool supply houses here use them as a *loss leader*, selling them at or near cost to lure you into the store.)

Figure 1. *Houses we frame are rarely on flat lots, and often exceed 3,200 square feet. This one has a two-story curved entry, 18-foot walls in open areas, and lots of roof complications.*

Figure 2. *Production framing crews prefer worm-drive saws because much of the cutting is done in place rather than on benches or horses. Dunkley requires his crew to bring their own saws, but backs those up with three Skil Model 77s of his own.*

Which blade to use is something that elicits strong feelings. For the last six months we've been using Black and Decker Piranha blades. These are thin-kerf, carbide blades that give a very smooth, clean cut. We use the 24-tooth version and have had excellent results. I spend about $9 per blade, and most have lasted three jobs even after slicing through the odd 16d nail. In fact, one of my guys clipped an anchor bolt with one not long ago, and it didn't even lose teeth.

When they get dull, I use them for roof sheathing. I love thumbing my nose at the guy who comes around to job sites to sharpen blades. I used to spend $40 a month with this guy and still end up with blades that didn't cut as well as the Piranhas do.

The last tool that comes out on that first day on site is my transit. The typical subfloor we build starts with concrete stem walls that vary in height from 3 feet to more than 12 feet. To get the subfloor level, we use a David White transit with an optical plummit. A year ago, it cost me about $700 with a tripod. I chose the optical plummit feature because it allows me to place the instrument right on the corner of a stem wall, swing 90 degrees, and shoot in a straight line up a series of stem walls even in a stiff wind that would play hell with a plumb bob. The only feature I wish I had added to it was the automatic leveler—a real time saver.

Floor Systems and Wall Prep

With my foundation walls snapped out, I drill the green (pressure-treated) plates for the anchor bolts with a 1/2-inch drill and a 13/16-inch auger bit. I currently use a Skil motor, but its plastic parts haven't withstood the punishment of the job site. I'll get a tougher model soon, probably a Milwaukee. To clinch down the 1/2-inch nuts, I use a Black and Decker electric impact wrench (they cost about $130). It's a tough little unit and has held up well.

If the floor is a girder system, I usually cut the 4x4 girder posts with a Skilsaw. If the floor is big—2,000 square feet or more—or if the girders are on 32-inch centers, then I cut the 4xs with a chop saw. I use my Makita

10-inch with a 60-tooth carbide blade. We get very smooth cuts with this setup, and the blade has remained sharp for nearly two years. My next chop saw will have a larger blade capacity; it'll probably be the Hitachi 15-incher. One of my guys owns one and I was impressed with what it did one day cutting rake backing.

Once the subfloor is out of the way, we drag out the rest of the support system, including the radial arm saw. This beast—a 12-inch Delta—weighs in at a herniating 300 pounds. Not surprisingly, when it's time to pull it out everybody gets real busy doing something else. I've thought about mounting it on a trailer, but the fact that I live on a hillside about 40 winding miles from most of my job sites has discouraged me.

I use a carbide blade on this saw; it cost me $110, but it's worth every penny. To make the saw setup more efficient, I haul around a set of job site tables to hold the saw and flank it on either side (see Figure 3). Made out of plywood and 2x6s, and equipped with rollers (like those on an outfeed table), the three sections are sized to fit in the back of my truck. But unlike the saw, they stay on site at night. The rollers are made of 1 1/4-inch Schedule 40 PVC pipe slipped over slightly longer lengths of 1-inch PVC so the outer pipe spins freely. I cut notches in the top rail of the tables and nailed the ends of the inner pipe into these slots so the "rollers" are all at the same height. For $2 worth of plastic pipe, it works remarkably well and has lasted for years.

Once the saw is set up, we cut all the headers, sills, cripples, trimmers, and special-length studs. I used to do all this with a Skilsaw, but not anymore. It's no longer unusual to have to whack out a hundred studs at a nonstandard length, and nothing beats a radial arm saw for doing this job when it comes to speed, accuracy, and safety.

While the frame package is being cut, I lay out the walls on the floor. The handiest tool I've found for squaring up the layout is the Construction Master II. This calculator deals in feet and inches, and by using the functions of rise, run, and diagonal, I can square up walls of any length. That means I can easily lay out 50 feet of walls to the nearest 1/16 inch. Although most of my work is done to 3/16 tolerances—and fudged to 3/8 inch sometimes—this is based on a layout that is as accurate as possible.

Figure 3. *A Delta 12-inch radial saw makes the trip to the job site every morning, but the site-built outfeed tables and saw stand stay put. The rollers are lengths of 1 1/4-inch PVC pipe that spin freely on fixed lengths of 1-inch stock notched into the table rails.*

For that accuracy to translate to three dimensions when you are plumbing and lining the walls later on, the plates have to be cut and detailed precisely. I use an aluminum layout stick with legs for 16-inch and 24-inch centers to mark my studs. For wall intersections and corners, I use an aluminum channel marker I picked up years ago (see Figure 4). These tools can easily be made on site or in the shop if you can't find them at your tool supply house. To mark doorways and window openings, we use colored *keel* (lumber crayon). But in the hot valley sun, the wax fades rapidly, and I have been thinking of using marker pens.

Walls

Today's framing demands nail guns. Staring at a 9-foot-high, 40-foot-long wall chock full of headers, cripples, sills, trimmers, and fire blocks can rekindle thoughts of finishing college. A nail gun will at least get you through the task with enough energy left to start the *other* 40-foot wall.

For air, I use an Emglo, gas-powered compressor (about $850). The front wheel of this wheelbarrow design means I can move it around on site so that its obnoxious sound and fumes are as far away as possible. I carry three, 100-foot air hoses (I pay about $35 each for these locally) to service the guns. Usually, one hose is sufficient for framing walls, but when it comes time to nail off the roof, two hoses are a must. Since air hoses have a tendency to be abused, I carry enough couplers and clamps to make immediate repairs.

On the receiving end of the air hoses are two Hitachi NR 83A guns (see Figure 5). These handle 8d through 16d headed, collated nails. The gun is the lightest weight of all framing guns, and that makes a big difference if you're handling one all day long.

The other important factor is how often they jam and breakdown. These Hitachis get few complaints from my guys and have held up for three years with no repairs. At $500 a pop, they're expensive, but well worth it as far as I'm concerned.

I had a flush nailer attachment put on the Hitachis so we can adjust nail depth. This comes in handy for shearwall—so the outer membrane isn't broken—and for exterior trim using galvanized nails.

I also have a Halstead 8d to 10d full-head nailer. This is the first gun I purchased. I bought it to nail shear

Figure 4. *These two production tools—an aluminum layout stick and a channel marker—can easily be fabricated on site, or with a little sheet metal in the shop. The layout stick is used to quickly mark 16-inch and 24-inch centers; a channel marker can scribe partition or corner intersections of 2x4 or 2x6 wall.*

Figure 5. *Dunkley and his crew like Hitachi framing guns for their lightness and durability. This gun and another like it have been through three seasons without repair. They take collated, headed nails from 8d to 16d.*

panel, floors, and roofs. It's a real workhorse, but I tend to use the Hitachis because they're so much lighter.

Once all the walls have been framed, and "plumb and line" braces have been attached, the 8-foot level comes out. Currently I use a Master level from Canada (about $145), and for over a year it has remained true. In the past, I have had Empire and MD 8-footers, but both had vials that would come out of adjustment frequently.

Although levels aren't supposed to get dropped and kicked about, in the real world they do. Finding one that will take some abuse is like searching for the Holy Grail. The best I've ever owned was an expensive, fixed-vial German import (Stabila), but these are hard to find in this area. Lately, I've been reading about a high-tech level by Wedge Innovations with a resetable digital readout. If the self-leveling feature works as they claim, it would be a great advantage.

Roof Framing and Interior Pick-up

Putting the roof together requires more thinking than any other part of the house. Although no job-site tool can relieve you of the need to conceptualize clearly in three dimensions, there are several that can make calculations go faster. For several years I have used the Construction Master II to figure my roofs. It helps me weave my way through the thicket of math calculations that are essential in putting together the typical roof where I build—a maze of hips, valleys, bays, turrets, multiple pitches, and the like.

Recently I was introduced to a calculator that takes the process one step further by supplying all of the lengths and cuts on complicated roofs with very little prompting. This program, called Roof Tech (Roof Tech, 5805 Morgan Place, Loomis, CA 95650; 916/652-5386), is wedded to a Hewlett-Packard scientific calculator and optional, infrared printer. It was written by Bill Rose, a framer in my area who is producing the programmed calculators in limited numbers. Although Roof Tech is expensive at $300 (the printer runs an extra $120), it can really knock down the amount of time spent in calculating members. But buyer beware: you must have a good foundation in roof cutting for the tool to do justice.

We do actual layout on the rafters with a Speed Square. Cutting these members is usually Skilsaw work, but big ridges and purlins can call for more firepower—a chainsaw. Although it's looked upon as the bastard child of the industry, I find it useful. Try cutting a 6x24 glulam without one. As long as a chainsaw is equipped with a nice sharp chain and in the hands of someone who knows what they're doing, it will leave a decent line. They do have to be used with great caution, however. I have an Echo with a 12-inch bar (about $180), and an electric grinding stone that keeps the chain very sharp.

Dogyu, A Framer's Best Friend

Probably the simplest, but most personal tool of all is the hammer. In my tool box are no less than seven—a testament to my changing age and style. In my younger days, the True Temper rigging axe was my favorite. Then I went to a 24-ounce Vaughan straight claw, and after that the Vaughan 24-ounce California framer.

Today my hammers are even lighter—all in the 20 ounce range, but their style varies with use. I like a 20-ounce Daluge straight claw for wall framing, but for anything requiring more finesse I use a very odd-looking Japanese hammer made by Dogyu (see photo). We order these from Woodline Japan Woodworker, 1731 Clement Avenue, Alameda, CA 94501; 415/521-1810, 800/537-7820. It has near perfect balance, an extended nose for tight spots, and very curved claws. The handle is a little short for wall framing, but it's great for pick-up. That leaves me holding a Vaughan 999 for most of my carpentry; I think of this model as the best all around hammer made.

My crew isn't as fanatical about their hammers as I am. One of my guys uses a Dogyu exclusively, while another won't touch one and swears by his 24-ounce Vaughan. — *D.D.*

The Dogyu hammer is guaranteed to draw comments the first time it's seen on a job site. This Japanese import has excellent balance, a standard millface, and weighs in at slightly over 20 ounces. It's a little light and a little short in the handle for some framers, but it has a lot of fans when it's time for pick-up work.

Figure 6. *This handy electric compressor from Emglo is used for interior pick-up work. Dunkley sets up a larger, gas-driven, wheelbarrow type when he is framing walls and sheathing.*

Unless the design gets in the way, I typically wait until after the roof is assembled and sheathed to do interior pick-up (detail work) and ceiling details. This buys us a little shade in the summer and some protection from weather in the winter.

Designers in my area seem to be locked into competition over who can incorporate the most coffered ceilings, corbeled soffits, and high arches in one house. This calls for more traditional carpentry tools, but I allow myself one exception. I use a $1^1/2$-horsepower electric Emglo compressor (about $350) to power my Hitachi nail gun in fastening all the pieces together (see Figure 6). Favored by many interior trim workers, this electric compressor contributes quiet, clean air. The unit can't handle the load needed for production framing, but it's just right for inside pick-up.

To build our arches, we use trammels to strike the arc, and a Bosch 1581-VS jigsaw to cut them out. On finished exterior arches we use my old one-horsepower Skil router to cut a nice smooth line. When it wears out, I'll replace it with a plunge router.

But when it comes to pick-up work, the true unsung hero of the job site is the Sawzall (mine, true to the name, is from Milwaukee and runs about $130). This tool is a lifesaver. It can cut a window or door header free in a minute, or cut off a misplaced anchor bolt that's sticking up in a doorway. I use it to trim shear panel on a regular basis, and to cut out the plates in doorways.

It also makes money. When the homeowner inevitably wants a larger window, a bigger door opening, or a wall moved, you don't have to invent a reason why the house would fall down if you were to make the change. You can just smile—with change order in hand—and say, "How wide did you want that, sir?" ■

Don Dunkley and his crew frame custom houses in California's Sacramento Valley.

How Two Guys in Two Weeks Build A Custom Shell

The formula: Good tools, good systems, and old-fashioned hard work

by Bill Walsh

These custom homes were framed, sheathed, sided, and roofed in 14 days or less by the author and his partner. A weathertight shell for a 24x28 ranch takes this two-man team about eight days to complete.

"Come on! Two guys can't build a 1600-square-foot house in 11 days."

That's what they all say when we pull up to look over the job. Eleven days later they ask, "Who were those masked men?" My partner Jerry and I live in Maine and were taking advantage of a recent building boom here. For the most part we subcontract the framing of custom homes by the square foot. We get our work from general contractors who are too busy to complete all the homes they sell. We carry our own insurance and supply only the labor at a fixed price.

Occasionally we finish a house, but it seems that one of the disadvantages of subwork is that the general usually saves the inside work for hourly employees. In this article we are talking about framing a watertight shell. My partner and I have been together only two years but we have developed a system (thank you, Henry Ford) that works extremely well. Let me just note that one important ingredient is old-fashioned hard work. We have no trouble sleeping nights.

We make a point of using plans and details, and stress the importance of good drawings with *rough-opening sizes and joist and rafter layouts.*

Okay now, let's look at equipment. We decided to buy "high-tech" building equipment rather than hire extra men and get locked into workers' comp, which is extra high in this state.

We use pneumatic nailers exclusively for nailing joists, studs, sheathing, rafters, siding, decking, roofing,

A radial-arm saw with 12-foot extension tables is set up on every job.

Using jigs and stops, the saw is used to pre-cut everything with speed and accuracy. But temporary electrical service, says the author, can create big problems.

soffit and fascia, and temporary bracing. Our compressor is two horsepower, with a 20-gallon tank and about 250 feet of 3/8-inch air hose. We usually operate two guns at once and always have plenty of pressure. Safety is a must with this equipment. These guns never get tired and pay for themselves very quickly. We oil the equipment once a day and that's it.

We also purchased a radial arm saw and set it up on every job with two extension tables, 12 feet long. This also saves time: No lines to square, just mark and cut, or set a stop and cut 200 pieces exactly the same. This is a very accurate way to cut and to maintain quality workmanship. We use this saw to pre-cut everything from floor joists to siding and soffit. This tool demands respect as it can eat fingers. The only maintenance with this saw is to sharpen blades.

How can two men put up a 40-foot wall? We use *wall jacks* because we are no longer young and stupid, and we value our backs.

We refuse to use scaffolding that takes more than 10 minutes to erect. Our motto is taken from my pastime of rock climbing: "fast and light." So, we use aluminum Type-I ladders and ladder jacks, with 2x10 planks, for anything over 12-feet high. We check out the planks carefully by testing them between two sawhorses before using them. We can cover approximately 9 feet of wall without moving the ladder jacks. Anything under 12 feet we use fiberglass stepladders and planks. Fiberglass ladders maintain their stability and last much longer than wood stepladders. We do own some pump-jack scaffolding but find we seldom if ever use it.

We both use circular saws—I prefer a worm drive and my partner uses a regular one. We also use a 4 1/2-inch circular saw for cutting fascia and trim as it is extremely light and accurate. The rest of our tools are like anybody else's in this business: rusty hammers and cats' paws, a sledge hammer for tight spots, crow bar—standard fare.

We use all grounded extension cords of #12 wire with a ground-fault receptacle. If one is not present on the site we bring one with us. A transit is also important on the first day, since you can't trust any foundation—particularly if the site is littered with beer cans. I have a tendency to ramble so I'll quit and get to business.

First Day

Set up transit and shoot foundation walls quickly to determine the extent of damage. Then we check all dimensions and mark them on the foundation. We usually mark the inside of our sill plate. Now check for square. This usually requires the use of many well-known four-letter words—we adjust our marks accordingly. Now we strike lines and pick a good spot for the saw and tables—hopefully not to be moved for ten more days. Next, set up the compressor and hoses.

My partner now cuts sill plates from dimensions on the foundation plan. We have already decided how to lay out joist and studs from the roof layout plan—so that trusses or rafters will bear over studs and joists.

Now I mark and drill holes for anchor bolts, while Jerry installs sill insulation, then bolts down the sill plates. I mark the layout for the stairwell opening and joists while Jerry cuts main beams out of 2x10 or 2x12 stock. I shoot the beams together and Jerry cuts the rim joist. I install the rim joists, then we both install the joists. We transfer the layout from the sill up onto the rim joists to ensure that joists are plumb when nailed. Now that some weight is on the carrying beams we can level them with the transit and set and plumb the posts under the beams. On a building with ten outside corners, three carrying beams, and 1,600 square feet of deck, this is about as far as we get the first day. This also includes some time for shimming the sill plate, which on a real bad foundation can take half a day. Our day is about nine hours long, which includes a half-hour lunch, a couple of breaks (and a sore back). Some of our breaks include discussions on the merits of a college education and guesses as to which Sears store sold the architect his license.

Second Day

A little stiff this morning but the plywood gets us moving and things go fast. We usually use 5/8-inch tongue-and-groove and install it with a 7-foot 2x4 and a ten-pound maul. Lay one course, shoot it down, lay the next course: Jerry's on the sledge and I work the T&G joint to assure a good fit. Every two courses of plywood we install bridging unless over a beam. I cut off plywood ends and Jerry nails off the deck.

I mark corners for wall plates and the center lines for windows and doors on the plywood deck and rim joists. We both strike the lines for walls and Jerry cuts all the plates (bottom and top), and makes a cut-list for headers, studs, jacks, and sills—all from the first-floor plan. Meanwhile, I am laying out the studs, jacks, and headers. This house is just one story so I lay out roof rafters and trusses on the top plates before the walls go up. We start to build walls even though it's getting late. We usually install sheathing, building paper, soffit nailers, and sometimes the windows before we stand the walls. Arrgh, time to go home. It gets dark early here in Maine!

Third Day

Stand up two walls first thing, then build the connecting walls to eliminate the need for bracing. Finish all the walls except the greenhouse, then line and brace all the walls. Next, we build center partitions or bearing walls and mark the ridge beam for a cathedral ceiling. We're beat—see you tomorrow.

Walsh and Fraser like to travel light and work fast. Staging is primarily ladder jacks and these Type-I aluminum ladders (top). Also in the truck bed is a two-horsepower compressor used to drive a variety of pneumatic nailers (bottom), with two typically running at once.

Fourth Day

We install ten 38-foot trusses, 4/12 pitch—yes, just Jerry and I. We move the trusses into the house upside down and prop up the center with the tails on the outside walls. Jerry uses a push stick to force the center of the truss up to me and I tack it to temporary bracing and go to the next truss. We use Truslock connectors for 16 feet, then install permanent bracing (pre-marked on the deck), and then continue until done.

For the cathedral portion of the roof, we install a 22-foot-long 4x14 ridge beam using wall jacks. We brace it off and set parallel-chord trusses with me on the ridge and Jerry on the side wall. We now begin to lay out and build two gable end walls. It's getting dark—see you in the morning.

Fifth Day

We sheathe in the gables—I usually cut on sawhorses and send to Jerry who nails in the gable studs and sheathing. There's no scaffolding—we work from inside on top of the walls.

Jerry cuts the flying rafters and we stack up plywood for the roof deck

Framing Tips: Fast and Accurate

- Make sure everything is square, level, and plumb right from the start.
- Try to have rough grade completed around the foundation.
- Have lumber delivered as close to the site as possible.
- Use all dimensions from plans and eliminate field measurements.
- Cut as many pieces as possible with the radial saw before each task.
- Try to picture each task and the following task to eliminate mistakes.
- Have one person do layout using master layout sticks or story poles.
- Check each other constantly for accuracy.
- Measure twice and cut once.
- Figure roof-deck cuts on the first floor and send cut and full sheets to the roof.
- When doing a task, complete it and don't leave things for later.
- Do as much as possible without scaffolding, but don't forget safety.
- Strike lines for everything, it saves time.
- Use air nailers—they save much time.
- Get a system of working together, perfect it, and stick with it.
- Set goals for each day: You'll be surprised how much you can accomplish.
- Fix anything that isn't perfect and keep quality your number one priority.
- If you go to work and actually work it will surprise you how fast the day goes. It will also surprise you how well you sleep.

where we can reach it from up top. We make approximate cuts ahead of time and send up full and cut sheets. We strike lines for the first course of plywood and sheathe the first courses together. Then we go for the rest as fast as we can. I trim any edges that have overhangs and Jerry nails off the roof deck. We move trim and miscellaneous lumber inside as we head for home.

Sixth Day

On any pitch of 6/12 or less, I will install all the fascia from the roof and Jerry will cut and help install if I can't do it alone. If I'm waiting for a cut piece of trim, I will change to a roofing gun and install a metal drip edge. Oh my aching back—time to haul shingles. We usually do it all at once to get it over with. Lunch time. Now it's time to paper while Jerry cuts starters for the entire roof. We install half the roofing this afternoon and go home early. We usually paper half the roof and strike lines every 20 inches or so and then shingle.

Seventh Day

Paper and shingle the other half and a couple of small shed roofs. Lunch time again. After that, we cut out the sill plate at exterior doors and install entry and slider doors. Finally, we finish installing windows that didn't go in when the walls were framed. Let's go home.

Eighth Day

Finish wrapping building paper around the corners and attach plywood strips to the rim joists. We then install corner posts to inside and outside corners, and use a story pole to lay out the siding courses. We install siding and soffits at the same time and finish each wall section completely as we go around the building.

Ninth Day

We install the 8x12-foot greenhouse (pre-cut unit). This takes a full day and causes us enough grief that we have to invent new four-letter words for the occasion.

Tenth Day

We can see the light at the end of the tunnel, but it's raining. So we cut and frame interior partitions, taking the rest of the day off.

Eleventh Day

Finish up siding, load up the truck, pick up the check and make like a big bird and fly.

Postscript

I'm getting tired just writing about this. But let me assure you that this is no rush job. It is a system of using plans, pre-cutting everything possible before each task, and then putting the parts together. We constantly check each other for accuracy. A square and plumb building is of utmost importance for things to go smoothly. If something is not just right, fix it immediately or pay for it later. Together the tools, plans, experience, and hard work help us to put up a quality house in a very short time.

Our prices vary, but here are a few examples of what we charge to contractors. We typically get $4 per square foot for a 24x48-foot ranch house, including vinyl siding, soffits, and asphalt shingles. However, we'll probably be adjusting up the base price by about 50 cents this year. With a truss roof system, we'll complete it in eight days. If the plan is more complicated than a simple rectangle, and the siding, soffits, and fascia are wood, we'll add 50 cents per square foot to our base price.

We'll also add 50 cents per square foot for any one of the following: vertical siding, over one story, cutting rafters on site (more for hips, less for partial trusses).

These prices include our own insurance and taxes. If we supply fasteners, we charge accordingly. If we contract with a private individual, we also charge more to cover the additional overhead for everything from selling the job to collecting the checks. ■

Bill Walsh and partner Jerry Fraser specialize in subcontracting custom home shells. Their company, Fraser & Walsh, is based in Aurora, Maine.

Section 2. **Floor Systems**

Courtesy: American Plywood Association

Taking the *Bounce* Out of Floors and Beams

Modest oversizing of joists and beams will minimize deflection and maximize customer satisfaction

by Harris Hyman

Half joking, people often ask of the architect or structural engineer, "Will the building fall down?" Well, when considering light-frame buildings, I can confidently answer, No! Such buildings almost never fall down. But they often have problems with sagging beams or floors: They don't break, but they might bend too much. This bending is called *deflection*.

A common example is an overly bouncy floor. You walk across this floor, and the dishes on the table rattle. The candles flicker and the TV changes channel. You really never will fall through the floor, but you might think that you will. Most important, you just don't feel good in this room.

Deflection Defects

Deflection becomes a real problem when the building is so deformed that it is not useful. For example, the gym roof on one New England school was a little too soft. When the snow piled up, the roof sagged a little—just enough to keep the great folding doors that divided the boys and girls from opening. After the first snow the gym was either divided or open, and stayed that way till the January thaw. Then the snow froze again until mid-March.

A passive solar building in New Hampshire had a sunroom with sloping south glazing on a low knee wall. The roof was a little too soft, and a good winter storm loaded it up, pressed it down, and pushed on the sloping windows. The wall bulged out, and the glazing panels twisted enough for the seals to break. By the time the snow had melted, most of the bulge had receded from the wall, but the double glazing had filled with condensate.

In another building, a contractor had failed to set sufficient fastenings into the major truss supporting the ridge in a fairly expensive and exotic Maine residence. The ridge sagged a little and pushed the rafters down and the walls out. Half a mil worth of new house took on a real Downeast seedy look.

Valley rafters are another problem area. They carry exceptionally heavy loads when, in a winter storm, snow collects in the valley. This rafter is usually a single stick, and it carries the weight of a group of jack rafters nailed into it. Consequently, valleys often sag and do funny things to interior finishes and to the inside corners to which they run.

Aspects of Deflection

There are four problems that can be caused by too much deflection. The designer should take into account: non-functionality of the building (the gym roof); the tendency to cause related damage (the greenhouse); unsightliness (sagging valleys or ridges); and human discomfort (bouncy floors).

Non-functional buildings. The first of the deflection problems is relatively straightforward to solve, but it requires a thorough look at a building's forces, not just the vertical bearing strength. The designer must assess all the possible ways the building can move out of the perfect square that is laid out on the drawings. This tactic requires some design sense.

For the most part, building dysfunction is limited to a few special situations. For example, I might be asked to design a room for a sensitive test device in a scientific lab where any bounce in the floor would cause the test device to malfunction.

Most likely, contractors will recognize the potential for this type of problem and hire an engineer to puzzle it out.

Related damage. The designer must also take a thorough look at the whole building to determine whether deflection in one part might cause damage to some other part. Most related damage from deflection is wreaked upon windows and doors, and analysis of those local areas is usually easy for the designer to handle.

The more difficult problems involve instability, where a part of the building moving out of place upsets an entire building. An example of an instability failure involves a post pushed out of place and a subsequent load bringing a porch down. Fortunately, this kind of problem is rare on light frame buildings.

Unsightliness and discomfort. The last two types of deflection failure are important to residential buildings but are extremely difficult to assess. It's perfectly clear when a building suffers dysfunction and damage. But what is unsightliness? What is discomfort? Deformations that can be detected by the eye alone might be called "unsightly," but we all have different eyes, and

Table 1. Floor Joist Spans

50 Pounds Per Square Foot Total Load

	Doug. Fir	Hem-Fir	S. Y. Pine	S-P-F
2x12 12" o.c.				
stress	24'-8"	22'-0"	25'-1"	20'-6"
deflection	20'-7"	19'-4"	20'-4"	18'-10"
2x12 16" o.c.				
stress	21'-5"	19'-0"	21'-9"	17'-9"
deflection	18'-9"	17'-7"	18'-4"	17'-2"
2x10 12" o.c.				
stress	20'-4"	18'-1"	20'-8"	16'-10"
deflection	16'-11"	15'-11"	16'-7"	15'-6"
2x10 16" o.c.				
stress	17'-7"	15'-8"	17'-10"	14'-7"
deflection	15'-5"	14'-5"	15'-1"	14'-1"
2x8 12" o.c.				
stress	15'-11"	14'-2"	16'-2"	13'-2"
deflection	13'-3"	12'-5"	13'-0"	12'-2"
2x8 16" o.c.				
stress	13'-9"	12'-3"	14'-0"	11'-5"
deflection	12'-1"	11'-4"	11'-10"	11'-0"

100 Pounds Per Square Foot Total Load

	Doug. Fir	Hem-Fir	S. Y. Pine	S-P-F
2x12 12" o.c.				
stress	17'-5"	15'-6"	17'-9"	14'-6"
deflection	16'-4"	15'-4"	16'-0"	15'-0"
2x12 16" o.c.				
stress	15'-1"	13'-5"	15'-4"	12'-6"
deflection	14'-10"	13'-11"	14'-7"	13'-7"
2x10 12" o.c.				
stress	14'-4"	12'-9"	14'-7"	11'-11"
deflection	13'-5"	12'-7"	13'-2"	12'-4"
2x10 16" o.c.				
stress	12'-5"	11'-1"	12'-8"	10'-4"
deflection	12'-3"	11'-5"	12'-0"	11'-2"
2x8 12" o.c.				
stress	11'-3"	10'-0"	11'-5"	9'-4"
deflection	10'-6"	9'-10"	10'-4"	9'-8"
2x8 16" o.c.				
stress	9'-9"	8'-8"	9'-11"	8'-1"
deflection	9'-7"	9'-0"	9'-5"	8'-9"

even a good eye needs a visual reference. On a sloping site, you can walk uphill and sight down a bulge in the eaves line that would never be noticeable from level ground.

Discomfort is even more difficult to gauge. A well-built floor feels good and solid; but how much bounce is tolerable? We do know that a couple of inches of hop is bad; but what about 3/8 inch? Research into comfort is limited. Since health and safety are not affected, this does not have a particularly high priority. No one is injured by bouncy floors; they just make life a little less pleasant.

Designing for Deflection

Despite the obvious and pervasive nature of the problem, we have only one useful standard for designing against deflection. Several model codes limit the allowable deflection of a span to $l/180$ of the overall span (or 1-inch deflection in a 15-foot span) for roofs over unplastered ceilings; $l/240$ for floors over unplastered ceilings; and $l/360$ for floors over plastered ceilings. This is based on an old design standard that suggests the plaster ceiling under the span will crack with any additional sag, and that the primary safety concern is falling chunks of plaster. Even though we don't use plaster ceilings much anymore, this criterion is still used and actually will give a reasonably comfortable floor design.

The best approach, I feel, is to calculate the deflection as well as the strength of a building. It's a little more work, but it's likely to produce a little better building—one that will (almost) never fall down.

The basic reason why buildings don't fall down is that a safety multiplier is applied to materials. Good clean spruce has a true bending strength of about 6,000 to 9,000 psi, but we normally use a design value of only 1,200 psi. This safety factor

Note: *All calculations were based upon design values for No. 2 Grade, 5 inches and wider. The deflection calculations assume a limit of $l/360$. This table illustrates the author's point that joists designed to limit deflection are usually of adequate size to meet the stress (bending strength) criteria.*

of about 5 covers the variation in the natural strength of spruce trees and quality of the wood. By comparison, steel has a safety factor of only 1.6 because it is much more regular in strength and quality.

There is no safety multiplier applied to the modulus of elasticity (E), which is a measure of the stiffness of wood of a particular species and grade. Deflection is computed with realistic values for the properties of wood of a given species and grade. Designing for deflection typically makes a conservative building design, because if a building is strong enough to resist bending, it is not likely to break.

Sizing Joists

With only instinct to justify it, I do use one design check for bounciness on floors: Will a load of 500 pounds distributed over four joists produce a deflection of more than 1/2 inch? The 500 pounds approximates a large man taking a strong step. A half inch just feels reasonable. In general, joists designed for deflection are covered for strength against failure and will usually meet the code requirements.

I also design the floors for living rooms, kitchens, and other public spaces in a house for 100 psf total load rather than the 40 psf live load required by some codes (see Table 1). This is a little overdesigned, but the added expense is relatively small, and it gives a house a good feeling. There is no need for this strength in bedrooms or bathrooms, unless the owners have some unusual proclivities.

Sizing Girders

The builder should also give some serious attention to the girder systems that hold up the joists. Carpenters and builders often *underdesign* girders. The typical triple 2x12 girder *looks* like a lot of wood, and certainly *feels* like a lot of wood. But it may not be enough if the span between posts is too long.

Let's take a typical situation—a 24x24-foot garage with a triple 2x12 girder down the center on a single post. This supports the floor joists for the apartment above. Depending on the species used, this may be a little light for a 40-psf floor load. It is very light for the 100 psf I suggest for residential public spaces. A second column wouldn't cost much and would stiffen the place a whole lot. For the post spacing I would use for 100 psf loads, see Table 2.

Table 2. Maximum Post Spacing for 100 psf Floor Loads

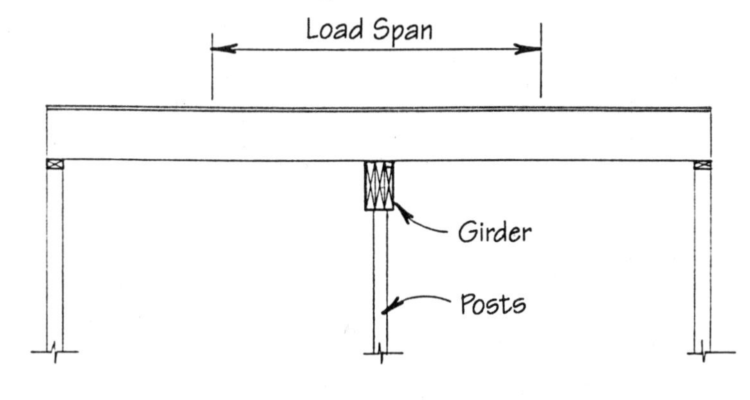

Load Span	Girders		
	Double 2x12	Triple 2x12	Quad 2x12
30'	4'-9"	5'-5"	6'-0"
24'	5'-2"	5'-11"	6'-6"
20'	5'-6"	6'-3"	6'-10"
16'	5'-11"	6'-9"	7'-5"
12'	6'-6"	7'-5"	8'-2"
10'	6'-11"	7'-11"	8'-9"

Note: *This table shows maximum post spacings for common girder sizes, where total floor loads are figured at 100 pounds per square foot (psf). The author uses the 100 psf value in designing girder systems for high-use residential spaces, such as living rooms and kitchens. If girders are built of 2x10s, reduce the post spacing in the chart by 20 percent.*

You don't usually have to worry about deflection in sizing girders. As it turns out, when a girder is designed to carry a load, it is almost always okay in deflection.

Creep

Creep occurs when, over time, deflection causes beams to be permanently deformed. But it just doesn't happen much. Constant loads that are very close to the design limits are necessary to produce creep. To limit deflection over time, the *Wood Handbook*, by the U.S. Forest Product Lab, Madison, Wis., recommends designing for about one-half the deflection ordinarily permitted for longtime deflection. The *Wood Handbook* is a good reference on the structural properties of wood.

A Soft Problem

You've noticed that I've used a variety of "soft" terms in discussing deflection: "instinct," "tolerable," "reasonably comfortable," etc. Well, the problem of deflection *is* soft, because it is often a question of judgment. Guidance by code is limited and in a single dimension, the vertical, while buildings actually move in four dimensions. Like most messy technical solutions to messy real-world problems, design for deflection really demands art and good sense in addition to number crunching. Good luck. ■

Harris Hyman is a civil engineer in Portland, Ore.

Out on a Limb With Cantilevers

How to get the dramatic effect without the bouncy floor

by Harris Hyman

It's taken years to finally admit it to myself, and then actually to others, but I belong to the Macho School of Structural Engineering. The basic principle of this school is, "Columns are for sissies." This approach, as long as I don't say it outright, pleases my architect clients who want the spaces the way they want them. It also frightens and irritates the builders, who sometimes worry about strength.

Cantilevers—beams and decks that hang out into space with no visible means of support—seem to fall into the Macho School. Their principal advantage is aesthetic. They are used in a variety of common situations, principally for decks, floor systems, and roof overhangs (Figure 1, next page). As long as they are quite short, they seem to work well. They are also subject to some unusual structural problems.

Bouncy Decks

Any beam can have three types of failure: deflection, bending, and shear. Of these, the most common is deflection, where the beam system is just too soft to do a proper job and either feels weak and bouncy or moves enough to interfere with something else. Shear failure occurs when the beam crushes or tears itself apart, usually down the centerline. Bending failure occurs when the beam breaks, and this is relatively rare, even with supported beams. Perversely, we usually design for strength in bending.

Cantilever beams are particularly bad with shear and deflection. Let's look at an 8-foot overhanging deck made of 2x8s, 16 inches on-center. If the deck is loaded with 50 pounds per square foot, it will have a deflection of 1 inch at the outer edge.

If the outer edge is supported with a post, then the sag in the deck is only 1/10 inch, about 10 percent of the sag with the cantilever. Try 2x12 joists instead of 2x8s: These reduce

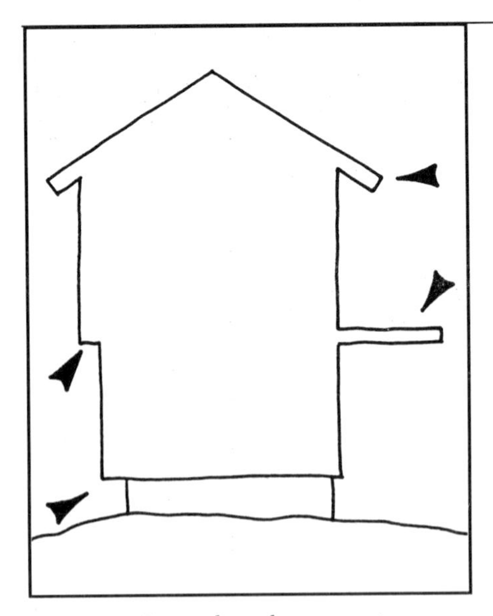

Figure 1. *In residential construction, cantilevers are used mostly for decks, floor systems, and roof overhangs.*

the outer-edge deflection to 1/4 inch. This is still not as good as supporting the outer edge, but it is probably good enough for a deck.

The width of the deck (length of the span) is quite important. The outer-edge deflection of cantilever decks increases with the fourth power of the overhang. This means that a deck that is twice as wide has 16 times the deflection. If the width of our deck were reduced from 8 feet to 6 feet, the outer-edge deflection would drop from 1 inch to 3/8 inch.

My personal criterion on deflections is that about 1/4 inch to 3/8 inch is perceptible, and that 1/2 inch is noticeable, but okay for exterior decks. This leads to the following table, which assumes joists 16 inches on-center, people walking around, and no outrageous loads like stoves, water beds, or hot tubs:

Maximum Cantilever Deck Span

Joist Size	Indoors	Outdoors
2x6	60 in.	66 in.
2x8	72 in.	81 in.
2x10	88 in.	96 in.
2x12	102 in.	114 in.

Deflection isn't the only problem we face, however, with cantilever decks. There is also shear. All of the load is concentrated at the root of the beam. This portion of the beam feels the force that the load exerts downward on the beam right next to the equal and opposite reaction force the building exerts up on the beam (Figure 2).

If the table above is used with ordinary construction-grade spruce, shear should not be a problem. But over the past few years, we've begun to rely more and more upon built-up systems, like Trus-Joists and rectangular floor trusses. Are these strong enough? In ordering rectangular floor trusses, the contractor should *always* submit plans to the truss fabricator and check the shop drawings they provide prior to building the trusses. Most of the truss shops have a computer and plotter on the premises. And if there is a time crunch, it is often possible to go directly to the fabricator and wait while the plotter draws out the plans. The point of checking the plans is to verify that there is plenty of vertical support at the root of the cantilever, so that an "off-the-rack" truss won't be crushed as in Figure 3.

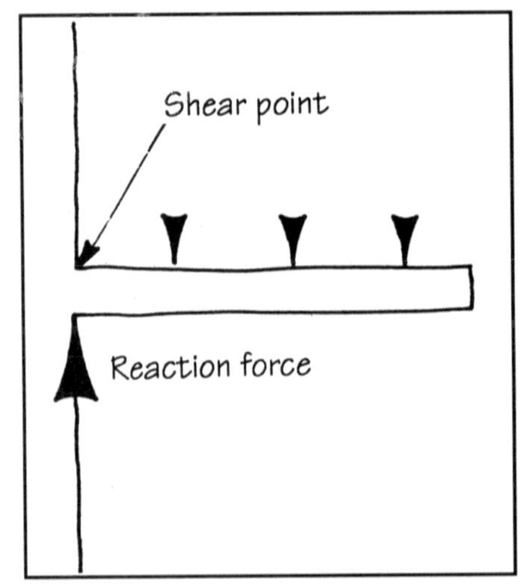

Figure 2. *All of the load on a cantilevered beam is concentrated at the root of the beam, which is loaded in shear. Excessive deflection at the outer end of the beam can also be a problem.*

I was unable to find the shear capacity of Trus-Joists, and from some cursory calculations I suspect that they should be reinforced with shear plates at the root of the cantilever. It would be wise to verify the shear capacity with the manufacturer, which is best done by sending in plans. This minimizes the opportunity for miscommunication.

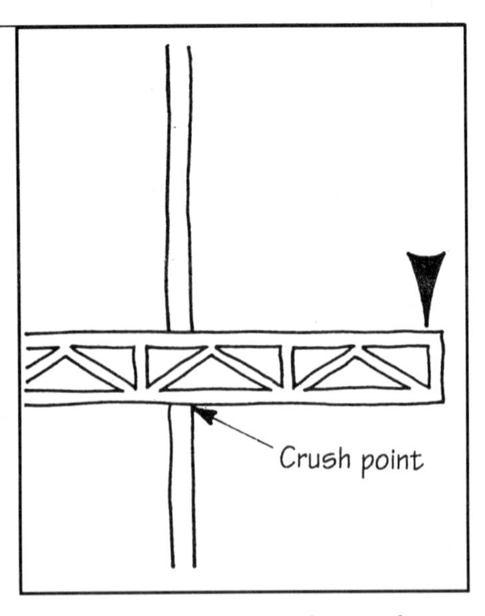

Figure 3. *Floor trusses often need extra vertical support at the root of a cantilever. This should be engineered by the truss fabricator.*

Hold Them Down

The next problem with cantilever beams is holding them straight. The roots of these beams must be embedded in the structure, and this is usually done by extending them across the building as floors as in Figure 4. An old builders' rule of thumb states that the cantilever extension should not exceed one quarter of the total length. This rule is somewhat conservative, so for most situations it will do the job. The conservative nature of the rule suggests that some careful engineering can push back this edge. The integration of the inside of the cantilever into the building structure can conceivably allow a much longer extension than the 25 percent allowed by the rule.

Insulation and Vapor Barriers

The last problem with cantilever decks is thermal protection. The beam structure extends from the inside of the building to the outside, making it difficult to get a continuous insulation layer, vapor barrier, or even an infiltration barrier. So, here is a

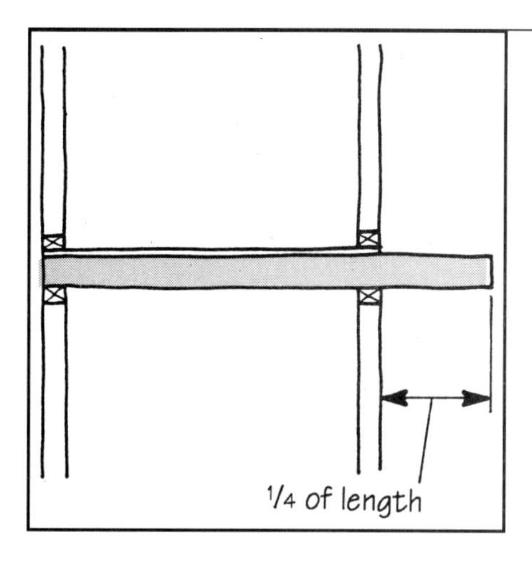

Figure 4. *For solid anchorage, a conservative rule of thumb states that the cantilevered section of a beam should not exceed one quarter of its total length.*

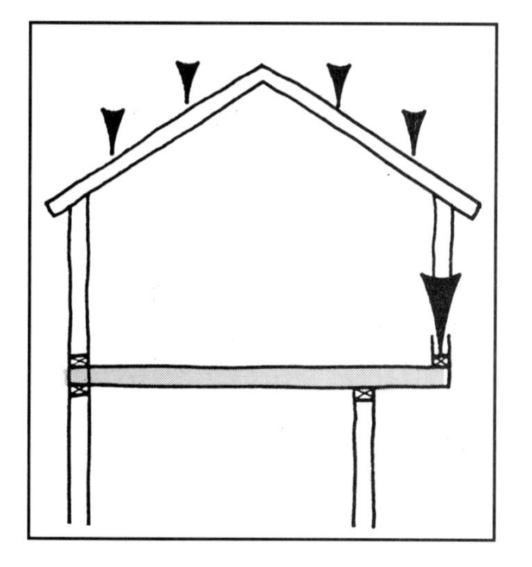

Figure 5. *In Garrison-type houses, the cantilever bears excessive shear loads, since it supports roof and wall loads. Use at least 2x10s, or use fabricated joists with caution.*

request to you builders and designers out there for some help in detailing an elegant insulation scheme for the root of a cantilever deck.

Garrisons, Bumps, and Beams

There are a couple of other cantilever situations that demand attention: building overhangs and beams. The style often called the "Garrison" is a typical example of overhang. It is also common to use cantilevers under bulges like bow windows, and even to extend the bow windows up to the second floor and the roof.

Deflection is rarely a factor for overhangs, but shear can be quite serious. In the Garrison (Figure 5), the cantilever bears not only the interior load of the building, but also the weight of the wall and half the load on the roof. Two-by-tens are the smallest joists that should be used on a Garrison or similar structure, with total building widths up to 24 feet. The joists on this type of building should be long and carefully tied into the rest of the structure. Here, even more than with decks, extreme care should be used in working with fabricated joists, since our experience with excessive shear loads has been gained with full dimensioned lumber—not with the new built-up products.

Bulges, unless they extend up to support the roof eaves, should present few problems for cantilevers. They are usually small, and generally they are lightly loaded. If the bulge requires joisting against the direction of the floor system, the bulge width should not be more than one third of the length of the cross joists (Figure 6).

Special beams, steel, and ply-beams are more complex in cantilever situations. Steel beams derive their economy from placing the weight of the steel out by the flange, leaving a relatively weak web. When there are high shear loads on a beam, they must be reinforced with welded (or sometimes bolted) shear plates at the support points. These plates have the effect of thickening the web for a short portion of the beam where the shear load is high. Most steel yards have some fabrication facilities, and can deliver a beam cut to length, with shear plates welded into place. So the use of such beams should be no problem for a contractor.

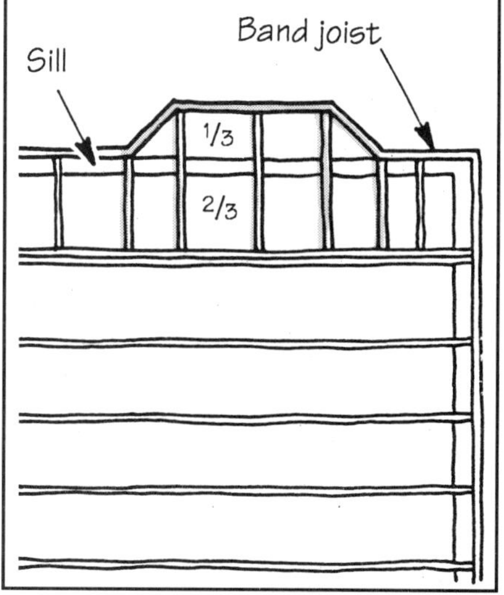

Figure 6. *Bow windows and other bumps are easily framed with cantilevers. In general, the overhang should not exceed one third the total length of the joists when framed perpendicular to the floor system.*

Ply-beams are similar, with a plywood web and flanges of dimensional lumber. At the support points, the flanges should have backing pieces (Figure 7) to support them in shear, strengthening the relatively thin plywood webs.

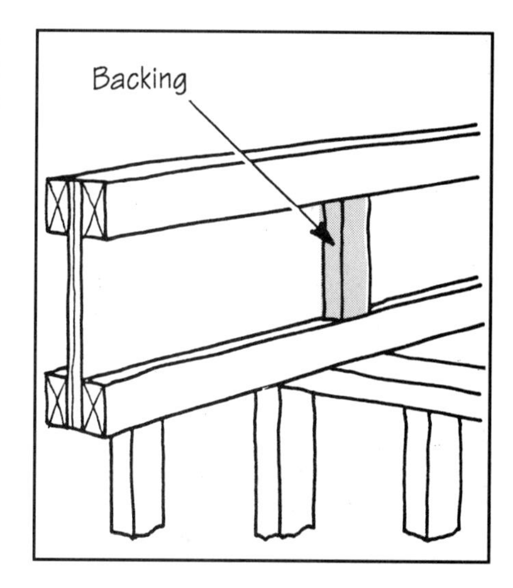

Figure 7. *When used in cantilevers, plywood I-beams must have backing pieces to strengthen the thin plywood web at the shear point. The same holds true for steel I-beams.*

Reinforced concrete beams, both straight and cantilever, are real specialty items. They are often prefabricated and require engineering, either privately or by the fabricator. These can be used with marvelous effects, as on George Howe's Clara Fargo House in Maine, which is one of the very best use of cantilevers, ever.

The basic question for the builder to ask with a cantilever is, "Is this really necessary?" If it is necessary—for economic, site related, or aesthetic reasons—think it through and work carefully. ■

Harris Hyman is a civil engineer in Portland, Ore.

Underlayments For Resilient Flooring

Resilient flooring needs the right underlayment installed the right way

by Paul Fisette

Installing floor underlayment is routine for most builders: Order thin particleboard or plywood and fasten it to the subfloor. But careful selection and fussy workmanship become more critical when the finish surface will be vinyl tiles or sheet goods. Resilient flooring is thin and subsurface irregularities are readily transmitted through thin vinyl. Nail-pops, swollen wood fibers, and deformed joints are the villains of the resilient underworld.

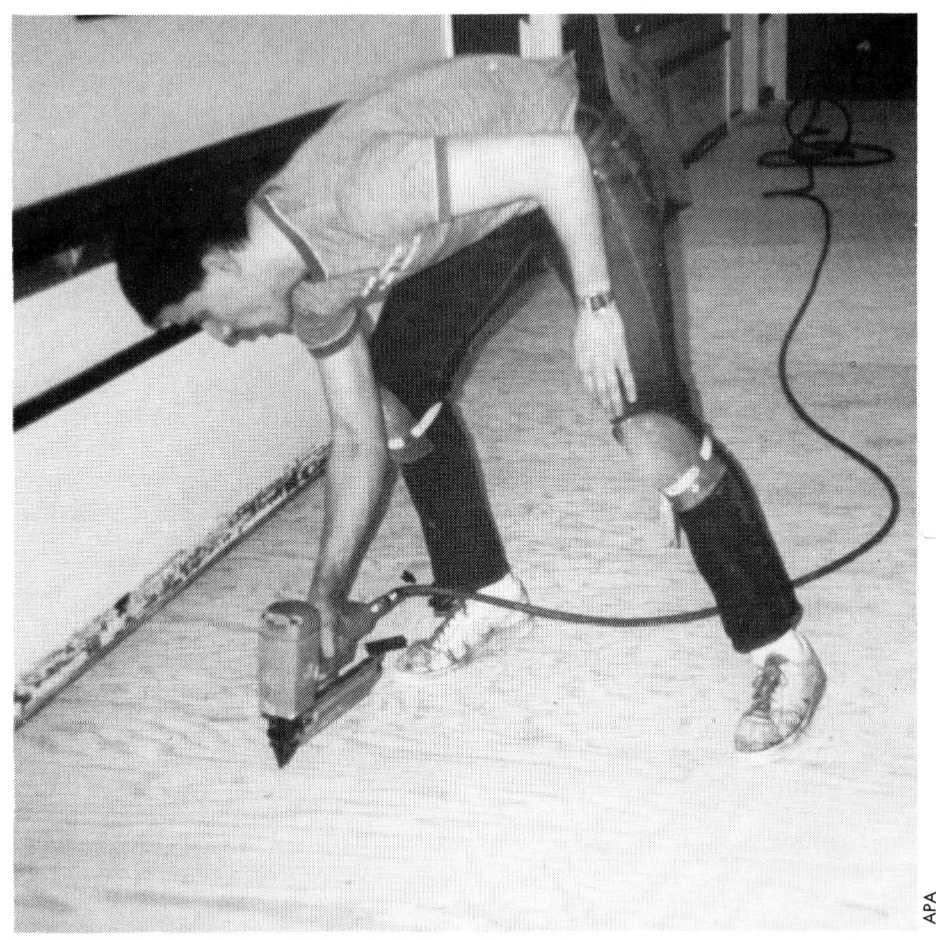

APA

Plywood is a good underlayment choice for resilient flooring. The right grade of plywood has a fully sanded face and no large voids in the inner plies.

Subfloor Vs. Underlayment

The terms "subfloor" and "underlayment" are often used interchangeably around the job site, but there is a world of difference between the two. A subfloor is a structural layer intended to provide support. It serves as the working platform and takes the abuse during construction. CDX plywood or oriented-strand board (OSB) are the most common subflooring materials.

Underlayment, on the other hand, is installed over a subfloor to create a smooth, durable surface to support the finish flooring. Underlayment should be installed just before the finish floor to avoid surface damage. Hardboard, particleboard, self-leveling concrete, plywood, and at least one OSB product offer the smooth, hard surface required for thin resilient flooring.

Some hybrid systems combine subflooring and underlayment functions in one layer, most notably American Plywood Association's (APA) Sturd-I-Floor.

Resilient flooring demands a lot of any underlayment. The underlayment must be very hard, smooth, dimensionally stable, and stiff. But few of the commonly used underlayment materials can live up to all these tasks in all situations.

Hardboard

Hardboard provides a thin, hard, smooth surface and, despite an increasing acceptance of newer materials, it is still a common choice for underlayment among remodelers. In fact, the Resilient Floor Covering Institute (RFCI), a Rockville, Md., association representing seven of the largest resilient flooring manufacturers, recommends hardboard as one of only two acceptable underlayments for resilient flooring (the other choice is an appropriately graded plywood).

The product the RFCI recommends, however—a .215-inch, Class 4, service-grade hardboard—is hard to get at the lumberyard. Few manufacturers make it any longer, so many builders may be using the wrong kind

Table 1: Recommended Underlayment Thicknesses and Nailing Schedule

Underlayment	Subfloor	Underlayment Thickness	Fastener (should be positioned $1/2$ in. from panel edges)	Nail Spacing (in.) along panel edges	Nail Spacing (in.) within field both directions
plywood	plywood ($1/2$ in. and thicker)	$1/4$ in.	$1\,1/4$ in. (3d) underlayment nails	3	6
plywood	plywood ($1/2$ in. and thicker)	$3/8$ to $1/2$ in.	$1\,1/4$ in. (3d) underlayment nails	6	8
plywood	plywood ($1/2$ in. and thicker)	$5/8$ in. to $3/4$ in.	$1\,1/2$ in. (4d) underlayment nails	6	8
plywood	plywood ($1/2$ in. and thicker)	$1/4$ in.	18 g staples x $3/16$ in. x $7/8$ in.	3	6
plywood	plywood ($1/2$ in. and thicker)	$3/8$ in. and thicker	16 g staples x $3/8$	3	6
plywood	boards up to 4 in. wide	$1/4$ in.	$1\,1/4$ in. (3d) underlayment nails	3	6
plywood	boards 4 in. and wider	$3/8$ in.	$1\,1/4$ in. (3d) underlayment nails	6	8
particle-board	plywood ($5/8$ in. and thicker)	less than $3/8$ in.	$1\,1/2$ in. (4d) underlayment nails	3	6
particle-board	plywood ($5/8$ in. and thicker)	$3/8$ in. to $5/8$ in.	2 in. (6d) underlayment nails	6	10
particle-board	plywood ($5/8$ in. and thicker)	$1/4$ in.	18 g staples x $3/16$ in. x $7/8$ in.	3	6
particle-board	plywood ($5/8$ in. and thicker)	$3/8$ in.	16 g staples x $3/8$ in. x $1\,1/8$ in.	3	6
particle-board	plywood ($5/8$ in. and thicker)	$1/2$ in. to $5/8$ in.	16 g staples x $3/8$ in. x $1\,5/8$ in.	3	6
particle-board	boards less than 8 in. wide	same as for $5/8$ in. plywood subfloor	16 g staples x $3/8$ in. x $1\,5/8$ in.	3	6
hardboard	all of the above subfloors	0.215 in.	$1\,1/4$ in. (3d) underlayment nails	3	6
OSB	plywood or boards, at least $5/8$ in. thick	$1/4$ in.	$1\,1/2$ in. (4d) underlayment nails	4	6

of hardboard underlayment. And, in spite of the RFCI's recommendation, many resilient flooring manufacturers flat out do not allow any hardboard under their fully adhered floors. They cite its inadequate uniformity, poor dimensional stability and variable surface porosity as reason for avoiding its use. Other manufacturers allow hardboard underlayment only for perimeter-bonded systems.

Hardboard is formed by densely packing wood fibers with heat and pressure. Because the material is dense and smooth, it affords great impact resistance which is needed in underlayment. Unfortunately, the tightly packed fibers also make it unsuitable in moist environments.

Product standards allow a 1/4-inch service-grade hardboard to swell as much as 30 percent (.075 inch) when subjected to a 24-hour soak test. By comparison, 1/2-inch particleboard is allowed to swell 12 percent (.063 inch). There is no product standard specifying allowable swell for plywood, but since the veneers are peeled from logs in a rotary fashion the thickness swelling of plywood mimics radial swelling of solid wood: about 2 percent to 4 percent. Since swollen panels can create unsightly bumps in the surface of the resilient flooring, hardboard should not be used in bathrooms where it might be exposed to high humidity and overflow from toilets and bathtubs.

Nailing can create problems with hardboard, too. When you drive a nail into a piece of lumber, the wood fibers around the nail are compressed into voids within the board. But because hardboard is very dense, when you drive a nail into it, the detached fibers will either be forced beneath the panel, causing a bump, or the fibers will mushroom out of the top near the nail head. These irregularities are easily telegraphed through thin resilient flooring. To prevent this, you need to use thin shank nails and pre-drill nail holes.

If you choose hardboard underlayment, the panels should be stood on their edges and spread around the room for 48 hours prior to application. This will give the panels a chance to adjust to ambient conditions. When installing the sheets, follow the nailing schedule shown in Table 1 and the placement specifications shown in Figure 1.

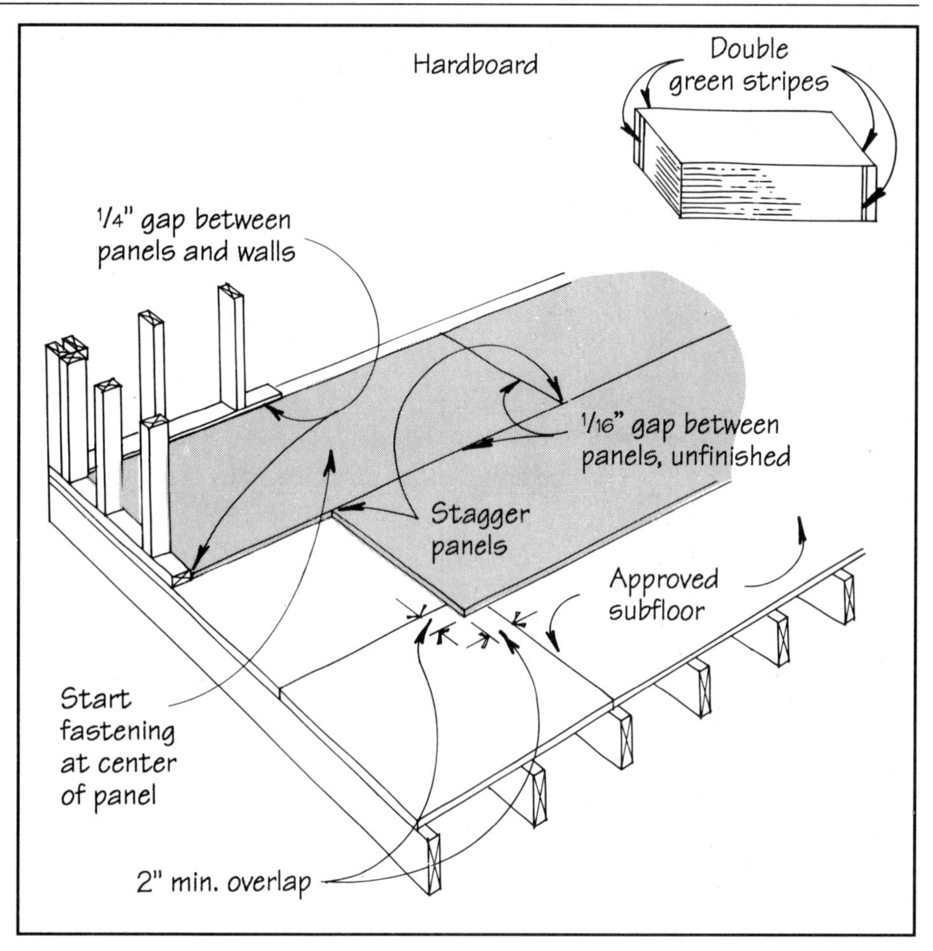

Figure 1. *The only hardboard underlayment approved by resilient floor manufacturers is a Class 4, service-grade panel. This grade is indicated by a double green stripe near the corners of a stack of panels (inset). To install, start nailing at the center of panels and work toward the edges. Leave a 1/16-inch gap at all joints.*

Particleboard

Particleboard is smooth, knot-free, and hard. It has no core voids and has great impact resistance. Because of this, many builders feel that particleboard is pretty darn hard to beat. In fact, about 1/2 billion square feet is installed each year as underlayment under resilient and non-resilient flooring. So why doesn't the RFCI recommend its use for fully adhered sheet vinyl or tile floors?

Thickness edge-swelling is the number-one complaint when it comes to particleboard installed under resilient flooring. Particleboard will take on moisture at its edge first, creating ridges in the finish flooring. Like hardboard, particleboard should be excluded as an underlayment option for bathrooms. And if vinyl tiles are chosen over sheet goods, particleboard should be excluded because the many seams in the tile floor could expose the underlayment to too much water.

Some sheet vinyl installers report another common problem with particleboard: during the "pull-back" operation, wood particles tear from the surface of particleboard. Normally adhesive is spread directly onto the underlayment except at the point where two adjacent vinyl sheets will meet, leaving a "dry zone." Once the flooring is positioned on the adhesive, the edges between adjacent sheets are "pulled-back" so that adhesive can be spread under the seam. At this point, wood particles are often pulled from the board, creating an irregular surface and a poor bond.

Rich Margosian, general manager with National Particleboard Association (NPA) claims, "The biggest problems are usually related to installation." Examples of this include laying the floor before the structure is enclosed or over unventilated crawlspaces without a vapor barrier. Once the particleboard gets wet or is installed over a wet subfloor, it's a losing battle. Improper on-site storage is another common problem, according to Margosian. Particleboard should be stored flat in a dry location. Panels should remain offsite until they are needed—this should be long after the concrete and plaster have dried.

If particleboard is used, a glue-nail

fastening system will produce the best results. White carpenters glue, not subfloor adhesive, is recommended by the NPA. Spread the glue onto the subfloor using a paint roller and then nail down the panels. Standard placement procedures are shown in Figure 2.

OSB

While there are over a dozen APA-approved oriented strand board (OSB) subfloor and sheathing panels, there is no approved OSB underlayment. Only one manufacturer, Weyerhauser, seems to be seeking APA approval for their 1/4-inch OSB underlayment, "*Structurwood.*" And it seems that a mere procedural technicality blocks the endorsement. The APA has not yet developed a standard for non-plywood underlayment, but hopes to have one in place by the end of the 1991.

Several of the large resilient flooring manufacturers have tested and approved Structurwood for use under fully adhered and perimeter-bonded floors. But perhaps it has been Weyerhauser's marketing savvy that has earned the product's widespread acceptance in the field. Weyerhauser installed and monitored test floors for four years before selling the first sheet of Structurwood. And since bringing it to market, the company has assumed responsibility for the performance of its product with a one-year warranty: They promise to pay for the repair or replacement of the damaged floor including underlayment, adhesive, floor covering, and labor—providing the product was professionally installed and handled in strict accordance with the installation guide (see Figure 3).

Surface smoothness can be a problem with OSB underlayment because strands lying next to each other may shrink and swell differently, causing an irregular surface that will telegraph through thin resilient flooring. Weyerhauser claims to have licked such problems with a proprietary stabilizing and conditioning process. They insist Structurwood uses a different surfacing process and greater fiber density so the panels remain smooth.

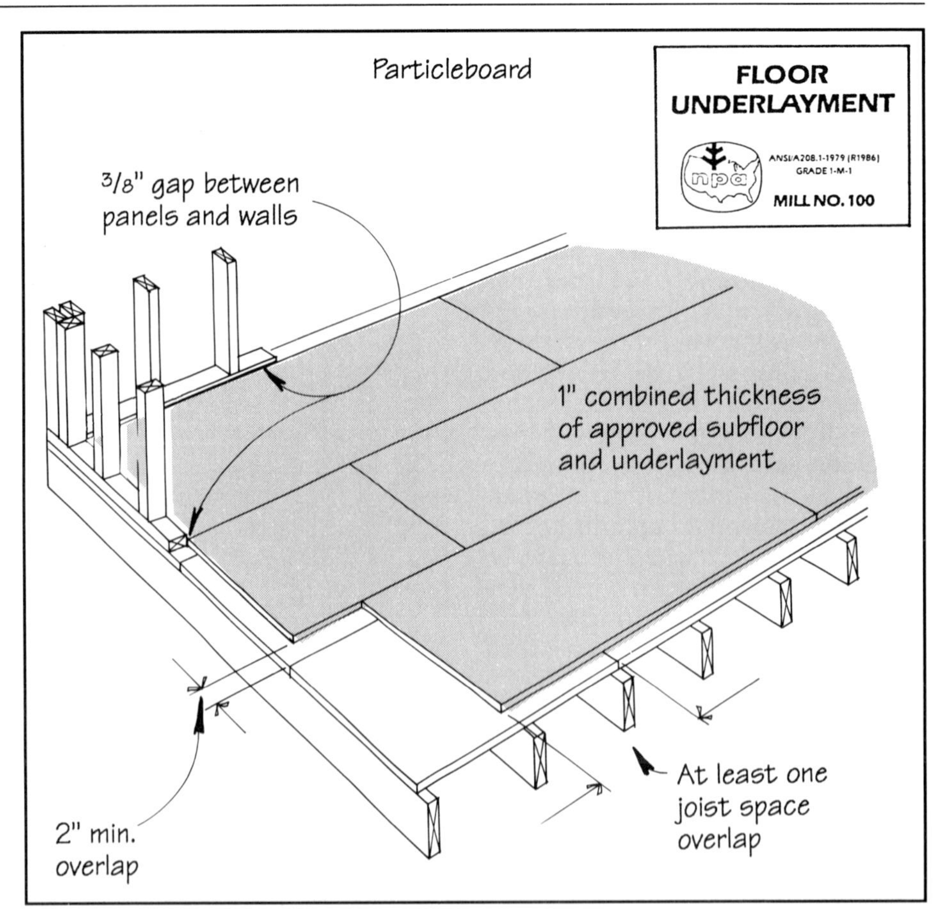

Figure 2. *Particleboard underlayment should carry the National Particleboard Association grade stamp (inset) on the face of the panels, and the combined thickness of the subfloor and underlayment should be one inch. Also be sure to protect the installed underlayment from moisture.*

APA Plywood

Plywood gets a clean bill-of-health from virtually everyone. As one builder put it, "I ask for premium-grade plywood underlayment. I want a fully sanded face and a plugged second ply. Why should I screw around?" Tried-and-true is appealing in an environment where everyone wants to blame the other guy for problems that might arise. All resilient flooring manufacturers approve the use of appropriately graded APA plywood under all types of resilient flooring, provided it is installed correctly (see Figure 4).

All APA approved underlayments are classified as either Exposure 1 or Exterior. Exposure 1 will survive limited exposure to the elements like those experienced during a construction project. Exterior indicates the use of an exterior phenolic resin. These panels can withstand repeated exposure to moisture and are the best choice for bathroom floors.

Approved underlayments for resilient flooring can be identified by a grade stamp which says either "underlayment" or "plugged crossbands" (see Table 2). The APA insists that underlayment plywood have a *fully* sanded face. PTS (plugged and touch-sanded) plywood is only sanded in spots, so it is not as desirable. In addition, the crossbands (the inner plys between the face and back veneers) must have plugged knot holes and voids, or knot holes limited to one inch. This specification is meant to eliminate the possibility of a concentrated load, such as a high heel, from puncturing both the finish floor and the face veneer.

Lauan Plywood

Quarter-inch lauan plywood is commonly used as an underlayment and it meets with fair success. All resilient floor manufacturers allow its

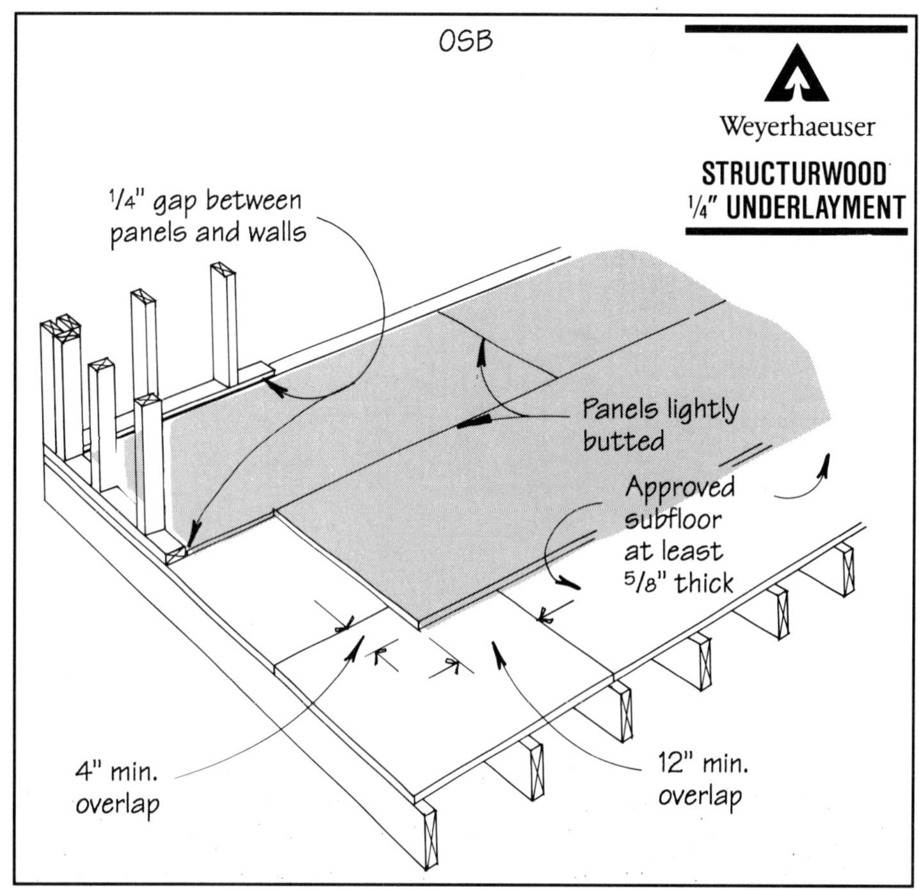

Figure 3. *Weyerhauser's "Structurwood" is the only OSB underlayment approved by resilient flooring manufacturers. The 1/4-inch panels should be laid over an approved subfloor that is at least 5/8 inch thick. Look for the stamp (inset).*

use under at least some resilient floor applications and some manufacturers even recommend it as an underlayment for all applications. Lauan owes its popularity to a solid track record and wide-spread availability. Lauan plywood should be installed following the guidelines issued for 1/4-inch APA plywood.

Still, builders should be wary of lauan. Lauan is made under a wall-panel specification of the International Hardwood Products Association (IHPA). No one accounts for its performance as an underlayment.

The most common problem with lauan as an underlayment arises from not specifying the correct type of material. There are two types of lauan: Type 1, with an exterior glue, and Type 2, with a water-resistant glue. Only Type 1 should be used under resilient flooring to prevent delamination of the plys from moisture, especially when using a water-based filler. The "type" is stamped on the very edge of the panel, usually on the 4-foot edge. But you will have to look closely since the panels are only 5.2 mm or 5.5 mm thick (1/4 inch = 6.35 mm). The panels are metric because they are imported.

Three face-grades are available—BB, CC, and OVL (overlay). Knot holes 3/4 inches in diameter are permitted in the face of the worst grade (OVL) if they are puttied. But take note: The putty might stain or react with the finish floor. Problems with yellowed vinyl have arisen in the past over puttied patches in domestic plywood panels. The APA claims these problems were related to only one patching compound which is no longer used with APA-approved underlayment. But in lauan plywood, which is not designed to serve as an underlayment, similar problems could arise. It is ironic that many resilient flooring manufacturers endorse lauan, even though no manufacturing specifications support its use as an underlayment.

Don't Fill Nail Holes

Whatever underlayment you use, all gouges, gaps, and chipped and sunken edges must be filled with a

Table 2: Recommended Plywood Grades for Underlayment

GRADE	EXPOSURE DURABILITY CLASSIFICATION	LOOK FOR THESE SPECIAL NOTATIONS IN PANEL TRADEMARK
APA Underlayment	Exposure 1	Sanded Face
APA C-C Plugged APA Underlayment C-C Plugged	Exterior	Sanded Face
APA A-C	Exterior	Plugged Crossbands Under Face
APA B-C	Exterior	Plugged Crossbands Under Face
APA A-D	Exposure 1	Plugged Crossbands Under Face
APA B-D	Exposure 1	Plugged Crossbands Under Face
APA Underlayment A-C APA Underlayment B-C	Exterior	Sanded Face

Adapted from ASTM F 499

Figure 4. *Plywood underlayment should be laid perpendicular to the floor joists. Space end joints 1/32 inch apart and don't locate them over joists.*

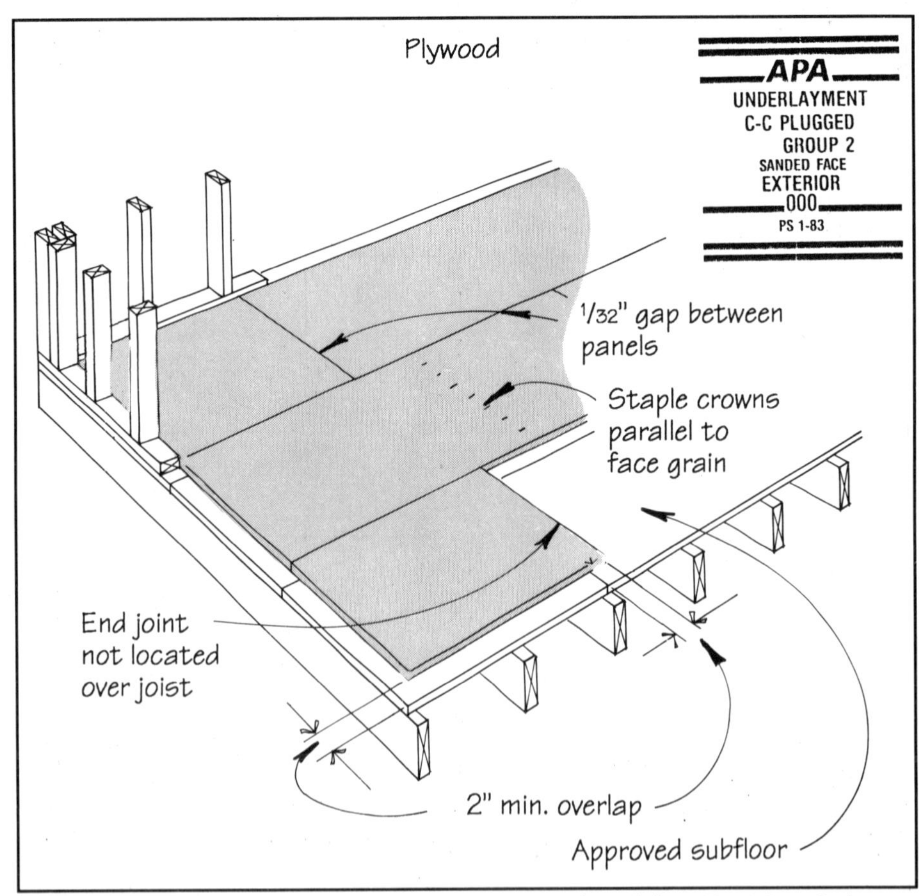

patching compound and sanded flat. Most resilient floor manufacturers specify a portland cement-based compound mixed with latex. Other fillers that are readily available but less effective include plaster compounds like calcium sulfate, plaster-of-Paris or gypsum mixed with a latex binder. With the exception of hardboard, all the joints between panels must be filled and leveled, too. *But don't fill nail holes.* The American Plywood Association, the National Particleboard Association, and the American Hardboard Association all agree that underlayment nails should not be set and filled. The reason: If a nail works loose, it can force the patching compound up and form a small bump in the finish floor. Instead they recommend driving the nail heads *just* below the surface of the plywood and leaving the holes unfilled. Obviously it's going to be tough to drive the nails just right every time. In the case of a grossly overdriven nail or deep rosette in the underlayment surface, the associations suggest it should be filled. This leaves a lot to the discretion of the builder.

Self-Leveling Concrete

One of the newest players in the underlayment game is self-leveling concrete. Self-leveling underlayments are an option for builders and remodelers working with out-of-level or bumpy subfloors (see Figure 5). These products are cementitious blends that are either gypsum-based or portland cement-based (see For More Information, below).

Gypsum-based materials can be pumped to depths ranging from 3/4 inch to 3 inches. Portland-based materials are poured at much thinner 1/8-inch to 1-inch depths. Portland-

For More Information

Trade Associations

American Hardwood Assoc.
1210 W. Northwest Hwy.
Palatine, IL 60067
708/934-8800

American Plywood Association
P.O. Box 11700
Tacoma, WA 98411
206/565-6600

Hardwood Plywood Manufacturers Assoc.
P.O. Box 2789
Reston, VA 22090
703/435-2900

International Hardwood Products Assoc.
P.O. Box 1308
Alexandria, VA 22313
703/836-6696

National Particleboard Association
18928 Premiere Court
Gaithersburg, MD 20879
301/670-0604

Resilient Floor Covering Institute
966 Hungerford Drive, Suite 12-B
Rockville, MD 20850
301/340-8580

Cementitious Underlayments:

Ardex, Inc.
630 Stoops Ferry Road
Coroapolis, PA 15108
412/264-4240

Gyp-Crete Corp.
P.O. Box 253
Hamel, MN 55340
612/478-6072

Hacker Industries
1501 Westcliffe Dr., Suite 310
Newport Beach, CA 92660
714/645-8891

Quikrete
1790 Century Circle
Atlanta, GA 30345
404/634-9100

Thoro System Products
7800 N.W. 38th St.
Miami, FL 33166
305/592-2081

Structurwood:
Weyerhauser Company
2000 Frontis Plaza Blvd., Suite 101
Winston Salem, NC 27103
919/768-5533

based materials are harder and more water-resistant than gypsum-based products. Typically the gypsum-based materials are less expensive but must be installed by licensed applicators. Both types of underlayments are lightweight so floor framing does not usually have to be strengthened.

Self-leveling underlayments cure to form a smooth, seamless surface that is ideal for resilient flooring. Moreover, they are inorganic, non-combustible materials that do not produce smoke or fuel a fire. In fact, they restrict the spread of fire and carry a one-hour fire rating. They also create good sound barriers by filling voids and cracks in the subfloor and beneath the wallboard and adding mass. Some manufacturers even claim that their products help control insect penetration and air infiltration. So finally, we have a product that resilient flooring manufacturers have to embrace—right? Not exactly. Several flooring manufacturers claim that they have had problems with self-leveling underlayments.

A technician in Tarkett's technical services department indicated that mold and mildew are associated with the use of these products. Tarkett also thinks that some gypsum-based products are not hard enough to support concentrated loads. They recommend a minimum density of 100 pounds per cubic foot (pcf), whereas many cementitious products are only rated at about 85 pcf. As a result, Tarkett does not recommend these products as underlayments for resilient flooring.

Congoleum's technical services representative, Bob Dempsey, also cautions against the use of lightweight or gypsum-base concrete as underlayments for resilient flooring. He has received complaints blaming cementitious underlayments for denting easily and damaging the flooring. Dempsey does not think the problems are inherent to the underlayment material, but rather a result of poor job-site control. Adequate drying time is critical so that moisture, which can soften the concrete and stain the vinyl, will not become trapped beneath the floor covering. Congoleum does not approve or disapprove of its use. Armstrong takes a similar stance, passing all liability on to the underlayment manufacturer or installer if it is used.

Figure 5. *Self-leveling concrete underlayment is a good choice for renovating bumpy, out-of-level, or damaged floors.*

Patrick Giles, quality assurance manager for Gyp-Crete Corporation, says, "The number-one problem associated with our product with regard to resilient flooring is that floor-goods mechanics do not run a moisture test before installing the vinyl flooring." In good climate conditions, it takes Gyp-Crete five to seven days to completely dry. After five days, Giles recommends that you tightly tape a 2x2-foot piece of polyethylene to the floor and leave it for 48 hours. If condensation collects on the underside of the poly, the underlayment needs to dry more.

Specifying and installing underlayment for resilient flooring seems like a game of dodge-the-bullet. Everyone wants to pass the buck. The safest course to follow is one paved by common sense. Know what you are asking your materials to do and understand how your materials will perform. And, oh yes—make sure you follow the manufacturer's recommendations so you won't get the buck passed to you if a problem does arise. ■

Paul Fisette is a wood technologist and director of the Building Materials Technology and Management program at the University of Massachusetts in Amherst, Mass.

Section 3. **Wall Framing**

Paul Spring

Fast & Accurate Wall Framing

Production speed doesn't have to mean a sloppy frame

by Don Dunkley

I frame high-end custom homes in California's suburbs, and I pride myself on turning out "clean" work. But I wouldn't be in business around here very long if I weren't fast. With my three-man crew, I'm expected to frame and sheathe a 3,000-square-foot custom home with lots of high rake walls and a complex roof in about one month.

That kind of speed comes from using production techniques. I learned them in the "tracts"—the West Coast's large, suburban developments that have been the breeding ground for these innovative techniques for more than 40 years. Contractors unfamiliar with these methods tend to think of them as inherently sloppy and out of control. On the contrary, each move is orchestrated to achieve efficiency.

The key is to learn the points in the process where exercising a little more care will allow you to really "crank" the rest of the time and still produce a tight frame. Here are some of the places where I insist on accuracy—and the techniques I use—in wall framing.

Following careful layout, walls are banged together and raised with assembly-line speed.

Laying It All Out

Production framing loads the "thinking" part of the job into the early stages—floor layout, plating, and detailing the plates—so that your crew can concentrate on pure speed later on. The first task is to establish a benchmark on the deck for square.

Snapping lines. Even if I built the floor, I no longer just hook my chalkline over the edge of a subfloor or slab and begin snapping out walls, compensating for square as I go.

Instead, I take the time to establish a square base—a giant 90 degree angle—with string. Then I use this as a reference for pulling my wall dimensions and establishing the layout on the floor with a chalkline. This lets me concentrate on the plans and how I want to build the walls when snapping out lines, rather than getting distracted with a lot of double-checking.

To establish square I pick the longest or least movable wall to use as a *base line*. I measure in $3^1/_2$ inches to the inside of the wall (or 7 inches if it's full of plumbing) and stretch a string the full length. With another "dry line," I establish the inside edge of a wall that's approximately 90 degrees to the first one (we'll call this the *adjustable line*). If this isn't a corner of the subfloor, I measure in the appropriate distance from the closest parallel wall.

I perfect my 90 degree angle by establishing arbitrary lengths for each line, marking these on the floor, calculating the exact length of the diagonal between them, and moving the adjustable line in or out to hit this dimension.

Most of us learned to do this with multiples of 3-4-5, but I find I have less chance of error and more flexibility in choosing dimensions (I want the longest lines the layout will allow) with a calculator. The one I use is a Construction Master II from Calculated Industries (22720 Savi Ranch, Yorba Linda, CA, 92687; 714/921-1800). By punching in *rise* and *run*, I get a diagonal length that I can measure from the end point of the base line to the end point of the adjustable line.

If the plan is complex or contains a step-down, I'll create several of these 90-degree angles. And if I have lots of 45-degree angles in a plan, I'll even bisect my 90-degree angle with a string so I can measure directly off it. Once I've established these reference angles, I can pull exact dimensions from anywhere on these strings, and trust them in snapping out the walls.

However, I always double-check for square any areas where tile will be used; its grid lines just don't offer any forgiveness.

Plating. Once the floor has been snapped out, my crew scatters plate stock. We tack together bottom and top plates for all the walls and cut them at the same time using the chalk lines of the floor layout as a template.

Here is a situation where I combine production techniques with old-fashioned care. In the tracts, I never got out a tape measure or a pencil for plating; instead, I'd cut everything by eyeballing it where it lay. But in my custom frames I measure and fit the wall plates as tightly as possible. If they don't fit just right, I redo them.

This way, when we nail the walls together and *plumb and line* them (brace the walls plumb and straight), the top plates can be trusted to be a duplicate of what's snapped out on the floor. Particularly when you're dealing with tall walls, angles, double walls, and other strange configurations, precise cutting will pay off.

With walls that exceed the stock

Figure 1. *A layout stick eliminates the occasional errors that come from stretching a tape and having to pick out stud centers. It also saves the extra steps of squaring off marks and making Xs.*

length, I tack plate stock on the subfloor till I get close to the end of the wall and either measure or scribe the remaining piece. However, I'll try to break my plate over a header, or second best, on a stud at least 4 feet from corners or partitions so the double top plate isn't the only thing keeping the wall stiff.

When plating for a rake wall, I only cut the bottom plate at this stage, leaving the top plates for actual framing later on.

Plate layout. Layout is a three-step process. First I mark all *corners* and *channels* (where one wall intersects another), then I detail all windows, doors, and *specials* (beam pockets, bearing posts, etc.), and finally I add the stud markings.

Every crew has its own philosophy and style of detailing. You want enough layout information that you don't have to think when you're nailing together walls, but you also want to keep things simple.

For window and door openings, I use *keel* (lumber crayon) to mark the location of the header along with its length. If it's a window, I also write down the length of the sill cripples. I make an X to the outside of these lines to indicate *king studs*. I don't show the *trimmers* (jacks) inside the king studs—it's a given—unless I want them doubled because the header is over 8 feet or because the wall is carrying a beam load from above.

I use one color keel for all my original layout marks. This allows me to use a second color to make changes or correct errors. My layout assumes what's standard; when stud length, header size, or anything else is unusual, I note these exceptions on the plates as concisely as possible.

I only use a tape measure at this stage to locate openings, partitions, and specials. For everything else I use a *channel marker* and a *layout stick*. These tools were born in the tracts, and make short work of stud and corner layout.

A channel marker is an L-shaped jig made of 1x4, metal, or plastic. It measures 3 inches on one side and 3½ inches on the other. It's simply a template that allows you to mark all corners and mid-wall intersections with a quick pencil scribe down each side.

A layout stick is usually 4 feet long with 1½-inch-wide fingers that extend down about 6 inches (see Figure 1). They're attached at 16 inches on-center. On exterior walls, I position the tool so the first finger is only half on the stock—this gives me the 15¼-inch or 47¼-inch offset (whichever way you do it) that allows the plywood to break on a stud. I then run my pencil against both sides of each finger, and move the stick down another 4 feet so the first finger is positioned directly over the preceding marks.

When using a layout stick on interior walls, I start a new layout whenever I run into a partition channel. This saves me the stud that would fall somewhere in the first 16 inches if I continued the original layout. That can add up to a bunch of studs over a whole house.

Building Walls

When it comes to banging together the walls, there's no need for finesse. Most of the details that bear on quality have been dealt with in layout and plating; all you're doing now is putting the pieces together. The main emphasis should be on grouping tasks so you're doing as much of one thing at a time as you can to increase your efficiency.

Establishing an order. I frame all the walls I can in the space available. After numbering them for location, I spread out the wall plates on the deck.

Experience is everybody's teacher when it comes to determining which walls should go up first. The trick is to make sure you leave room to build the next set after the first ones are up. Long hallway walls can really be a bear if you box yourself in. It helps if you keep reading the floor layout in front of you and imagine the walls going up.

Studs and specials. After all the plates are spread, I stock them with studs, headers, trimmers, sills, cripples, corners, channels, and specials like 4x bearing posts. All of these are cut on the radial-arm saw by one of my crew while I'm laying out, so when we start framing our rhythm isn't interrupted by missing pieces.

Although I cull out badly bowed or warped studs when stocking the walls, I don't bother to crown them. I've found with the green framing lumber we use in this area that I get better results by waiting until we do *pick up*. (These are the small framing details or corrections we do after the walls are up so they don't break the flow of the work.) This way, the studs that are going to go sour will have as much time as possible, and I can fix or replace them.

Figure 2. *To speed up the process of cutting in sway (let-in) braces, framers hold the 1x in place with one foot and run the saw along each side. This short-cut takes practice and caution.*

Nailing. With the plates spread and stocked, the standard wall banging chore can begin. We nail together all the walls we can accommodate before cutting double top plates, let-in braces, and blocks. The process of doing one step at a time really shows its worth here.

Double top plates. I typically install the double top-plate and blocking without ever using a tape measure. This saves time and eliminates the transfer of measurements that causes so many cutting errors. To top plate a wall fast and accurately, lay your stock down along the existing top plate and use it as a template to mark where it should break for partitions.

To produce the "tab" that will lap the neighboring wall at a corner, hold the end of your double top plate stock to the *far side* of the nearest partition channel and cut the other end flush with the end of the wall. Then shift the double top plate back to the *near side* of the partition before nailing it. If the wall has no channel, I use the 3½-inch side of the table on my worm-drive saw or a scrap piece of 2x4 as a gauge.

Let-in braces. We use "let-ins" a lot where I build. These are 1x4 sway braces that are notched into the studs and plates on the outside face of a wall. Their ultimate purpose is to provide shear strength. They should run from the top corner down to the bottom plate at a 45-degree angle, and be repeated every 25 feet or so on long walls.

Let-in braces also make plumbing the walls a good deal easier by giving us a way to quickly lock the wall in position when it's plumb. We cut the braces into the wall while it's still down, but only nail the brace at the bottom plate and first stud. However, we start nails at every other stud and at the top plates so there's no fumbling around when other crew members are straining to rack the wall into a plumb position.

I even use let-ins on exterior walls that will eventually be shear paneled. This allows me to plumb and brace those walls along with the rest, and then put one of my crew on shear panel while I'm laying out floor or ceiling joists.

The reason I use let-ins so freely is the speed we've developed when installing them. The procedure most of us around here use takes some practice and won't win any awards for safety, but it allows you to let in a brace in two or three minutes.

First, lay the 1x4 down in position; it should extend slightly beyond the bottom and top plates. Then, while you hold the brace in place with your foot, trim it off flush at the bottom of the wall and run the blade alongside the brace, cutting into the bottom plate (see Figure 2). You will need to set your saw at a little more than $1^1/_2$ inches deep so that it can ride on the brace and still notch the plate by at least $^3/_4$ inch.

Then keeping one foot on the brace, run the blade up both sides of the 1x4 cutting into each stud as you go until you reach the top plate. Keeping your elbow locked against your knee will help guard against kickback. Then trim off the top of the brace about $^3/_4$ inch shy so it can't run above the double top plate when the wall is racked plumb.

Now kick the brace out of the way and use your saw parallel to the deck (the table will actually be resting on the face, not the edge, of the stud) to cut the bottom of each notch. This "pocket" cut is a little awkward at first, but you can get very fast with practice. You will have to overcut the kerfs that form the sides of the notch on the face of the stud so the bottom of the blade makes the full cut on the other side.

Now you can clear the chunks of waste in between the cut lines with your hammer, nestle the 1x in place, nail it at the bottom, and bend a nail over the brace at the top plate so it won't flop around too much. It ain't pretty, but it's very fast and effective.

Figure 3. *These tight-fitting blocks were cut in place without measuring by laying the stock on the wall and eyeballing the saw blade.*

Blocking. Most exterior walls require fire/draft stop blocking. This can be a tedious job if you measure each one to fit, and it won't contribute a bit to real quality. Instead, lay a piece of stock alongside the bottom or top plate (not in the middle of the wall where bowed studs can throw you off layout), and cut the blocks for each stud bay by eyeballing your sawblade. With a little practice, you can churn out tight fitting blocks in no time (see Figure 3).

When nailing the blocking, it's a good practice to leave the last block out, or better yet, tack it to the stud bay for later installation. This will avoid those situations where the block bows out the last stud enough to make it hang up on a neighboring wall while you're struggling to raise it.

Plumb and Line

"Plumbing" refers to racking the walls side to side to make them plumb. "Lining" means pushing the tops of the walls in and out to straighten them. "Plumb and line" begins when the walls have all been framed, stood up, and tied off. This includes nailing off (or with a slab, shooting in) bottom plates to the snap lines on the floor, and nailing together intersecting walls at corners, channels, and double plates. Remember to drive these together tightly or you will be wasting the accuracy you achieved with careful layout and plating.

The first step in plumb and line is to scatter the stock for the braces throughout the rooms and tack them where they'll be needed. Then you'll need to systematically rack each wall until it's plumb, and nail off the braces to keep it there. Finally, you'll need to make sure the tops of your walls are straight and parallel by forcing them in or out using *line braces* and nailing them off.

Stocking braces and cleats. To keep the walls plumb, I rely largely on the let-ins, but you can also use temporary 2x4 diagonal braces on the inside face of walls at the same general spacing and angle as let-ins. Long interior walls, or ones that are largely independent of other walls, should be diagonally braced as well. Nail diagonal braces at the top, but just start nails on the rest of the brace for now.

Line braces are set perpendicular to the walls, and run from the top of the wall down to the floor at approximately 45 degrees. We spike our line braces to the face of a stud just under the top plate (see Figure 4). If we need one at a header, we'll nail a 2x4 cleat to the top plate and header, and nail the line brace into the side of the cleat.

On fairly short walls, I generally use a line brace right in the middle. On long runs, I take a peep at the wall to spot obvious bows and put line braces there. If I have a long wall that will remain open for some time, such as a vaulted ceiling wall, I use a few extra braces so it can't move around. I also drive all my nails home.

We nail 2x4 cleats to the floor where the line braces fall. Centering a 2-footer on the end of the line brace usually allows enough adjustment back or forth to line the wall and still get good nailing. We also supply each room with a few 3- to 4-foot lengths of 2x to use as *kickers*. These are jammed between the floor and the underside of flat line braces and act as fulcrums that help pull the wall in.

Plumbing. I rack all the exterior walls plumb using one of my guys on the level, another on a push stick, and a third nailing the let-ins or temporary diagonal braces. In my piecework days in the tracts, there were just two of us doing plumb and line, but it's pretty hard to keep your eye on a level and nail off braces at the same time.

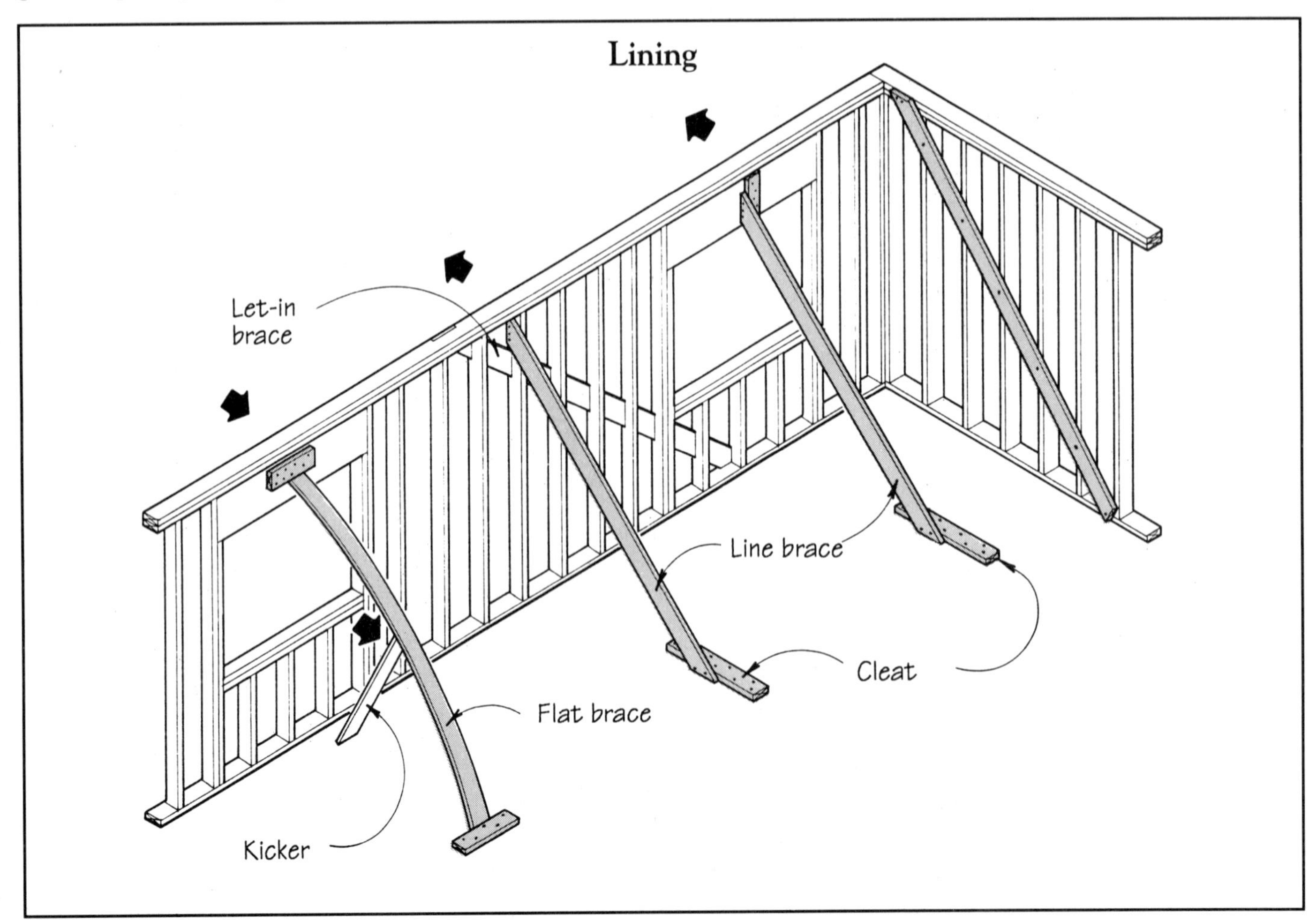

Figure 4. *Bringing the tops of your walls into line requires standard line braces for pushing walls out and keeping them there, and flat braces with kickers to bring them in. The author uses his eye rather than a string for lining most walls because, with practice, it's fast and accurate.*

On 8-foot-high walls, a 10-foot push stick made from two face-nailed 1x4s works well. It is wedged between the floor and the intersection of a stud and the top plate at a 45-degree angle and then flexed to rack the wall. But most of the custom homes I build now include tall walls with two or more rows of blocking that require something with more leverage.

The answer is a scissor-action push stick (see Figure 5). This consists of two studs nailed together so they overlap by at least a few feet. The amount of overlap depends on the height of the walls. I use three 16d nails grouped in a tight circle to allow the top "lever" to rotate. A notch at the top helps to engage the top plate of the wall I'm trying to move, and an angle on the bottom keeps it in one place on the subfloor.

You can use a scissor stick to rack a wall plumb, or to push it into line. To plumb a wall with it, place the notch end at the intersection of a stud and the top plate (a king stud is best so you don't blow out a regular layout stud) and with the bottom leg in a near vertical position, push down on the tail of the top leg. This "stretcher" can really tweak some walls.

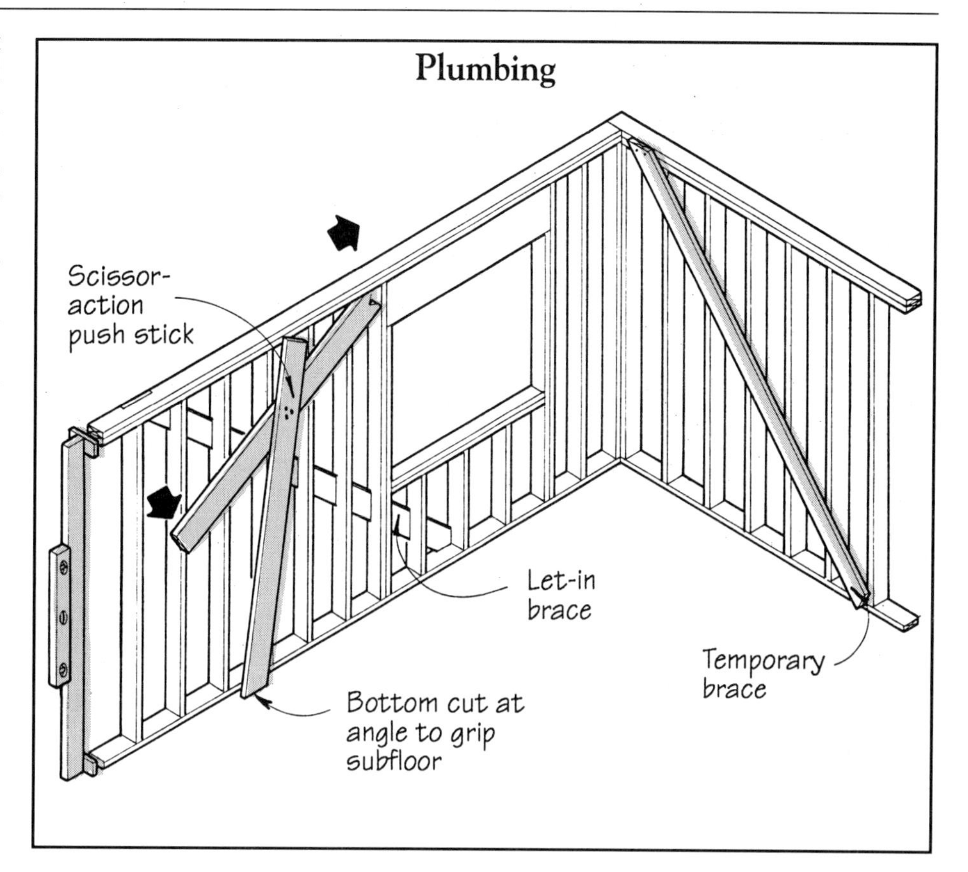

Figure 5. *The scissor-action push stick shown here is a good site-made tool for racking tall walls with lots of blocking. It can also be used to persuade walls that bow in at the top during the "lining" procedure.*

Another problem in a house with walls of various heights is how to rig up a level to accommodate them all. Like everybody else, I've taped or nailed levels to long 2x4s, but it's a clumsy arrangement at best. Lately, I've been using a level that will adjust in height from 5 to 13 feet (Plumb-It, 3045 North Dodge Blvd., Tucson, AZ, 85716; 800/759-9925), and I've been amazed at how quickly and accurately it's allowed me to work on some of the crazy custom designs I'm asked to build.

Armed with the level and a push stick, we start at an exterior corner and methodically work our way through the house. The crew member on the level should keep the guy on the push stick aware of what's needed, and ask him to keep a steady tension on the stick once the bubble looks good. Then he needs to quickly check the other end of the wall to make sure it also reads plumb. If it doesn't, they'll have to split the difference. When the wall is where they want it, they'll give the third guy the go ahead to drive home the nails that have been started in the braces while they check to make sure the wall is holding steady.

I do all the exterior walls and those that intersect them first. If there are two intersecting walls close together, such as a closet, I plumb one and catch the other later. After exteriors are done, I work my way through the interior walls from one end of the house to the other.

Lining. After all walls have been satisfactorily plumbed, I line all the exteriors and then any of the longer interior walls. Although a dry line held out by plywood spacers is the ultimate in accuracy, I find my eye is sufficient most of the time and a great deal faster. I steady myself on a 6-foot stepladder and just eyeball the top exterior corner of the double top plate, while talking with the guy nailing the bottoms of the line braces.

Walls that are bowed in are pretty simple; just lever up on the bottom of the line brace — the scissor push stick can be used on really stubborn walls —and nail to the cleat when it looks good. Line braces can also be run back to the bottom plates of interior walls. If the wall bows out, you'll have to nail your line brace flat to the wall with a cleat above it to keep it there, and use a kicker to create an upward bow on the line brace.

If you've taken some care with layout, framing, and plumb and line, you'll end up with top plates that are straight, square, and rock solid. And that means less measuring and adjusting when you add joists or rafters, which in turn, saves time, effort, and error. ■

Don Dunkley is a framing contractor and the owner of Village Homes. He lives in Cool, Calif., on the edge of the Sacramento Valley.

Time-Saving Framing Tips

by Marshall Gross

Misplaced hammers and disappearing chalk lines got you down? Here are a few ideas you may not have thought of to help you save time (or at least keep you from losing your cool) on your next framing job:

• That pretty blue chalk you've been using has no place on the framing site. Use red cement dye and fill your chalk box 7/8 full. Lines snapped with this stuff won't wash away when you sweep the floor or when it rains (or when you drop your coffee).

• If your framing stock is larger than 2x6 (say 4x6 or 6x6), use a framing square instead of a combination square when drawing lines. Next time you're at the hardware store, check all the combination squares against a framing square, and you'll see why.

• As a general rule, when measuring timbers, do it twice. Measure first from the left and then from the right. Always begin with the larger size on the list. If a mistake is made you can still use it for one of the shorter sizes and thereby save the day.

• Use 16d sinkers with a waffle head for framing. The hammer should weigh about 24 ounces. If the hammer is new or you've just had it rewaffled at the saw shop, beat it for a while on steel or concrete to take off the razor-sharp edges.

With one swing, the nail should be set into the plate; the next blow should drive the nail home. (This is called "two-licking." Three-licking will get you a check within your first hour on the job, as it did with me on my first framing job.)

When you're ready to start two-licking, open the keg and grab a handful of nails. Gather them like sticks, dropping the ends against your palm. Now there are heads at both ends, but not for long.

With the thumb and forefinger of each hand, you can easily separate them into two groups (heads up and heads down). *Put them all in one direction in your pouch.* Repeat this a couple of times until your pouch is almost full, then get one more handful and move on to the studs.

• When you run out of sinkers, leave your hammer at that spot with the handle sticking up. Then go get more nails. You won't have to look for your place (or your hammer) on the way back, or miss some nailing.

• Everyone makes mistakes (as in nails placed off center, for which you use your cats'-paw or "gooney spoon" to pull them out). Stanley Tools makes an excellent nail puller (no. 55-035) with an offset claw on either end—one at 90 degrees and one at 30 degrees. You'll really appreciate the claw at the long end when the stud spacing is close and you can't get at a nail with the short end.

• There should be one let-in brace for every 25 linear feet of wall; 45 degrees is their minimum angle, while 60 degrees is the maximum. With 16-inch spaces, the brace should cover five spaces and go from the rafter plate to the sole plate, slanting up and out toward the upper corners of the wall.

To cut a let-in brace, lay a piece of 1x4 in the correct position on the wall (after you've checked that the wall is square, of course). Set your saw to 1½ inches deep and cut along one edge of the brace, then the other. The sole plate of the saw will ride along on top of the 1x4. Cut the ends off to the correct shape as you pass with the saw. If you work carefully, the brace will fit nice and tight.

Now move the brace to the side to make the bottom cut (¾ inch deep). This is a dangerous cut that requires a lot of practice; get someone to show you how it's done the first time around.

Place the let-in brace into its channel and nail it to the sole plate with 8d sinkers. At each stud and the top plate, start (just *start*) two nails into the 1x4. Drive a nail toward the upper part of the brace and bend it over to hold the 1x4 in place while raising the wall. (The brace will be nailed in later.)

• Before you can lift a wall, you have to place a block under the rafter plate about every 10 feet so that you can get your fingers underneath it. To do this, place a 2x4 scrap on its edge against the rafter plate, then drive your hammer claws into the plate and lift up the wall. The scrap piece will fall under the rafter plate and hold up the wall.

• Just because you need an eight-foot level doesn't mean you have to buy one of those expensive jobs that you carry in a glass case and certainly don't want to drop.

Instead, go to the scrap yard and get a 20-foot piece of I-beam or channel aluminum, and cut it into an eight-foot level, a four-foot level and a 6'6" level for door trimmers. Use a torpedo placed against its side, and you've got it made. (Don't buy a torpedo with an exposed bubble, though—lines wear off.)

The aluminum I've been kicking around on jobs for the past ten years has held up well. Occasionally I buy a new torpedo and suddenly have four brand new levels to use—and at great prices, too. And you can always make a pouch for the torpedo out of scrap leather and rivets so you can keep it by your side. ■

Carpenter and teacher Marshall Gross lives in Cottonwood, Idaho. He is the author of Roof Framing *(Craftsman Book Co., 1984).*

Bracing Walls Against Racking

Plywood sheathing is tops. But shear panels and let-in bracing do well, too.

by Harris Hyman, P.E.

A wood-frame house is a surprisingly safe building. It is almost impossible to construct one that will fall down. Most builders instinctively work with lumber that is heavy enough to do the job. In fact, the wood is usually much, much stronger than the bare minimum that is needed to hold things up. Failures are generally water-related — rotting or leaking.

Structural problems in a house almost always relate to deflection, that is, unwanted movement of the framing. The most annoying of these unwanted movements is a bouncy floor, but the one that is most noticeable and causes the most deterioration in the house's value is racking. In this condition, walls that presumably were plumb and square shear off into parallelograms (see Figure 1).

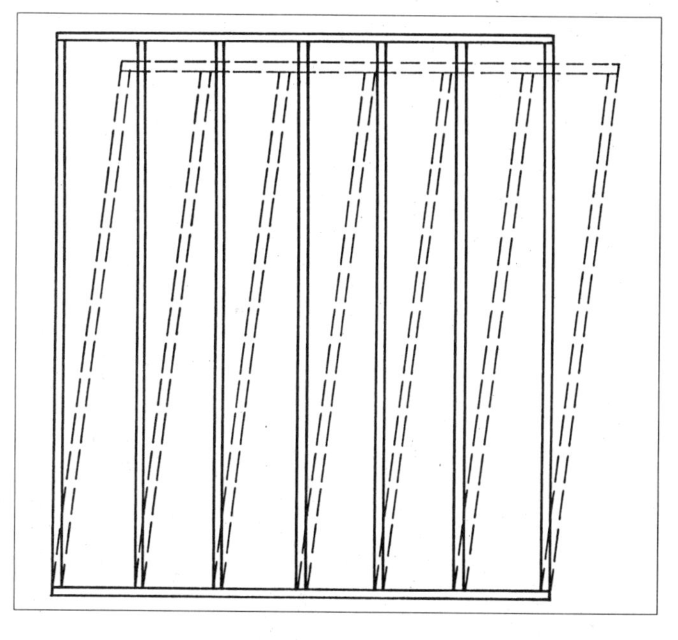

Figure 1. *When a wall racks, what was square becomes a parallelogram.*

Racking also drives carpenters crazy. They set up the job carefully, cut the studs to within 1/16 inch, measure and align everything twice, and even use a plumb bob. Then, a couple of weeks later, it's time to set the doors and windows, and the rough openings are 1/8 inch out of plumb. It's enough to drive a carpenter to toss a hammer through the window —which I've seen.

So what happened? Simple—the building moved. The upper northeast corner went a little farther northeast while the sill stayed in place. The whole thing was put together properly, but it sheared and twisted. (Unwind the plumb bob, or measure the wall diagonals, and you'll see.)

It's a lot worse when a building racks *after* the owner moves in. A contractor can pretty easily calm down a carpenter, but an unhappy client is another matter entirely. If the building shifts a little and the casement windows suddenly won't close and the drywall screws pop, it can take some real doing to appease the client's attorney.

Another racking problem is the building that sways in the wind. The engineer may have stamped the plans and assured the owner that nothing would happen, but every time a nor'easter blows, the house shudders —and the owner is beginning to wonder. Even I worry in a building that shudders. My inner ear can contradict a head full of calculations and statistics.

This tendency toward flimsiness and movement is going to occur more and more as exotic glazings and green houses become integral parts of a structure. Holes in the skin weaken a building quite a bit, but with R-6 and R-9 windows you want to cut a lot more holes. A greenhouse is *all* holes, with hardly any structure around it, and when a house is half greenhouse, a lot of the normally expected strength is gone.

There are several ways to strengthen a frame in shear and prevent racking. The traditional method is let-in diagonal bracing. Ideally, diagonal bracing should zigzag at 60 degrees or so from the sill to the top plate and back down again. Where windows and other openings interrupt the pattern, you should still try to place the braces so they go unbroken from sill to top plate.

After World War II, plywood became a common skin material. It proved to be much stronger and easier to do than let-ins, which were gradually abandoned. Over the past few years, however, a number of contractors have stopped using plywood sheathing and returned to boarding.

Boarding has a couple of things going for it. It gives a generally flatter and smoother appearance than 1/2-inch plywood; it can be less expensive and much easier to handle than 5/8-inch plywood; and it provides a structure that is consistent with Lotz's Law, which recommends a relative permeability between the outer and inner skins of 10:1 for moisture control in the walls. But the return to boarding makes a weaker building.

Another new practice is the use of a one-inch foam insulation on the exterior. The foam sheathing is placed between the framing and the skin, where it acts as both an infiltration barrier and as a thermal break for the studs. If plywood is installed on the exterior side of the foam to provide a nail base, it does little to improve the strength of the building because the plywood hangs out on one inch of unsupported nails. Additional bracing would be needed to prevent racking. (My solution is to use the foam insulation on the interior under the drywall.)

I have an old copy of Albert Dietz's classic, *Dwelling House Construction*. He gives sketches and cites NFPA strength data, which is summarized in

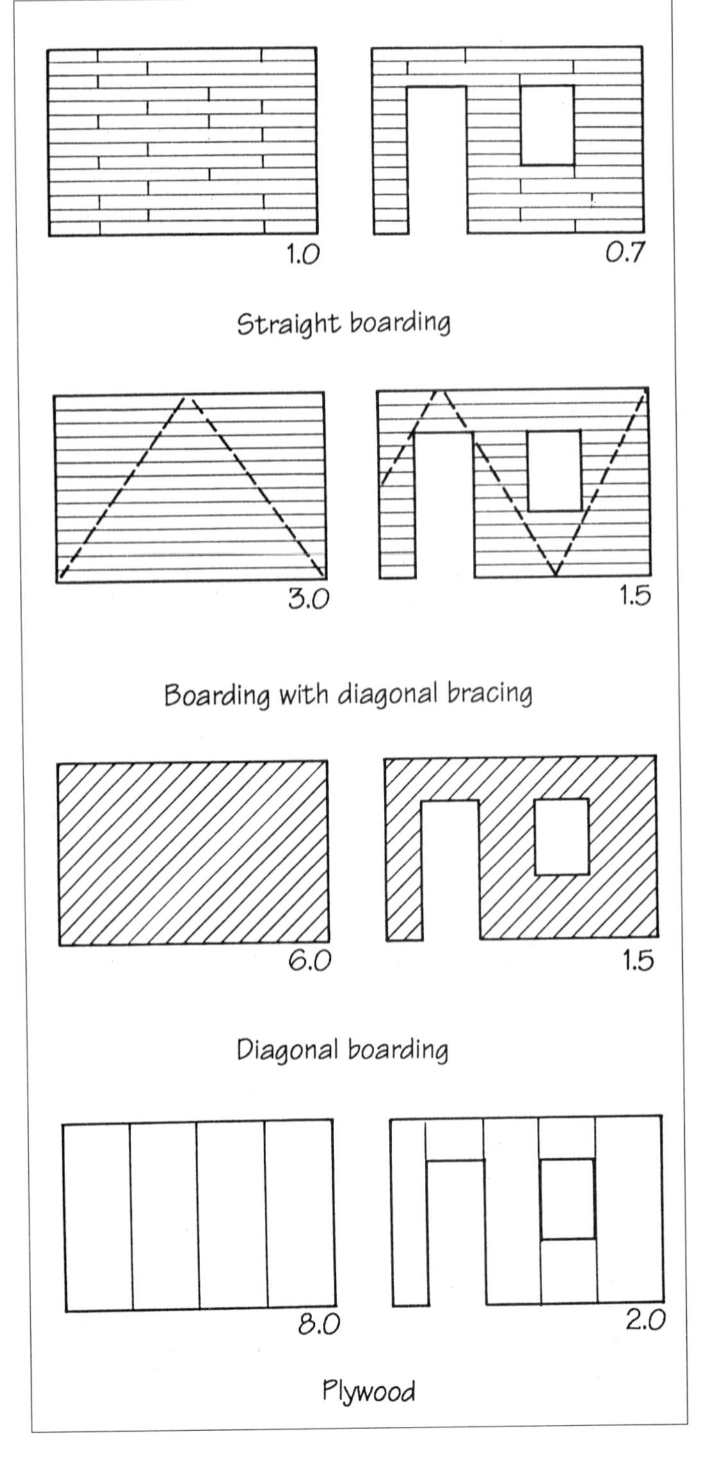

Figure 2. *If a boarded frame wall has a strength of 1.0, diagonal bracing would have a value of 3.0, but only 1.5 with door and window openings. Plywood is tops with eight times the strength of a boarded wall.*

Figure 2. A boarded frame wall is given a baseline strength of 1.0. With diagonal bracing, it is about three times as stiff. With two door and window openings, the wall is only about three-quarters as stiff. Covering the wall with plywood makes it five times as stiff ,and if the plywood is glued on, the wall has a relative stiffness of 10 to 20. With doors and windows, the plywood wall is still twice as strong as the plain boarded wall.

Diagonal sheathing is my personal favorite. It has the advantage of a smooth, flat skin (and the permeability) and is somewhere between diagonal bracing and plywood in strength. But it is tedious and expensive to lay up, and about 20% is waste.

Another anti-racking system (see Figure 3) uses a shear panel in the wall. This is basically a boarded wall construction with one or two sheets of plywood for stiffness. These are usually placed at the two corners of each wall where they provide the greatest rigidity, even with "broken" plates. This system seems to have all the advantages of diagonal boarding, but is much less expensive. I've used it and it appears to work well—the buildings feel good—but I don't have specific stiffness figures. Preliminary calculations suggest a ratio of 20% plywood to 80% boarding.

Dow Chemical, which makes Styrofoam, recommends a hybrid plywood/boarding system for use with foam on the exterior. Here, the plywood at the corner 4 feet of each wall is 1/4 inch and the foam over the plywood is 3/4 inch. This system *looks* good, but may be difficult to actually put together. For one thing, the quality of the 1/4-inch, construction-grade fir plywood that is available—at least in eastern Maine—is not good, although lauan and Masonite seem like pretty inexpensive and reasonable alternatives. Furthermore, 3/4-inch foam sheathing is never available, unless you are willing to buy a full pallet.

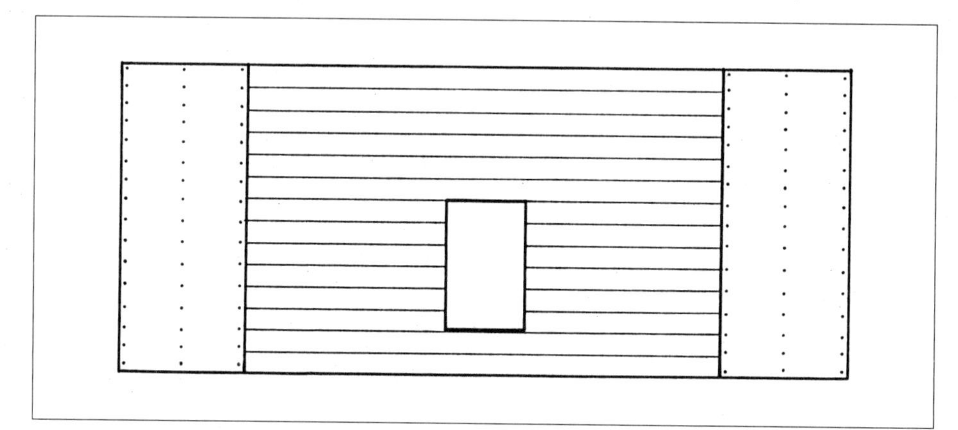

Figure 3. *Plywood in the corners (shear panels) seems to have all the advantages of diagonal boarding with much less expense.*

There are two other stiffening systems: diagonal flat metal strapping, and diagonal T-strapping. The flat metal strapping is all right in tension, but against a compressive load it's a wet noodle (see Figure 4). Using it corner-to-corner in an X pattern might work, but with nothing to go on except instinct and experience, I don't particularly like it. It seems as though it will "work" around, enlarge the nail holes, and then not provide any more stiffness.

The T-section (see Figure 5) is a little more promising since it will support compression. You install it by cutting a diagonal kerf one inch deep and nailing it up. All I've seen are the ads, and it isn't stocked by any of my local suppliers. It might be worth some experimenting, but the reviews I hear through the grapevine are mixed.

Certain types of buildings are structural horrors that are particularly susceptible to racking. Foremost is the garage-apartment, where an apartment is constructed over a garage. It's wonderful for a rural

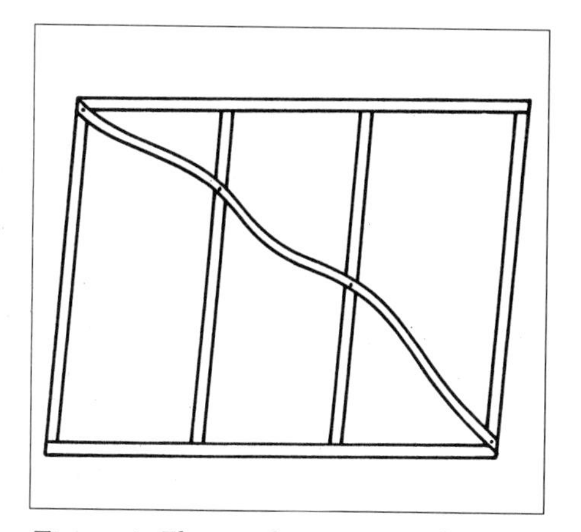

Figure 4. *Flat metal strapping is all right in tension, but is a wet noodle in compression. An X pattern might work.*

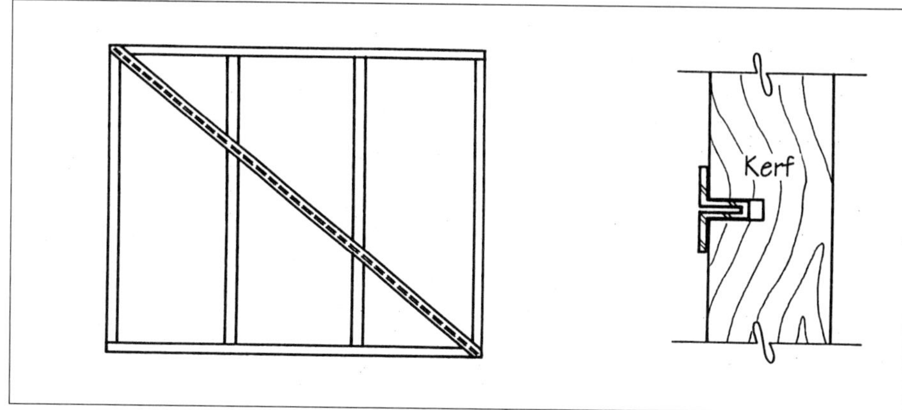

Figure 5. *T-section should outperform flat metal strapping.*

rental unit—and can make a solid contribution to the mortgage payments—but it's not physically strong. The lower walls have huge cutouts for the garage doors, the garage space must be empty of interior partitions, which will add stiffness, the apartment floor is heavily loaded, and the whole thing tends to shake. These structures often require a double, staggered skin of plywood on the garage-door wall, and careful control of the size and placement of the door openings. And even then, they shake.

Another monster is the building that has a battery of sliders or other windows. Structurally, this is similar to the garage, but not as bad. There are usually interior walls that can be stiffened with plywood sheathing to provide the rigidity that the exterior skin lacks. The bad part is that the multiple-window configuration usually occurs on an expensive house, and the owner will be especially intolerant of shaking.

So, as usual, be careful.■

Harris Hyman is a civil engineer in Portland, Ore.

When Walls Go Round

Simple techniques to help builders when designers throw a curve

by Neil Momb

After 40 years of building whatever the architects can draw—from simple circular walls to S-shaped and flared walls—I now think of myself as an accomplished radius wall framer. Still, I remember how easy it is to be intimidated by a set of drawings showing curved walls.

Curved walls are simple once you begin to think "round." A round wall is just a straight wall with curved plates. But details such as window and door openings and fireblocking can get tricky when things go round. This article will describe some of the standard ways I frame radius walls.

First, let's look at the essentials of drawing and cutting curved plates.

Drawing Curves

There are a couple of different ways to draw curves full scale. When the radius is fairly short (under 5 feet or so), I drive an eight-penny nail partway in at the center of the circle and hook the end of my tape measure over the head. Then, holding a pencil on my tape, I draw the curve. This method is fast, but with large-radius curves the tape can get hung up.

For big curves, I cut a piece of strapping to the length of the outside radius of the wall. I then measure back the width of the plates and cut a notch in the side of the strapping for a pencil. With one end of the strapping nailed to the center point, I hold my pencil in the notch to draw the inside plate line, and hold it on the end of the board for the outside line.

I draw the plate sections on plywood. A plate built up out of two layers of 3/4-inch plywood is plenty strong and will match the nominal thickness of 2x plate stock, which

The author sits atop the completed portion of a two-and-a-half-story tower wall. The 22-foot-high wall was tipped up in 8-foot-long sections and the double-layer plywood plates spliced together.

Figure 1. *The author uses a plywood jig to hold a router for cutting curved plates. The radius can be adjusted by screwing on any length of strapping.*

allows you the use of a common stud length throughout. Half-wall sections with an outside radius of less than 4 feet can be drawn on one sheet of plywood. Larger radius curves, of course, require more sections. But as you will quickly discover, it takes more material to cut small-radius plates than larger ones.

Cutting Curves

I use a heavy-duty, 3-hp plunge router and a pilot-panel bit to get the fastest, cleanest radius cut possible. I use a Makita plunge router; it's proven to be a very tough machine. The panel bit has a sharpened end for plunging into the wood, and either one or two cutters (flutes) along the shaft. With a 3-hp router, you can cut through a 3/4-inch piece of plywood in one pass, so you need a bit with at least a 1-inch cutting edge. Carbide cutters are a must because the plywood glue quickly dulls any other. I use 1/2-inch diameter shafts, as 1/4-inch bits tend to chatter with heavy cutting and break easily.

To cut a curved line with a router, I make a jig out of a piece of plywood which is a few inches wider than the base of my router (see Figure 1). The piece is routed out for the base and has a hole for the bit to stick through. I can screw this jig to any board that gives me the correct length of the radius. When you measure to cut the radius, remember which side of the bit the cut is made on.

When you start cutting, you'll notice that you can only swing through the cut in one direction—that is, against the rotation of the router spindle.

Once you've cut one plate section, you can use it as a template for cutting the rest. I temporarily fasten the finished plate to the remaining uncut plywood with 1 1/4-inch drywall screws, and then turn the plywood over. The pilot tip of the bit acts as a guide along the original plate.

Layout and Assembly

At this point the plates haven't been cut to length. Lay the rough plates where they actually go, overlapping the ends. I first lay out for stud spacing, and then mark openings, intersecting walls, and special bearing points for loads above.

The rule of thumb for on-center spacing is 1 inch per foot of diameter measured along the outside of the curve. For example, on a 4-foot diameter wall (with a 2-foot radius), the studs should be 4 inches on-center. On curves over 12 feet in diameter, stick with 12 inches on-center. This number of studs is needed to define a smooth curve, more than for carrying loads. When measuring along the outside of the plates, use a flexible nylon or fiberglass measuring tape. It will wrap around the radius much better than a steel tape.

With a straightedge or chalkline stretched along the radius line, I mark the end cuts on the overlapping ends of each plate section, cutting the plate stock so the splices fall on the center of a stud.

I nail together the two layers of plywood for each plate section with 1 1/4-inch joist-hanger nails (available from Teco, Simpson, Kant-Sag, etc.). The overlap between layers should be at least 16 inches. After assembling the double-layer plates, I turn them curve up to nail in the studs.

Two people working together, one at the top plates and one at the bottom, speeds the task of nailing the studs in place. I prefer to use air nailers for this because they don't knock things out of alignment. It's best to leave the first stud off each section since it falls on the splice.

Curved walls are naturally stable, so depending on the radius, you can get away with a lot less temporary bracing to stand the walls up. The first stud must be braced in two directions. After that, a 2x4 brace between the wall and the floor deck is needed only every 8 feet or so to hold the studs plumb. I plumb the sections as I go, and this alignment telegraphs around. On a completely round tower, the last section just pops right into place.

Pre-cut plate sections are used for the second top plate after the wall sections are lifted into place. Use the standard rules for double-plating—that is, at least 32-inch overlaps at plate joints and header ends.

Fireblocking

In a round wall, only the center of a stud lies on a true radius line, but the angle at each end of a fireblock is slightly different. To find the correct angle, I locate and mark the centerline for two adjacent studs on a section of radius plate stock. I then measure in half of the stud width on

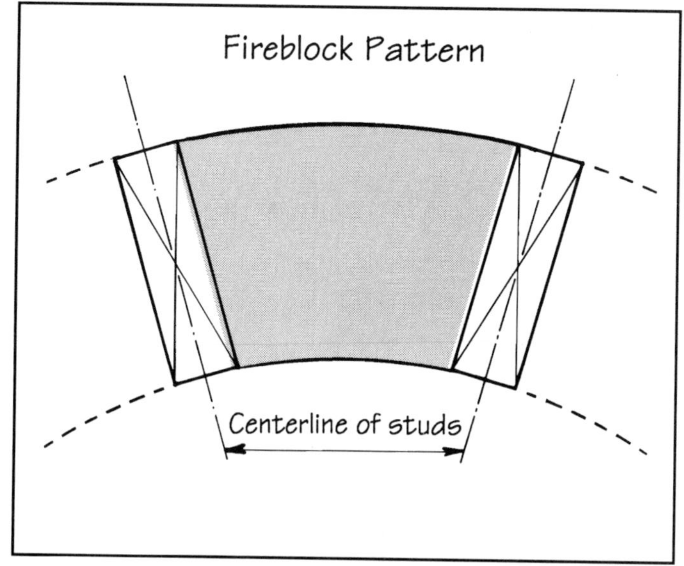

Figure 2. *To find the correct angle and length for fireblocking, mark the centerlines of two adjacent studs on a section of curved plate stock and measure in half the stud width from each end.*

each end, as shown in Figure 2. This gives me both the angle and length for fireblocks. I cut a fireblock pattern and duplicate it as many times as needed. To keep the wall straight after the blocks are installed, you have to take a lot of care to ensure that each fireblock is exactly alike.

Most of the round walls I build are linked to straight walls, and dead space is almost always created at the intersections (see Figure 3). To firestop the dead space at the ceiling, I substitute scrap pieces of 3/4-inch plywood for double top-plates. When all the walls are plumbed and braced, I nail an oversized plywood scrap over the intersection, overlapping all the joints to tie the walls together. Then, I trim around the perimeter with a router, and add a second layer (pieces are okay), trimming this the same way.

Skinning the Walls

When sheathing exterior walls, I run the plywood vertically. It must be ripped to splice on studs, and all horizontal joints need to be blocked. Half-inch CDX will bend around large curves (8-foot radius and up). For tighter curves I use either one layer of 3/8-inch CDX or two layers of 1/4-inch, staggering the splices between each layer. The double thickness is only needed for holding siding nails. Bent into a curve, one layer of 1/4-inch plywood is remarkably strong.

Be generous when nailing off the skin. I generally use a 3- to 4-inch spacing along the edge of the plywood, and about a 6-inch spacing in

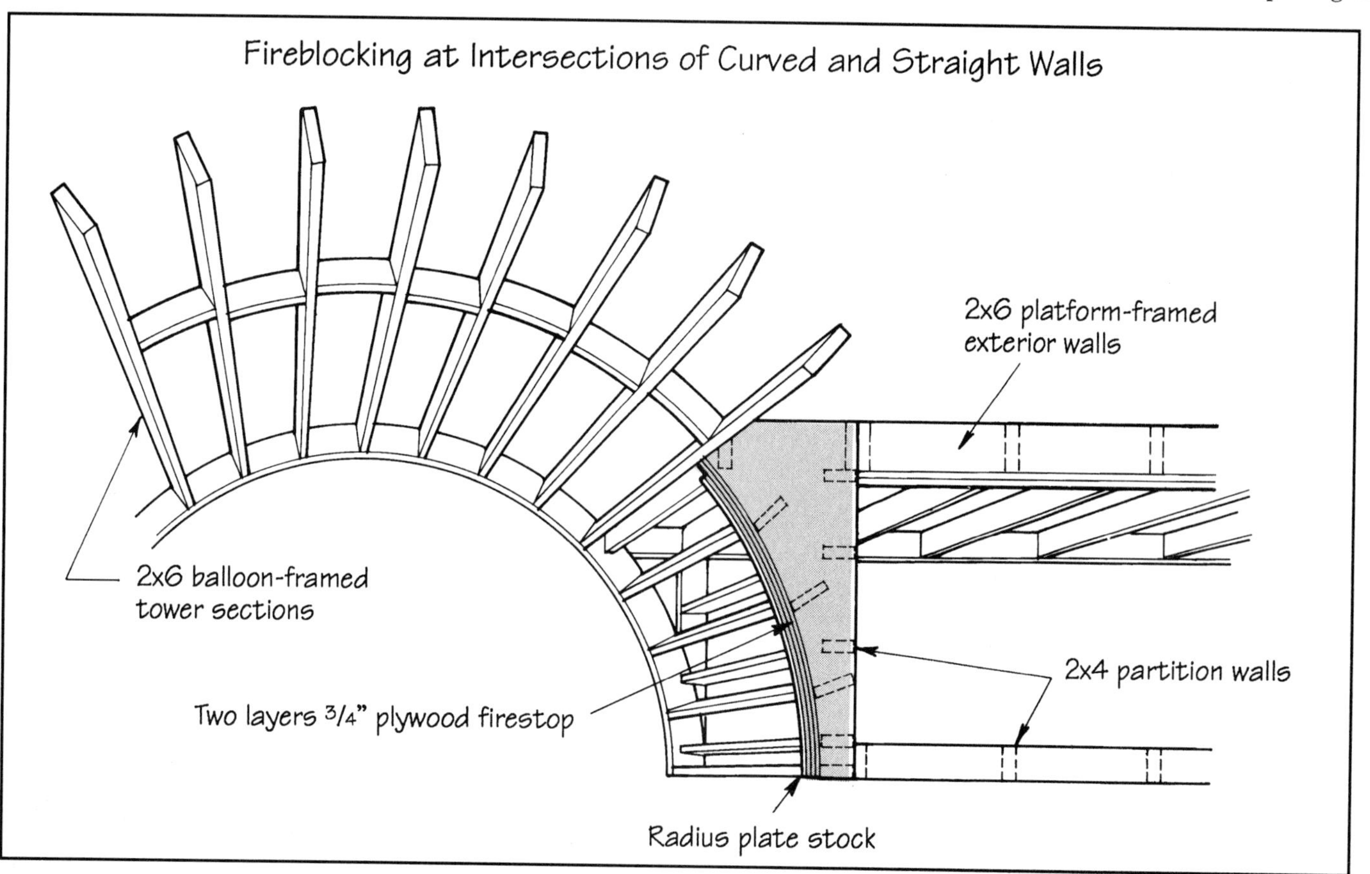

Figure 3. *When a curved wall intersects a straight wall, the dead space must be fireblocked. The author uses scrap pieces of 3/4-inch plywood (shown in grey) in place of double top-plates. The scraps are nailed over the cavity and then trimmed to the contour of the walls with a router.*

the field, to flatten out all the bulges. The way you nail the sheets is important for getting them aligned properly. On the first sheet, always work from the center towards the edges. On the next sheets, nail the edge nearest the first sheet of plywood, starting in the middle of this line, and working your way up and down. Then move over to the next stud, starting in the middle of the line again, and so forth. If you don't follow this pattern, the plywood sheets will tend to wander out of alignment as you wrap them around the frame. By the time you discover this you're really in trouble.

Openings in a Curved Surface

Windows in a round wall are an opportunity for the designer to have some fun, but the builder ought to exert some influence. If the windows are too wide, the corners will protrude inside. And since curved glass units are super expensive, narrow windows are usually the choice. In the stair towers, I installed tall, arrow windows between the studs with very satisfactory results. The studs are 12 inches on-center, and the windows are 10 inches wide by 4 feet high. These fit in nicely above the treads inside.

You don't always have to go so skinny, though. On the top of the same tower, I built an observation room where wider windows were needed. Using two stud-spaces I was able to fit a 20-inch-wide window in the 2x6 wall.

The rough sill for the window opening is cut from radius plate stock. I then add another sill on top of this which is straight on the outside to accommodate the window, and curved on the inside to fit the wall (see Figure 4). This arrangement is reversed for the top of the opening. I can then run a straight structural header between this top sill and the top plates.

The finish sill is cut from a 2-inch-thick cedar plank. I cut the outside curve with a jigsaw and then bevel the top face of the board with a power plane. Before setting the windows and the finish sill, I lay down a heavy-duty flexible vinyl flashing.

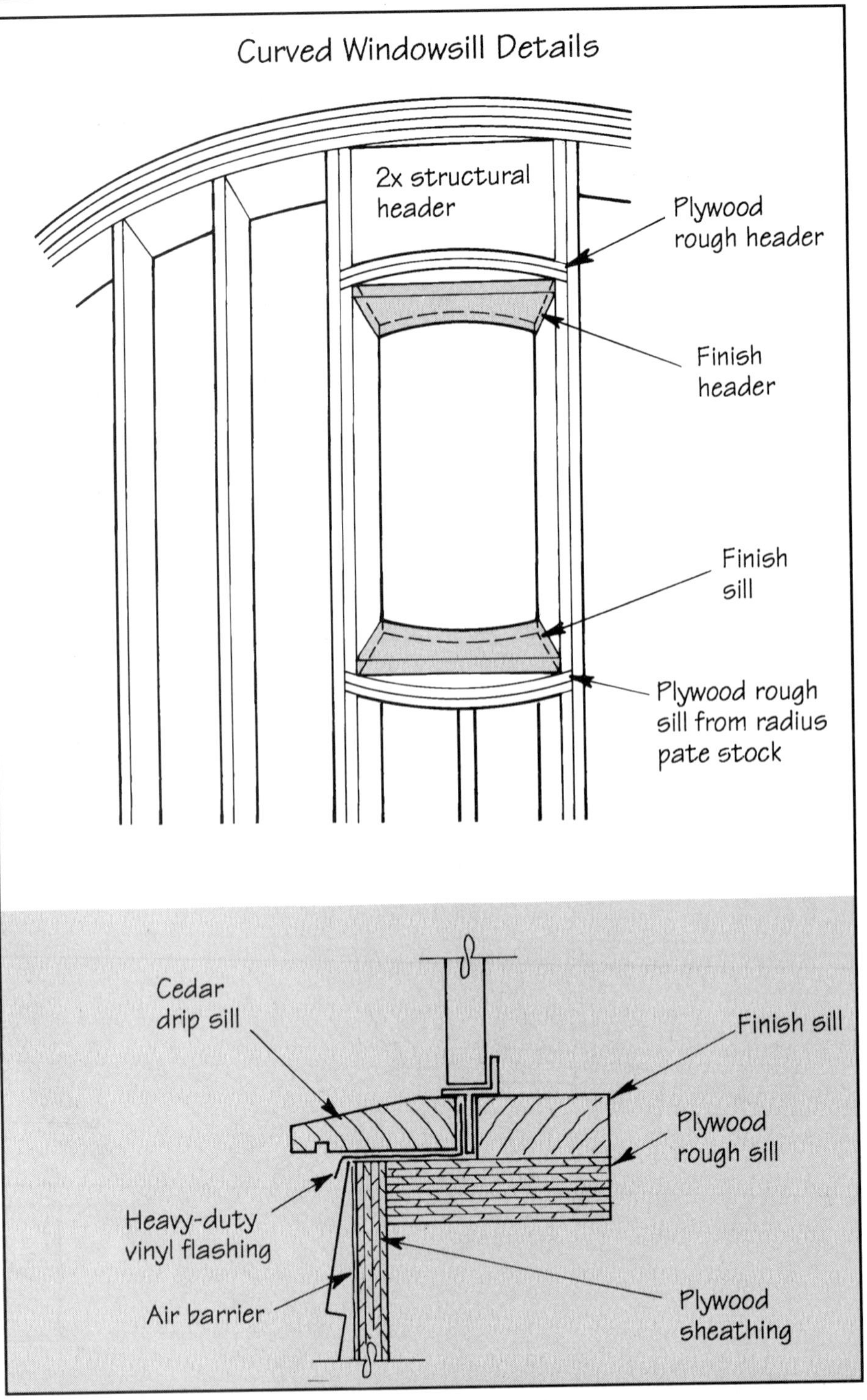

Curved Windowsill Details

Figure 4. *The rough sill is made of a piece of curved plywood plate stock, with finish sill cut from 2x lumber nailed on top. The finish sill is straight on the outside to match the window and curved on the inside to fit the wall (top). A cedar drip sill—cut with a jigsaw and beveled with a power plane—trims out the window (above).*

For wider windows using curved-glass units, I fashion a laminated header from several ½-inch plywood strips, building it to the thickness of the stud width. To build the header, you must first make a curved pattern-frame with an outside radius equal to the inside of the curved wall. The first layer is screwed to the frame so the screws can be removed later on. Then using a resorcinol glue or construction adhesive, I bend the plywood strips together, pulling the pieces in with staples or small nails until the glue sets. At this point, I leave the ends long and I don't nail where the header will be trimmed to length.

If the opening is very wide, I extend the header beyond the rough opening, at least to the next stud. This means that the header cantilevers over an extra jack stud on either side of the rough opening (see Figure 5).

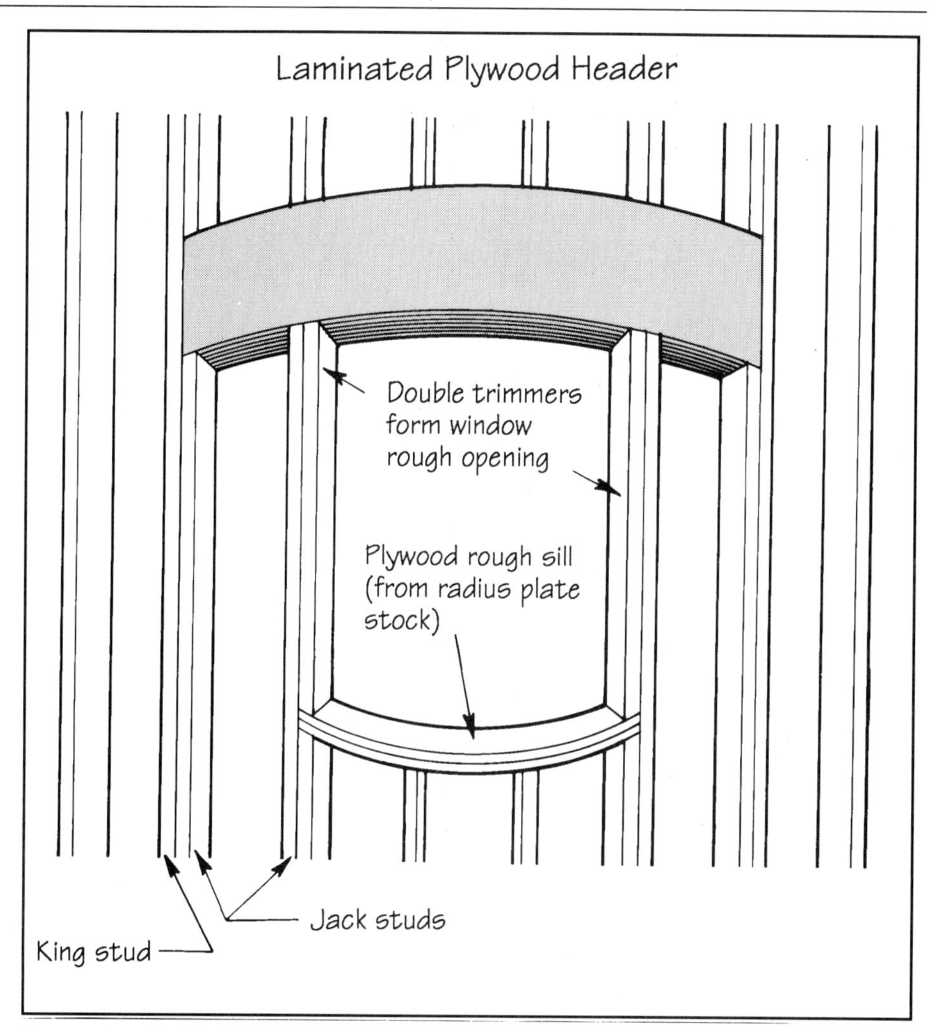

Figure 5. *For wide window openings, the author builds a laminated header from plywood strips (shown in gray). To keep the curved header from tipping outward, it must be cantilevered over the jack stud on either side of the rough opening.*

Finishing Up

Siding on exterior walls requires special consideration, and here is another place you can influence the designer if the budget matters. Clearly, the simplest, least expensive siding is stucco. Cedar shingles also work well. Stone and brick do nicely to accent the curve, but can be expensive. If wood is preferred, I try to recommend narrow vertical siding, such as 1x4 T&G. It's beautiful, but be careful to pre-stain so you don't get any unfinished joints when the wood shrinks. Avoid horizontal clapboards if you can. On one large radius curve (24 feet), I used 1/2x4-inch bevel cedar siding with great success, but I would discourage using anything wider or thicker, or trying it on tight curves.

On interior walls that have a 2-foot or smaller radius, I use one layer of 1/4-inch shop plywood under one layer of 1/4-inch drywall. This gives a truer curve than wetted drywall alone. On concave surfaces, I sometimes go with a thinner plywood and use 3/8-inch drywall, since it's harder to bend the plywood into the curve.

Round walls add a lot to a design, and are quite fun to build. After a couple of projects, you'll find yourself looking for a place to include them. ■

Neil Momb is a framing contractor in Issaquah, Wash.

Seismic Bracing

Retrofitting structural supports will help wood-framed homes survive earthquakes and storm damage

by David Benaroya Helfant

The unbraced cripple walls of this West Oakland, Calif., home succumbed to the 7.1 quake of Oct. 17, 1989, leading to their collapse at the rear of the house. Shearwall bracing and substructure diagonal braces would have given this home a better chance.

Editor's Note: *In many ways, 1989's 7.1 Loma Prieta earthquake confirmed—with frightening realism—what seismologists, engineers, and builders already knew: One- to two-story, wood-framed structures are fairly resilient if they are on solid ground, bolted to their foundations and braced to withstand the lateral racking forces that come with a strong or prolonged tremor.*

Unfortunately, most homes built prior to 1940 aren't bolted, and few if any, are adequately braced. The evidence appears in countless photos: stately Victorians and solid-looking bungalows literally on their knees. The reasons for these losses are evident at first glance: some houses simply left their foundations during those 15 seconds, while others bear the telltale bulges low on the siding that mean collapsed cripple walls.

Although a small number of contractors in earthquake country have been trying to eke out a living doing residential seismic retrofitting in the last few years, this moderately strong earthquake brought a new sense of urgency to Northern California residents and millions of others with older unsecured homes who live along active faults.

Unfortunately, there's a lot more eagerness than information surrounding seismic retrofitting. Foundation bolting seems pretty straightforward until you hit the first cripple wall that is too short to accommodate an upright hammer drill, or a foundation that is suspiciously easy to punch holes in. (The first condition requires a steel plate lag-screwed to the edge of the sill and bolted to the inside face of the stem wall; the latter condition can require a new foundation.)

However, information both for professionals and homeowners is becoming available. One such source is Earthquake Safe: A Hazard Reduction Manual for Homes *from which the following article and illustrations are adapted. This 54-page booklet covers seismic basics including descriptions of mudsill bolting, bracing, and foundation drainage.*

We chose seismic bracing schemes to concentrate on here. (Bracing follows foundation bolting in a typical retrofit but often gets less attention.) But keep in mind that these procedures are only part of a complicated whole that includes siting considerations, soil and foundation types, building shape and mass, structural details, and drainage. To be thorough in reducing the hazards of living where the earth shifts, you'll need to take all of these factors into account.

Frightening and destructive earthquakes rocked the San Francisco Bay Area, Coalinga, Calif., Alaska, Mexico City, Armenia, and Japan in the 1980s. Earthquakes were also felt in the Midwest, the South, and in New England. Recent information suggests that there are potentially active fault systems in New York City as well as other parts of the mid-Atlantic region.

In fact recently revised earthquake hazard maps show that in the United States alone over 50 million people live in areas of greater-than-average seismic risk, while perhaps a half a billion share the risk world wide.

While the idea of earthquake *proofing* is naive, the need to protect the structures we design and build is valid and real. Applying what engineers have learned about the effects of quakes can go a long way toward *reducing* the risk of serious structural failure.

Reducing the Hazards

Earthquakes are typically experienced as lateral motion: waves traveling through the earth that produce a roller coaster effect. But sudden upthrusts can also result. The intensity of a quake and its duration are major factors in how destructive it is.

Soil conditions and subsurface soil structure also dramatically affect the amount of vibration that is transferred to a building during an earthquake. For example, bedrock or stiff soil transfers much less vibration up through the foundation to a structure than do other soils. And the siting of the building can have a major effect.

Then there's the building itself. Four kinds of structural elements have to be examined on any house to determine what strengthening strategies are necessary. The first is the foundation that supports the building. Second are the horizontal members, such as the floors, that support and transfer the weight of the building and its contents to the walls or vertical members. The third are the columns, support walls, pilasters, posts and other vertical members which transfer the weight down to the foundation. The fourth structural element is all the points of connection between members. It is these points of connection that require special attention in earthquake country (see Figure 1).

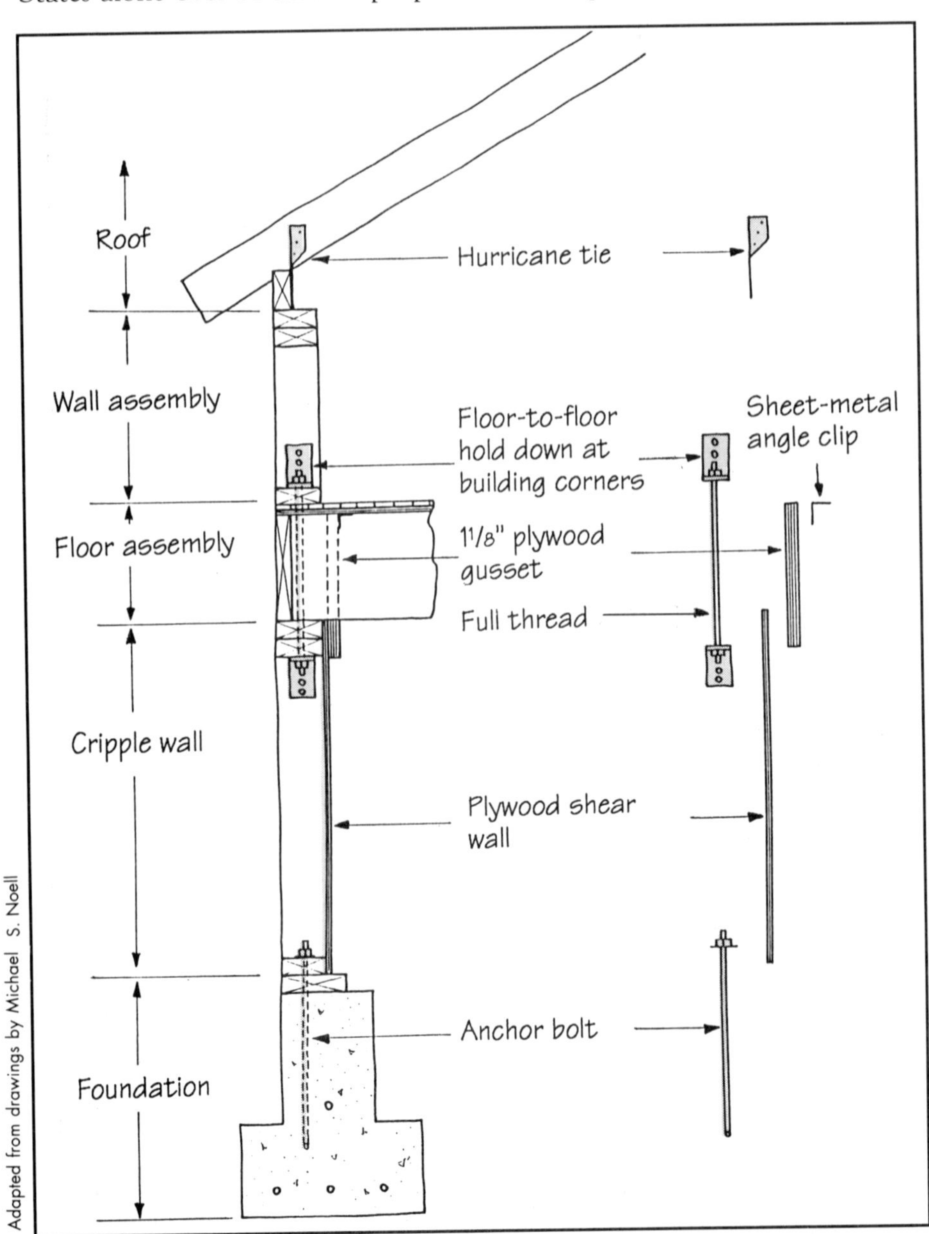

Figure 1. *Tools of the seismic retrofitter:* *Hurricane ties connect roof rafters and top plates; hold-downs connect between floors; plywood gussets and angle clips go in joist spaces; plywood shear panels strengthen cripple walls; and anchor bolts keep the mudsill tight to stemwalls.*

Although I've seen newspaper ads that promise foundation anchoring in a day for $5 a bolt, generally it takes a skilled three-man crew 100 to 200 labor hours to complete a comprehensive substructural seismic reinforcing job for a typical 2,000-square-foot, two-story house. This work is based upon a thorough inspection report that yields a program of bolting and bracing that leaves few vital connections without some form of strengthening.

This brings up the question of codes and standards. To date, code bodies and state governments have not dealt with existing homes. As a result, many municipalities don't have procedures in place for dealing with seismic retrofitting. In our work, we use the new construction requirements of the 1988 Uniform Building Code (Chapter 23) as the standard.

Reinforcing a building to reduce earthquake hazard should begin at the foundation with a careful examination of the concrete, and the connection between it and the mudsill. If the sill isn't bolted to the foundation, that has to be remedied.

The next concern is with the structure immediately above the mudsill. In many cases, this is a short cripple wall. Severe racking and moment failure of cripple walls are the major sources of structural damage to homes in serious earthquakes. The remedy for these stresses is the installation of bracing.

Existing Bracing

Most wood-frame residences have some kind of bracing nailed to the framing. One-by-six sheathing, either under exterior siding or standing alone, is common in older homes. However, board sheathing is the weakest form of wall bracing. One-by-six *let-in braces*, which run a 45 degree angle between top and bottom plates and are cut into the edges of the plates and studs, lend greater strength. But plywood *shearwall* that connects mudsill, cripple studs, and plates is the strongest wood-frame construction technique.

Mesh-reinforced stucco contains some shear strength, but these values vary a lot with the type and amount of mesh reinforcement, the way it's attached to the frame, and the thickness and quality of the stucco itself. Also, most of stucco's shear value is lost when it cracks. It shouldn't be thought of as a substitute for shear panels.

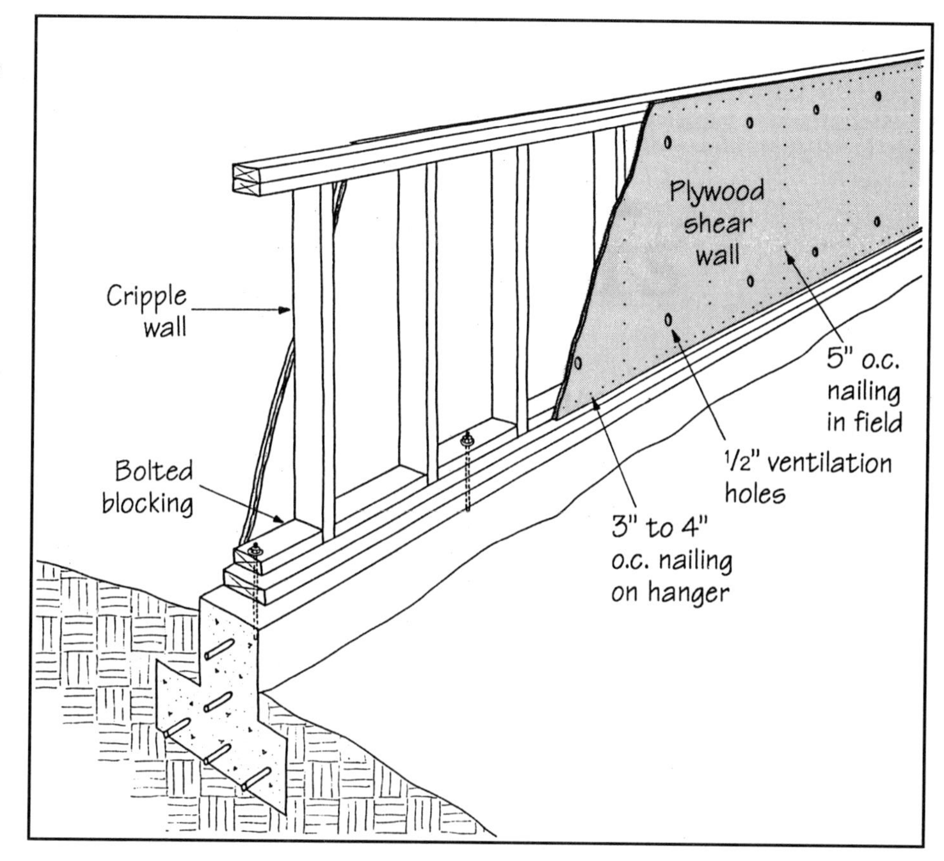

Figure 2. *The primary form of seismic bracing on cripple walls is plywood shear panel with a very close nailing schedule. When the mudsill is wider than the wall framing, blocks have to be anchor bolted down through the sill to provide the vital connection to the foundation.*

Shearwall Basics

In seismic construction, 1/2-inch CDX plywood is the standard for one story homes. Houses that are two-stories or more typically require 5/8-inch plywood, with 3/4-inch plywood used for three-story or taller homes.

In an existing building, you can often retrofit the shearwall onto the cripple studs from the inside of the crawlspace. This saves the expense of tearing off existing exterior siding and replacing it with new once the plywood has been nailed up. However, you won't be able to avoid this procedure on houses that are built on slabs or that don't have a crawlspace large enough to let you handle plywood.

The walls that require the greatest amount of shearwall are those that bear the greatest portion of the structural load from floor joists and roof rafters. However, you should start with each corner of the building. The standard rule of thumb is to install shearwall 6 to 8 feet, measured horizontally, in each direction from each corner of a one-story house of 1,000 square feet or less. This will account for about 40% to 50% of the substructure wall area. For two-story houses, you'll need 8 to 16 feet in each direction (50% to 70%). Plan to shear-wall the entire substructural area in homes that are three stories or larger.

To calculate the shearwall needs of a home more precisely, you can turn to the UBC design criteria for new construction. It's based on a number of factors: soil conditions (rates S1 to S4), seismic risk zone (0 to 4), occupancy use "importance," total building weight, and architectural and framing details. If you are familiar with calculating loads to find building weight and can make some educated guesses about general soil conditions, it's not difficult to apply these criteria.

However, there are some conditions that pose special risks, and require a foundation or structural engineer. These include:

- Questionable soil (fill, high water table, history of slides, etc.);
- Steep sites;
- Brick foundations, or concrete that is badly deteriorated, severely cracked, or lacking reinforcement;
- Pier-type foundations;
- Unusual distribution of building mass or large openings in the structure, such as garages that have living space above (these are called *soft stories*).

A Simple Installation

Shearwall has to be nailed to a fully blocked, flush surface (see Figure 2). The plates above the cripple studs, the edges of the studs themselves, and the mudsill must all be in the same plane, and the edges of the plywood must be supported continuously by this framing.

That raises a common problem. In many older homes, the mudsill is wider than the studs and plates that are nailed to it. To compensate for this, you have to install flat blocking on top of the mudsill that is flush with the wall framing. This should be bolted through the mudsill into the concrete foundation with an 8½ or 10-inch anchor bolt (depending on the number of stories). You should bolt at least every other block, making sure that you don't skip the ones that fall at the intersection of two panels. At the corners, install a hold-down rather than an anchor bolt. This will bolt the block down to the foundation and tie in the corner post as well.

Make sure all your bolting—including all hold-downs—is complete before cutting and nailing up shearwall. Once you cut a panel, snap chalk lines to mark studs, diagonal bracing, and any blocking in the wall. Before installing the plywood, it is a good idea to paint the mudsill and the studs with a wood preservative in wet or very moist environments to help guard against termites and dry rot. Also coat the plywood on both sides at least 6 inches up from the bottom edge.

Although nobody is likely to see your work, a flat, tightly nailed panel is most effective. Begin by tacking one corner of the plywood, and then working diagonally down the sheet, smoothing out any kinks, bumps, or warps ahead of you as you go. When you're finished nailing, the panel should resonate like a drum when you slap it between studs.

The best nailing schedule for the perimeter of the plywood is 4 inches on-center; you can reduce this to 6 inches for intermediate studs. While this may seem like overkill, the more nails you drive, the stronger the bracing becomes because the strength of the plywood shearwall is, in part, a function of the shear value of the nails used. We recommend 10d common nails.

You want the nail to penetrate approximately 50% of the depth of the studs, so a 6d, 8d, or 10d nail is appropriate using ½-inch shearwall over 3½-inch studs. Shear values are greater for 10d common nails. Nail guns are fine, but some municipalities do not allow clipped head nails. In general, full-headed nails are better.

Once you've finished nailing off the plywood, go back with an electric drill and punch a row of ½-inch-diameter vent holes 5 inches down from the top plate and the same distance up from the mudsill. Lay these out 16 inches on-center between framing members. This will provide sufficient air circulation for any moisture that builds up providing there is good storm drainage along the exterior of the footings and gutters and downspouts are properly drained. If good site drainage isn't in place, you can have more than a moisture problem in the crawlspace, since water-saturated soil exerts tremendous additional pressure on foundations during an earthquake.

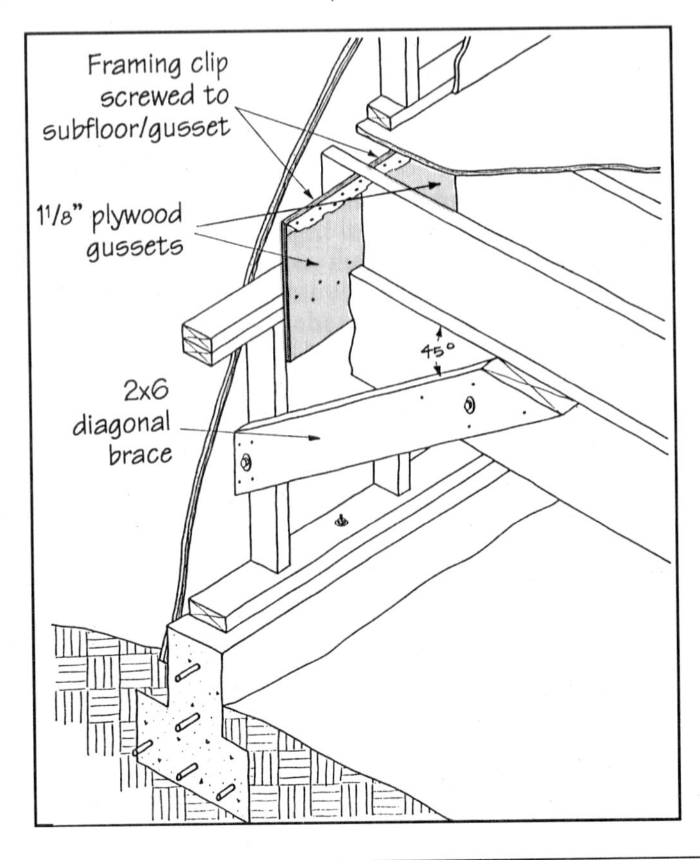

Figure 3. *If the floor system doesn't have solid end blocking or a rim joist, installing plywood "blocks" will keep joists from tipping, and will act as gussets to connect the floor and cripple wall. Diagonal braces provide moment resistance along the length of the joists.*

Bracing Floors

After the cripple walls have been shear-walled, you'll want to look at the first-floor framing. Good floor diaphragms have blocking between the joists at all plates and at 8-foot intervals in their spans. Joists that lack blocking are subject to tipping and can collapse during a serious earthquake. If the joist span is 10 feet or less, a line of blocking at mid-span and one at each end is sufficient.

Now turn to the floor-to-plate connection. First, make sure the joists are properly blocked at the perimeter with a rim joist or end blocking. If they are, install a sheet metal angle clip—a Simpson L90 or an equivalent connector—on top of the cripple wall in each bay. One leg of this connector nails to the top plate; the other nails to the rim joist. Use another, smaller clip in a vertical attitude that connects the joist end and the rim joist.

If blocking is needed, you can use it to make a solid connection between the floor system and the framing below. Cut a block from 1⅛-inch plywood that fits snugly between the joists. The top of it should butt the underside of the subfloor, and the bottom should lap the top plates of the cripple wall by at least 4 inches. Face nail this block-gusset to the cripple wall top plates at the bottom, and toenail the sides into the joists. At the top, screw a sheet-metal angle clip to the face of the plywood and into the subfloor above using #10 screws (see Figure 3).

If the floor joists sit directly on top of the mudsill—no cripple walls—the connection you use will again depend on blocking. If joist blocking at the sill is sufficient, use sheet metal angle clips to tie the mudsill and joist ends to the rim joist, and angle iron to connect the joists to the foundation (see Figure 4).

This 2¼x2¼x¼-inch angle stock should be machine bolted to the joist faces, and anchor bolted to the foundation. (If you are using this system with 4x girders, use angle iron on both sides of the beams.) You'll have to get the steel from a fabricator, but that gives you an opportunity to order it pre-drilled. Recently, I assisted the Simpson Strong Tie company in designing a mass-produced foundation-to-joist strap anchor now available in their FJA and FSA lines.

If you need to add blocking where there are no cripple walls, cut 1⅛-inch plywood gussets that extend from the bottom of the subfloor to about 10 inches below the mudsill. You're going to combine this with a ¼-inch steel plate that covers much of the plywood and attaches to the mudsill and the concrete.

Install the gussets with an angle clip at the top and toenails into the joists as shown in Figure 3. Use a layer of 30-pound asphalt felt where the plywood laps the concrete. Then, lag screw the pre-drilled, steel plate, which should measure about 12 inches square, through the plywood gusset into the edge of the mudsill, and anchor bolt it into the face of the concrete foundation.

Other Bracing

Installing diagonal braces that run from the joists to the cripple walls provides moment resistance against the hinging action that long runs of joists are prone to in prolonged earthquakes. A typical diagonal brace is a 2x6 nailed and bolted to a cripple stud about halfway between the mudsill and top plate. A holddown is recommended to reinforce the stud to sill connection for diagonally braced studs. It extends up on a 45 degree angle to just under the subfloor and is nailed and bolted to the joist face it parallels, as shown in Figure 3.

Diagonal braces are typically placed in open bays, but can be used on shear-walled sections of cripple wall by installing the braces first, blocking around them to support the edges of the shear panels, and then cutting slots in the plywood so the sheets can be lowered down over the braces.

Another extremely weak intersection in wood-frame buildings is post-to-beam connections. These should be reinforced with T-straps. Plywood gussets may also be used along with diagonal 2x4 braces from the main post up to the beam that it supports. Posts should be secured to their piers with steel straps bolted to both the concrete and the wood as shown in Figure 5.

Although it requires removing and replacing some finish wall material on at least one level, bolting in holddowns and connecting them with *all-thread* between floors at building corners is a way of ensuring that platform-framed floors remain attached.

Roofs and Chimneys

Roof systems are triangulated wood structures, and tend to be rigid if carefully framed. However, where rafters and ceiling joists rest on the top plate they should be reinforced with sheetmetal "hurricane ties." Adding sheetmetal angle clips in each bay to tie the blocking to the double plate further increases the attachment between the roof and the

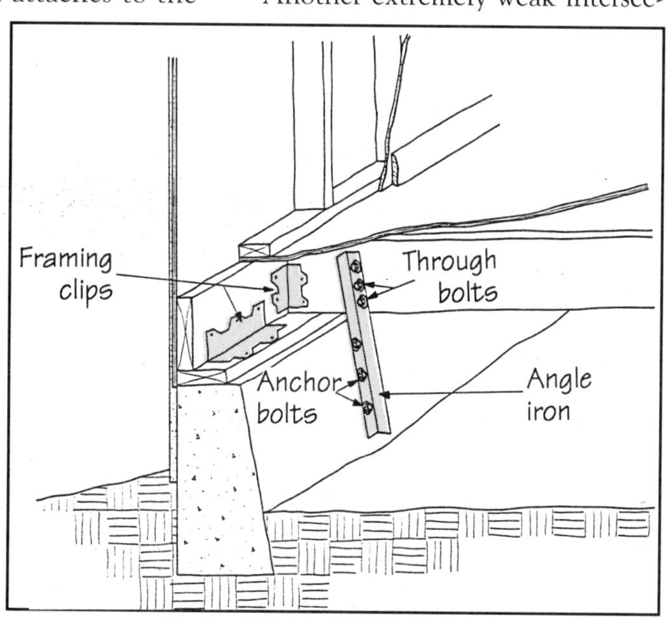

Figure 4. *This bracing scheme—for floor joists that bear directly on the mudsill—relies on ¼-inch angle iron bolted to the joist and the stemwall, and sheet-metal framing clips nailed to the joist, mudsill, and rim joist.*

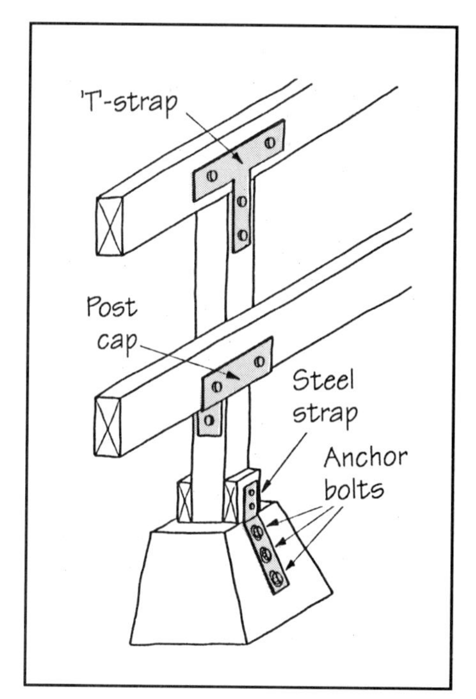

Figure 5. *Post-to-beam and post-to-pier connections are reinforced with steel brackets and machine bolts. Pier systems—rather than grade beams or stem walls—are very vulnerable in earthquakes and difficult to retrofit effectively.*

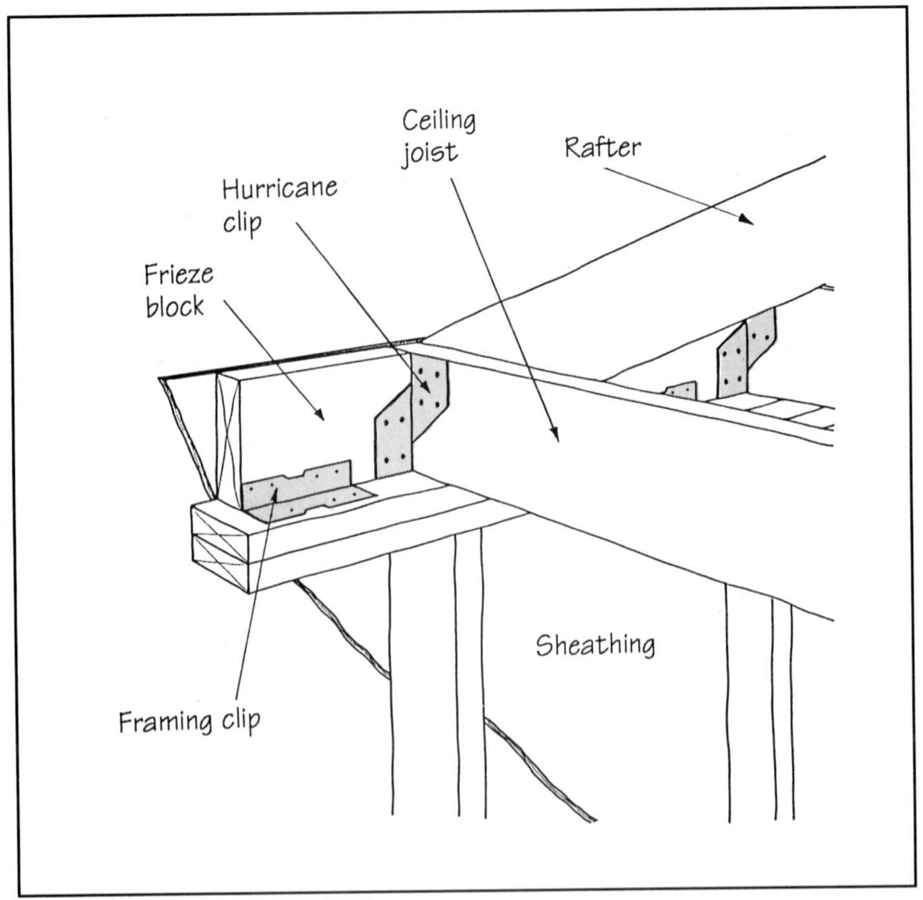

Figure 6. *Although most roofs are well triangulated by sheathing, nailing hurricane clips to plates and rafters or ceiling joists, and framing clips to plates and frieze blocks, ensures a connection between the roof system and the remaining framing in the house.*

vertical framing (see Figure 6).

A cautionary note regarding reroofing: Although most municipal building codes allow three layers of roofing on an existing roof frame, that represents a great deal of weight. Stripping the roof will give you a chance to add the sheetmetal connectors more easily and add more nails to the existing plywood to increase its ability to resist racking.

Brick chimneys often suffer in earthquake country from two general design limitations. First, they are typically built on their own inadequate foundations. Second, they are often not structurally connected to the building.

If there is bad drainage around the foundation, masonry fireplaces invariably sink. As they settle, they tend to rotate and pull away from the building. Proper drainage, a deepened, steel-reinforced foundation, and bracing back to the frame of the building can help.

During earthquakes, however, chimneys still have a tendency to break at the roofline. One solution for this is to take off the brick to the roof line and build a wood chase with a lightweight flue system, or a new, steel-reinforced masonry one. Another precaution well worth taking is to install plywood in the attic adjacent to the chimney so that a collapse won't send bricks through the ceiling and into living space below.

Chimneys that are structurally attached to the eaves of the roof tend to buckle near the center of their vertical run if bad drainage and a shallow foundation have resulted in some subsidence. To protect against this, you can typically brace these chimneys with structural steel and tie them back to the framing at points where they are prone to buckle. ■

David Helfant, Ph.D., a general engineering and building consultant, is a principal of SEISCO, Structural Engineering Inspection Services Co., Emeryville, Calif., and manages Bay Area Structural, Inc. an Oakland, Calif., engineering/foundation contracting company.

For More Information:

Earthquake Safe: A Hazard Reduction Manual for Homes *is published by Builders Booksource, 1817 Fourth Street, Berkeley, CA 94710; 415/845-6874, or 800/843-2028.*

Section 4. Cutting Roofs

Clayton DeKorne

Production Roof Cutting

Streamlined layout and cutting techniques can speed up gable, hip, and valley framing

by Will Holladay

I started cutting roofs on California tract housing, and while I only work on custom houses now, I still use the production techniques I learned back then because they increase my efficiency on the job site. Here are some of the tricks I have collected for laying out and cutting simple gables, hips, and valleys. All of the examples shown in this article are figured for a roof with a 6/12 pitch. Using the tables that accompany this article you can adapt the system to other roofs.

Setup

The first step in roof cutting is to set up racks so the rafters can be gang cut. Two styles of rafter racks are shown in Figure 1. A good set of racks doesn't need to be perfectly level, but the top edges must be parallel. You can check this by eyeballing across the tops to see if they lie in the same plane. Brace the middle of a long rack and wax the top edge with a block of paraffin to

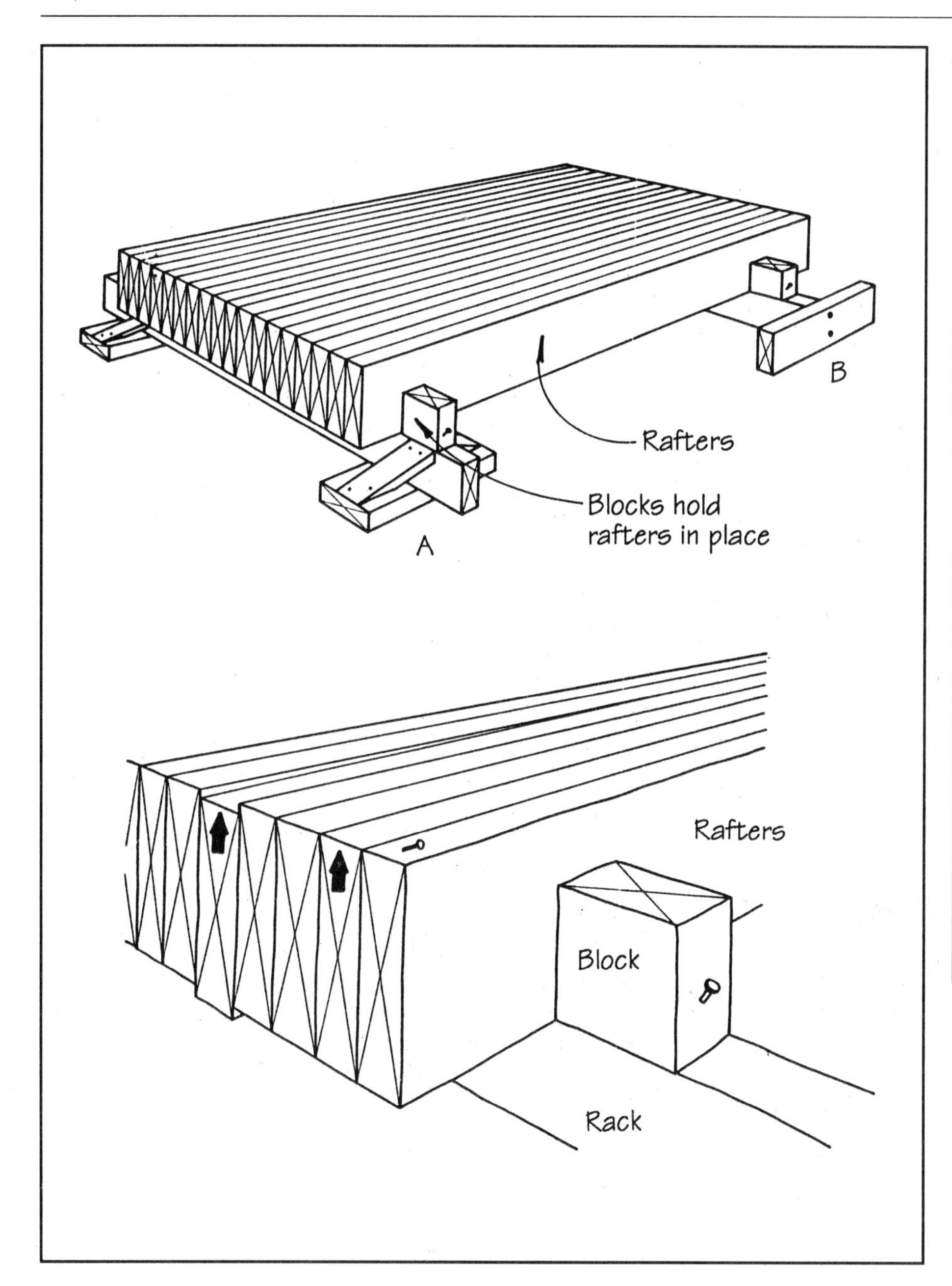

Figure 1. *For gang-cutting, lay out the rafters on racks and hold them in place with scrap blocks (top). A and B show two rack styles that work. Where boards are uneven due to extreme crowns or inconsistencies in the milling, raise the low boards and toenail them to hold them flush at the cut lines (lower).*

TABLE 1
Rafter LL Ratio = LL/Rafter Run

Pitch	Common	Hip/Valley
1	1.0034	1.4166
1½	1.0077	1.4196
2	1.0137	1.4239
2½	1.0214	1.4294
3	1.0307	1.4360
3½	1.0416	1.4439
4	1.0540	1.4529
4½	1.0680	1.4630
5	1.0833	1.4742
5⅓	1.1000	1.4865
6	1.1180	1.4999
6½	1.1372	1.5143
7	1.1577	1.5297
7½	1.1792	1.5461
8	1.2018	1.5634
8½	1.2254	1.5816
9	1.2500	1.6007
9½	1.2750	1.6206
10	1.3017	1.6414
10½	1.3287	1.6629
11	1.3567	1.6852
11½	1.3850	1.7083
12	1.4142	1.7320
14	1.5366	1.8333
16	1.6666	1.9436
18	1.8028	2.0615
24	2.2361	2.4495

make it easier to slide the rafters around.

Stack the rafters crown down and leave about 12 inches of the rafters hanging over one rack. The second rack should be placed approximately under the birdsmouth cuts. Nail end blocks on the rack to keep the rafters together. Rafters that are low can be raised and toenailed to the adjacent rafters to keep them flush on top, as shown in Figure 1.

Gable Roofs

Laying out common rafters. To lay out the ridge cut and birdsmouth, I determine the rafter's *line length* (LL). *The LL equals the length from the ridge cut to the heel cut of the birdsmouth.* This length is measured along the bottom edge of the rafter. The LL for a straight gable run is equal to the *actual rafter run* multiplied by the *common LL ratio* for a specific pitch. Most people are used to using a theoretical rafter run. To find the actual rafter run, subtract the thickness of the ridge from the span of the building, and divide by two (see Figure 2, next page). The common LL ratio is, in trigonometric terms, the secant of the roof pitch angle. Rather than using rafter tables or running through the Pythagorean theorem, I use the LL ratio to figure all my roofs. Table 1 lists the common LL ratios for most pitches

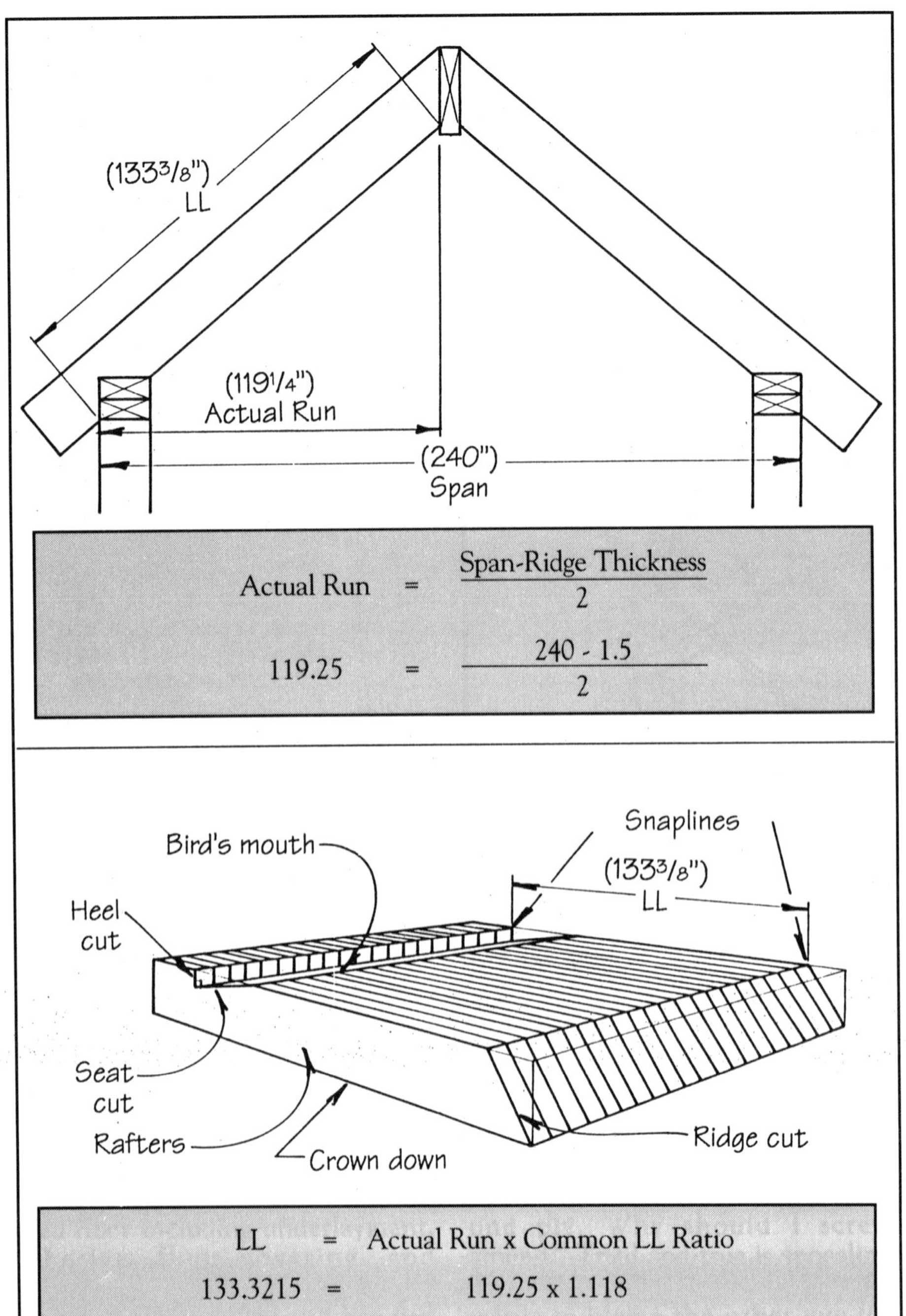

Figure 2. *To find the rafter run, subtract the thickness of the ridge from the span and divide by two (top). This is multiplied by the LL ratio to find the length of the rafter (LL). On the racks (lower), the ridge cut line is snapped on one end of the pile. The LL is measured from here to find the heel cut line.*

TABLE 2

Common Run = 12 Roof Pitch Degrees

Rise	Angle
1	4½
1½	7
2	9½
2½	11¾
3	14
3½	16¼
4	18½
4½	20½
5	22½
5½	24½
6	26½
6½	28¼
7	30¼
7½	32
8	33¾
8½	35¼
9	37
9½	38½
10	40
10½	41¼
11	42½
11½	43¾
12	45
14	49½
16	53¼
18	56¼
24	63½

between 1 and 24.

Snap a line perpendicular to the rafters for the short point of the ridge cut near one end of the rafters as shown in Figure 2. Make sure you come in far enough from the end so that the long point of this cut will stay on the board. Measure the LL distance down the rafters at each end for this snapped line and snap another line between these points. This marks the heel cut of the birdsmouth. To make a crisp cut for exposed rafter beams, check diagonals across the pile from the seat-cut line to the ridge-cut line to be sure the cuts will be square.

Gang cutting commons. The birdsmouth can be cut with either a dado saw or a swing table saw. The dado saw makes the seat cut in one pass, whereas the swing table saw requires two. Use a board as a guide when cutting exposed birdsmouth cuts.

When the birdsmouth cut is finished, make a starter pass on the

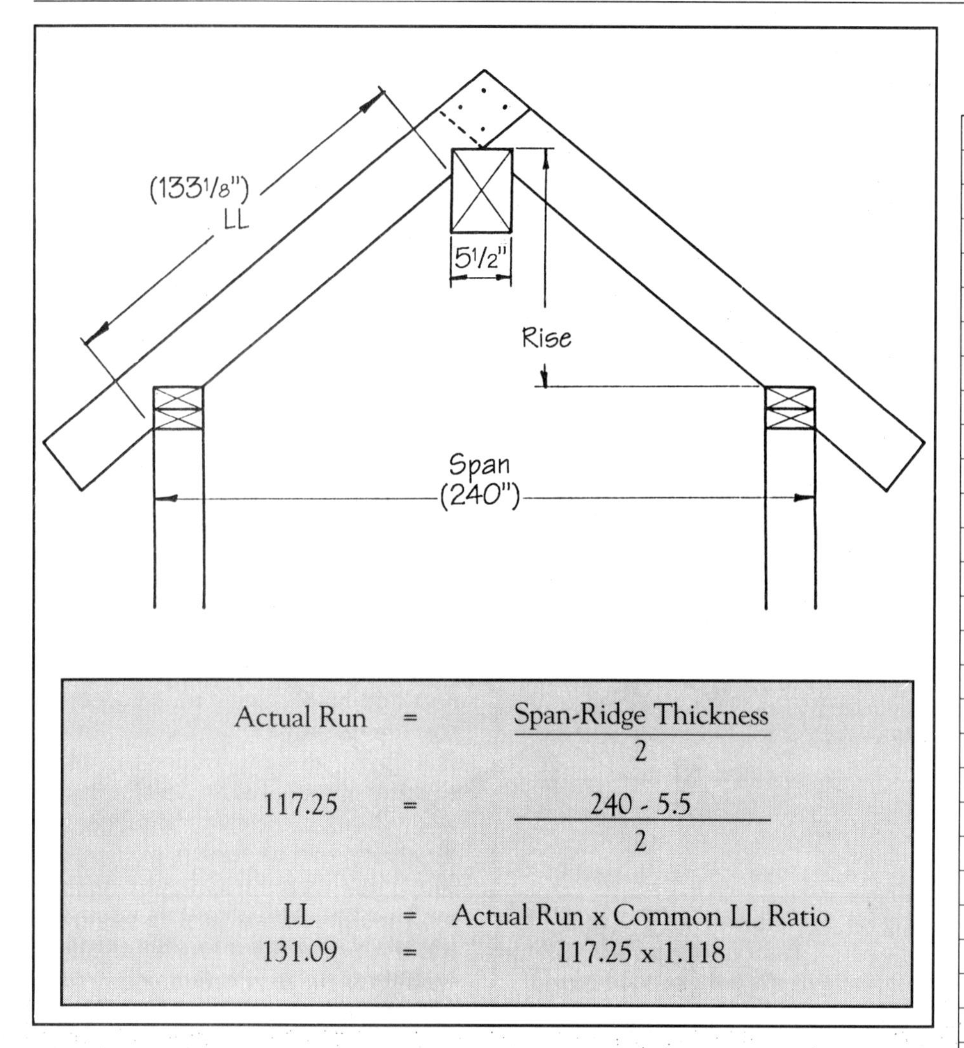

Figure 3. *For cathedral ceilings, the rafters sometimes lap over an exposed ridge beam. The rise is figured to the top of the ridge, and the rafter length (LL) is figured between the two birdsmouths.*

TABLE 3

Roof Rise Ratio = Rise/Run

Pitch	Common
1	.0833
1½	.1250
2	.1666
2½	.2083
3	.2500
3½	.2916
4	.3333
4½	.3750
5	.4166
5½	.4583
6	.5000
6½	.5416
7	.5833
7½	.6250
8	.6666
8½	.7083
9	.7500
9½	.7916
10	.8333
10½	.8750
11	.9166
11½	.9583
12	1.0000
14	1.1667
16	1.3334
18	1.5000
24	2.0000

ridge cut line with a circular saw set at the proper roof pitch angle, as shown in Table 2. Return the saw to square and change the depth to 1¾ inches. On one side of the pile draw a plumb cut line from the bottom of the starter pass with an adjustable bevel square. (I use a Squangle—Mayes Tool Co., #2 Claremont Dr., Johnson City, TN 37601; 615/926-6171—which has both common and hip valley cuts already marked out on it and a long enough blade to reach across the width of most framing lumber.) Saw in a vertical fashion to finish the cut. The blade will leave a kerf line on the second board. Push the cut board back out of the way of the saw table. Follow the kerf to cut the next board, and so forth. Check the angle of the cuts every few rafters with the Squangle to make sure the cut is not wandering off. If the roof pitch is very steep you will have to lay each board flat and cut horizontally.

Lapping over an exposed ridge beam. For cathedral ceilings, I'm often called upon to run the rafters of a gable roof over an exposed ridge beam. In this situation, I find it easiest to offset the rafters and lap them on top of the ridge beam. I lay out the on-center spacing on the ridge and the plates with an X on the long side of the layout mark for one side of the roof. On the short side of the layout mark, I set the rafters for the other side of the roof. Since each side of the roof is sheathed separately it doesn't matter if the rafters are offset.

There is no plumb ridge cut here, but there is a birdsmouth for the ridge beam. The rise for the ridge beam and the LL for the rafter are both figured from the actual run. This run is found by taking the span, subtracting the thickness of the ridge beam and dividing by two (see Figure 3). The rise is figured from the top of the outside plates to the *top* of the ridge. (The rise can always be found by multiplying the actual run by the *roof-rise ratio*. The roof-rise ratio is, in trigonometric terms, the tangent of the roof pitch angle. Here too, I figure these numbers out for common pitches and they are included in Table 3. To find the LL between the top and bottom heel

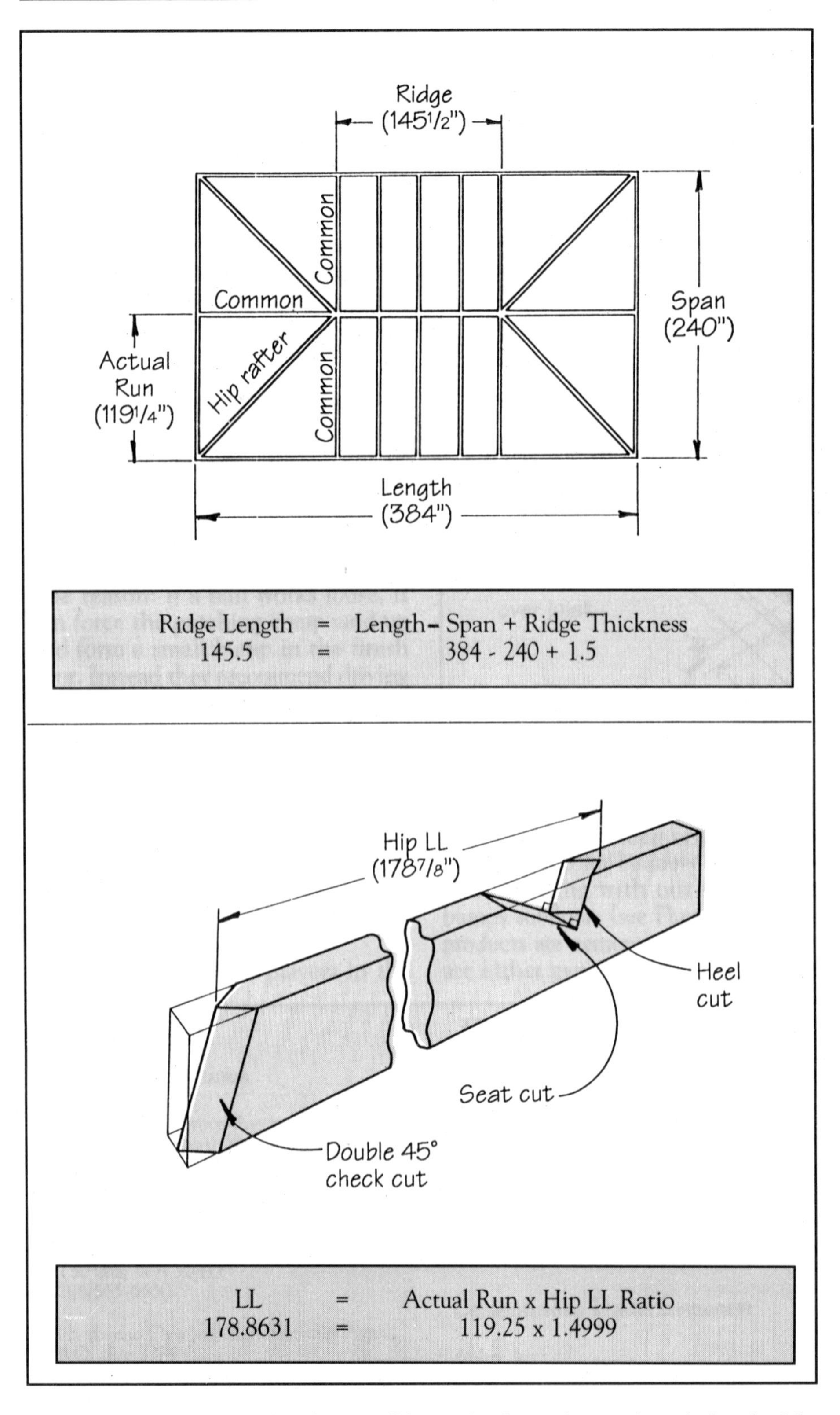

Ridge Length	=	Length – Span + Ridge Thickness
145.5	=	384 - 240 + 1.5

LL	=	Actual Run x Hip LL Ratio
178.8631	=	119.25 x 1.4999

Figure 4. *To find the ridge length for a full hip roof, subtract the span from the length of the house and add the thickness of the ridge (top). To find the hip run, subtract the ridge thickness from the span and divide by two (lower). This run is then multiplied by the hip LL ratio to find the LL of the hip rafter.*

cuts, multiply the run by the common LL ratio.

I usually cut the overlap on the ridge ends in the air to save time. Or, if this is hard to reach, you can lay out a pattern by scribing the cuts on a pair of rafters that have been tacked onto the ridge after the birdsmouths have been cut.

Hip Roofs

Hip ridge and commons. To figure the ridge length for a full hip roof, take the length of the building, subtract the width of the building and add the thickness of the ridge (see Figure 4). Hip roof commons are identical to gable roof commons. I put a common rafter at each end of the ridge, as shown. This helps steady the ridge when stacking the roof and provides a strong corner for the hip rafter to tie into.

Hip rafters. To figure the LL of the hip rafter, take the span, subtract the thickness of the ridge, and divide by two as shown in Figure 4. Multiply this distance by the hip/valley LL ratio in Table 1. The resulting LL is measured on the bottom of the board from the top double 45° cheek cut to the point of the V-notch at the birdsmouth. To find the cut lines for the top double 45° angle, measure downhill (towards the tail) half the thickness of the hip.

Make sure this distance is measured square to the hip/valley LL plumb line. When drawing the plumb line, don't forget to set the Squangle for the hip/valley plumb cut. I keep two Squangles painted separate colors for hip and common rafters so I won't confuse them.

To determine the depth of the seat cut, measure the plumb distance above the birdsmouth on a common rafter. Then transfer this distance to the hip rafter along a plumb line that is drawn parallel to the LL mark but uphill (towards the ridge) at a distance of half the thickness of the hip.

Remember to measure this half thickness square to the LL line. Cut the seat through this point perpendicular to the plumb line as shown

Backing Hips and Valleys

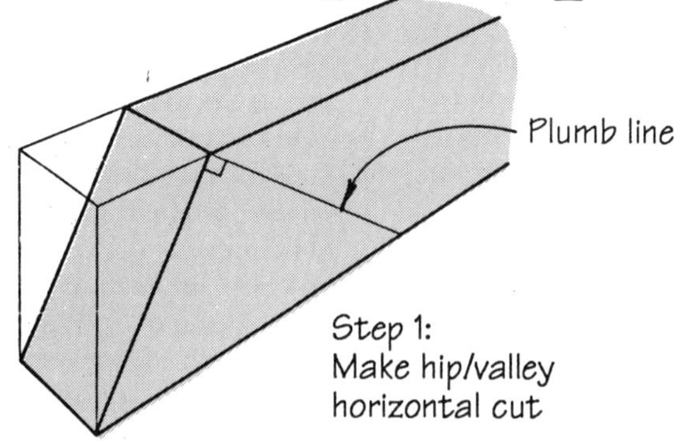

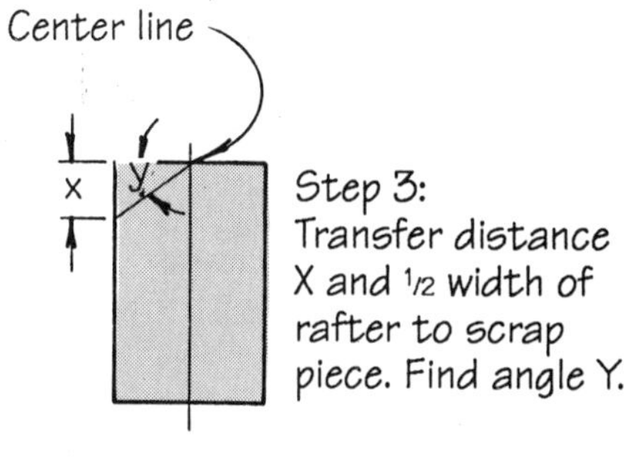

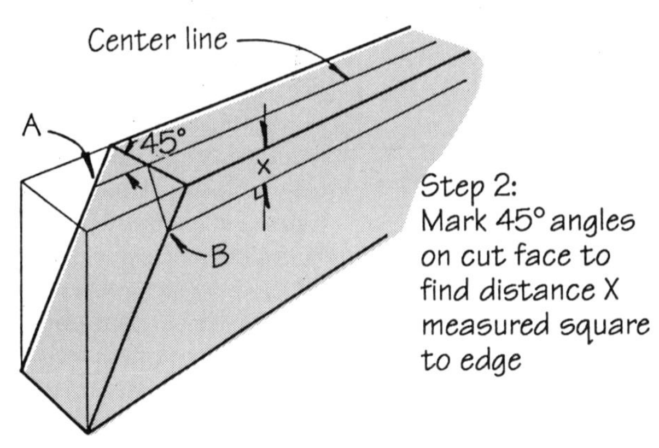

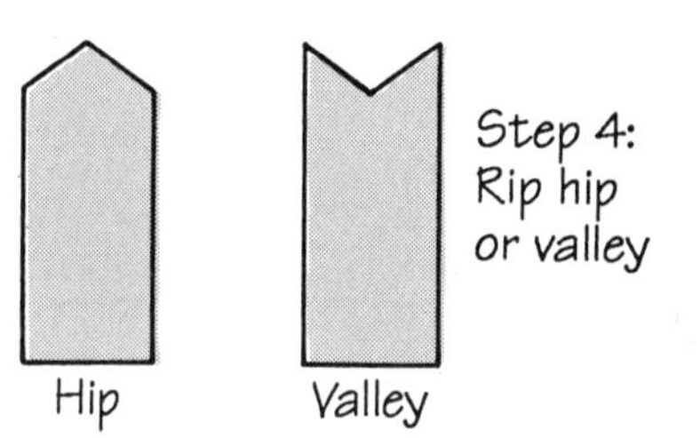

Wide hip and valley rafters can be ripped along the top edges so the two roof planes meet in the center of the rafters. The photo at right shows a valley rafter that has been "backed," or beveled, along the top edge.

The first step in finding the angle of this rip is to find the *backing distance*. This is the distance "X" shown in the drawing above. On a scrap that's the same width as the hip or valley rafter, draw a line perpendicular to the hip/valley plumb line and make this cut. On the face of this cut, draw a 90° angle that bisects the centerline of the rafter as shown. (For polygonal roof ends that meet at an angle other than 90° or bastard intersections, use the angle at which the wall plates meet.) Extend the ends of this angle to the edges of the cut, and from the end points A and B, draw two lines parallel to the top edge of the rafter. The distance from the top edge to this line will be the backing distance (shown as X).

To find the angle to set your saw at for the rip, draw a small right triangle with one leg equal to the backing distance and the other leg equal to half the thickness of the rafter. I use a Squangle to find the backing angle (shown as Y). To back a valley, set the depth of the saw to the long side of the triangle and rip along the edge of the rafter towards the center. To back a hip, rip along the scribed line X. –W.H.

A wide valley rafter can be "backed" so the roof planes on either side meet in the center of the rafter.

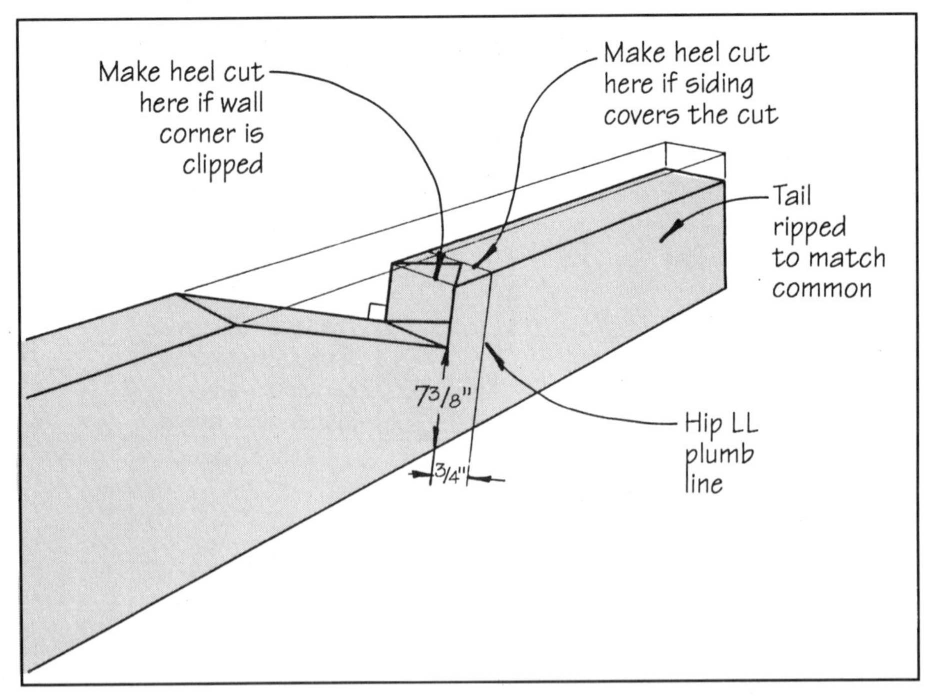

Figure 5. *The depth of the hip rafter's birdsmouth can be found by measuring the length of the plumb line above the birdsmouth on a common rafter. This distance, in this case 7 3/8 inches, is transferred to a plumb line on the hip rafter, which is drawn 3/4 inch (half the rafter thickness) uphill from the LL plumb line. If the heel cut doesn't show, this line can be cut square at the LL plumb line. Or, you can cut at the uphill line and nip off the corner of the top plate.*

With the help of a crane, the author lifts a full set of gang-cut rafters into position.

in Figure 5.

For a 2x hip, square cut the heel cut of the hip birdsmouth at the LL mark if the siding will cover the heel cut of the birdsmouth. Or you can also clip the outside corner of the top plate. For exposed beam tails wider than 2x, the V-notch has to be chiseled or chainsawed. For wide hips you might have to *back* the hip, especially if the tails are exposed (see "Backing Hips and Valleys," previous page).

Hip jacks. I lay out hip jacks on the racks using the rafter material that is the same length as the commons. I mark the LL for the longest jack on one side of the hip and step down to the shortest jack. I mark the other half on the cut-offs. I pair the shortest jack of one set with the longest of the other set, and so forth. The *step* is the length each jack is successively shorter. It is found by multiplying the on-center spacing of the commons by the common LL ratio. As shown in Figure 6 for 16 inches on-center, this is 16x1.118=17 7/8 inches. I usually take a lumber scrap and cut it to the length of this step to help mark off the jacks.

To start, determine the number of jacks needed for one side of the hip. To do this, take half the span (in inches), divide by the on-center spacing, and round up to the nearest whole number. For example, in Figures 4 and 6, this would be 120÷16=7.5, which rounds up to 8.

Next, find the LL of the longest hip jack and step down the remaining jacks. The LL of a jack is measured from the heel cut line to the short point of the cheek cut along the bottom edge of the board. The longest jack is shorter than the common LL by one step measurement and the LL of half the thickness of the hip rafter measured on the diagonal.

Lay out the hip jacks starting at the tail. Measure up for the overhang, and snap a line for the birdsmouth. From the birdsmouth line, measure up the length of the longest hip jack you just found, and place a mark at A, as shown in Figure 6. Make a keel mark going in the direction of the cheek cut to remind you that this is the short-point measurement. From A, mark off successive steps on each board using the step pattern until you reach the birdsmouth line. This completes one set of jacks.

To lay out the other side, set up another tail and birdsmouth at the other end of the pile. Mark off another long jack on the other side of the pile, and step down to the new birdsmouth line as shown. You should end up with two sets of opposite cheek cuts. Make the marks for each set with a different colored keel so the two sets don't get mixed up. I usually add an extra rafter to the pile and start the step off for the second set on this one, so the long points of the cheek cuts don't run into each other.

Make the birdsmouth cuts on each end before making the cheek cuts in the middle of the board. And check as you cut each line to see that they are going in the right direction.

Valleys

The LL and the ridge cut of a valley rafter is identical to a hip rafter shown in Figure 4. The only difference between the two is that the V-notch in the birdsmouth is reversed. To reverse the cut, measure half the thickness of the valley rafter perpendicular to the LL plumb line towards the tail. Draw a second plumb line here. Then measure the plumb height of the common rafter above the birdsmouth and transfer this measurement to the LL plumb line on the valley rafter. Then draw a perpendicular line through this point to mark the seat cut. This automatically drops the valley to the ridge height. Again, if the heel cut will be covered by siding it can be cut square at the second plumb line.

Valley jacks. Valley jacks are "stepped off" similar to hip jacks. Here, however, you don't cut a birdsmouth. Instead you lay out two parallel plumb cuts; the end that intersects the valley is a single cheek cut and the other end is a regular ridge cut.

There are four sets of jacks in a valley intersection—two sets for the

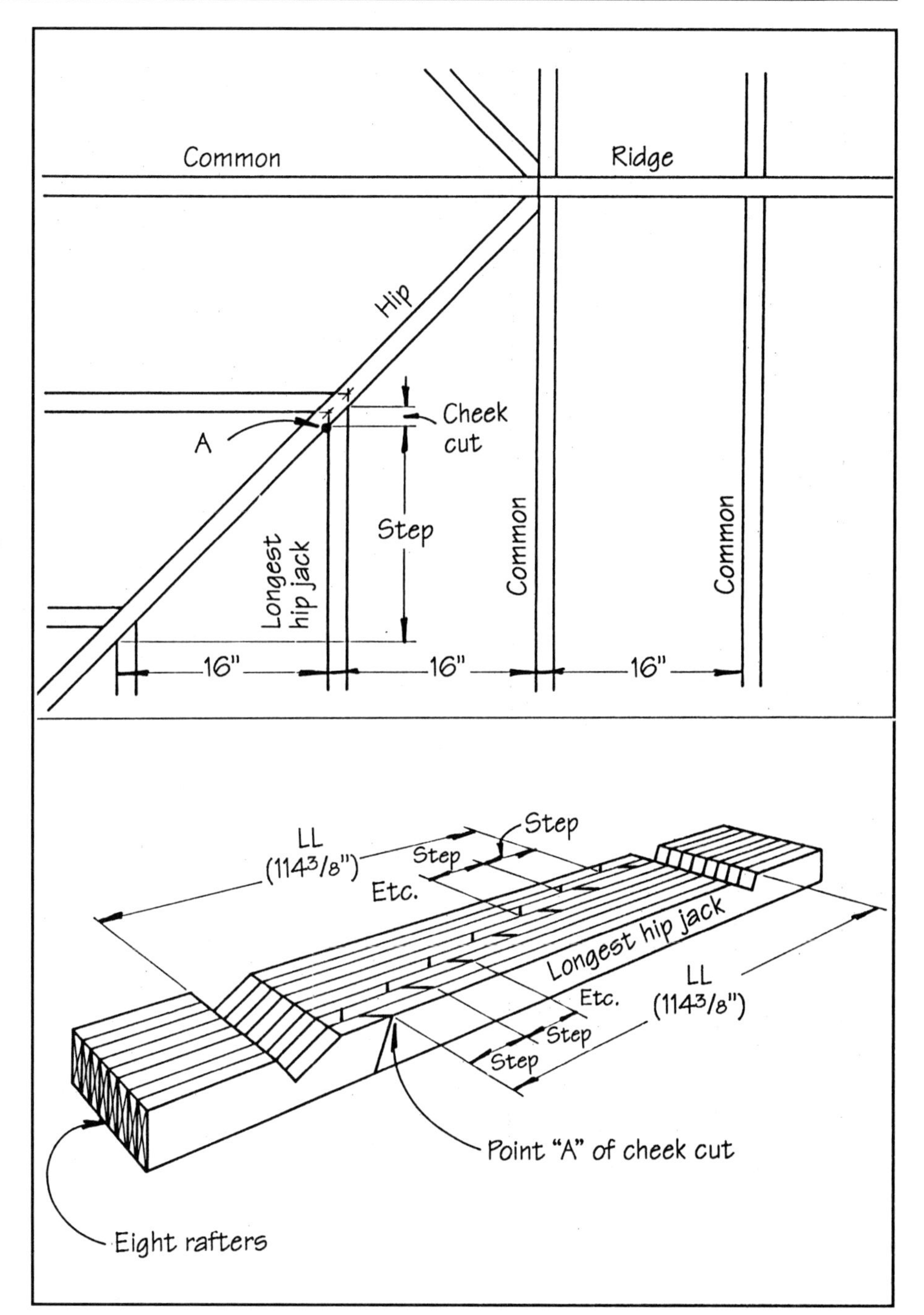

Figure 6. *To figure the length of the longest hip jack (top), subtract the step distance and the LL of half the 45° thickness of the hip rafter from the length of a common rafter. The "step" is found by multiplying the on-center spacing of the commons by the common LL ratio. Note that the LL of the jack is measured here to the short point of the cheek cut, labeled A.*

On the racks (lower), one set of hip jacks is stepped off on common rafter stock. The other set is stepped off in the opposite direction on the offcuts. Notice that the keel marks, which indicate the direction of the cheek cut, are made in opposite directions.

span of the main roof and two sets for the span of the intersecting roof—as shown in Figure 7. I lay them all out and cut them one at a time.

The longest valley jacks on each side have the same LL as a common rafter for the same run. I start the layout by snapping a ridge-cut line, and then I measure down the LL of the common rafter for each roof span. Next, I "step off" the jacks to lay out sets A and D as shown. The step marks indicate the long points of the cheek cuts.

To mark the alternative sets B and C, begin by calculating the LL of a 45° cheek cut. This is the thickness of the jack multiplied by the common LL ratio. Measure this distance towards the ridge cut line on the shortest jack in sets A and D, and mark the short point of the cheek cut on the opposite edge of the board. The short point will be the long point of the longest jack on the alternate set for each span. Measure down the LL of the common rafter to find the ridge cut lines for these alternate sets. I make the ridge cuts for two jacks in one pass.

More Complex Roofs

These techniques can be used to lay out and cut more complicated roofs. If you can get used to thinking with the LL ratio, you can easily grasp the relationship among all the roof's parts. I use the LL ratio in place of the Pythagorean theorem for figuring out *any* roof cut. ■

Will Holladay is a framer from Santa Barbara, Calif. He is the author of A Roof Cutter's Secret To Framing The Custom House, *from which this article is adapted.*

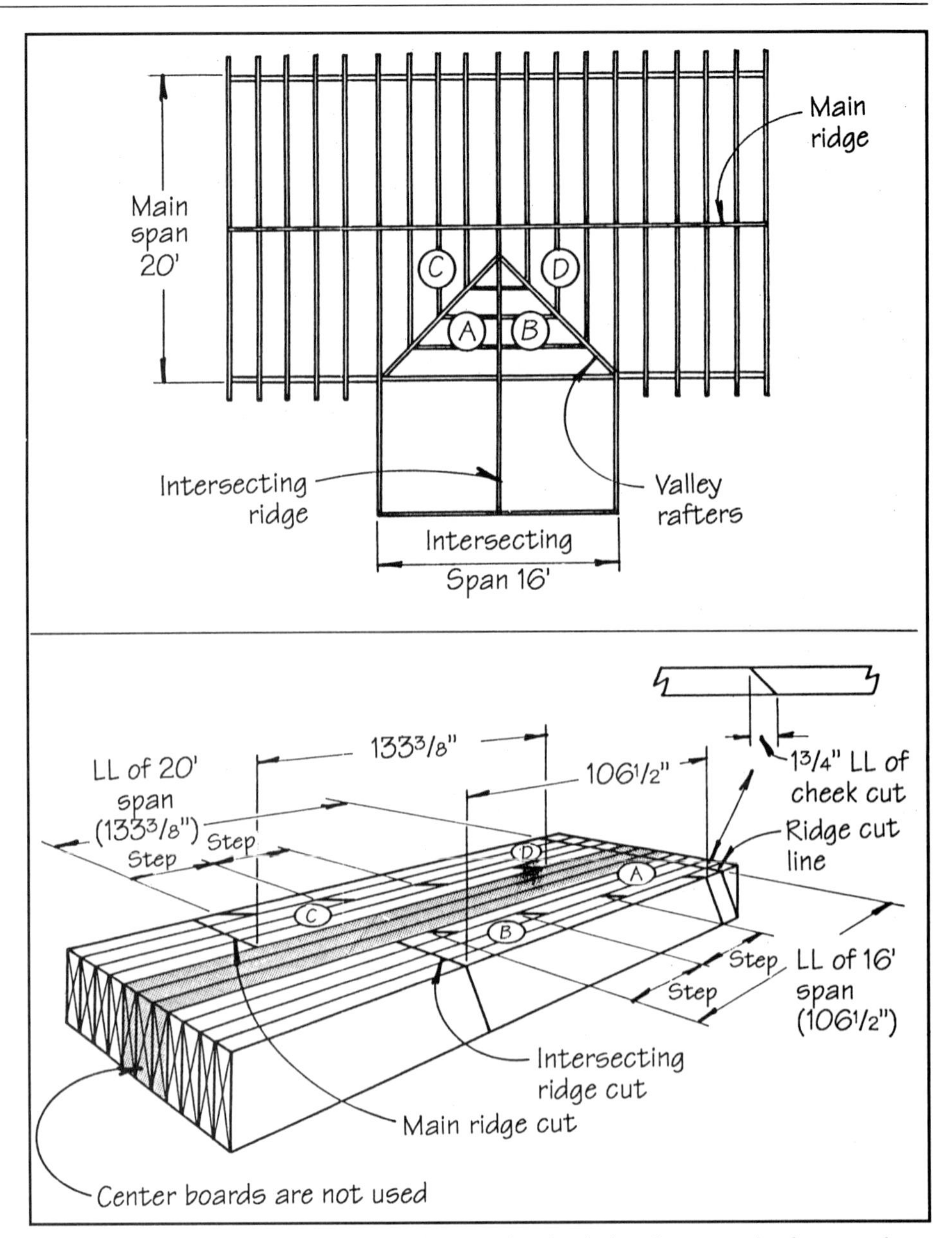

Figure 7. *A valley intersection, shown here in plan (top), has four sets of jacks—two for each roof span.*

All four sets can be laid out on the racks (lower). The long points of one set from each span are stepped off first. Then the short points of the other sets are found using the LL of the cheek cut. Note the ridge cut and the cheek cuts go in the same direction.

Cutting Jack Rafters

After figuring the first one, it's as easy as measuring, setting your saw to 45 degrees, and cutting

by Robert Adam

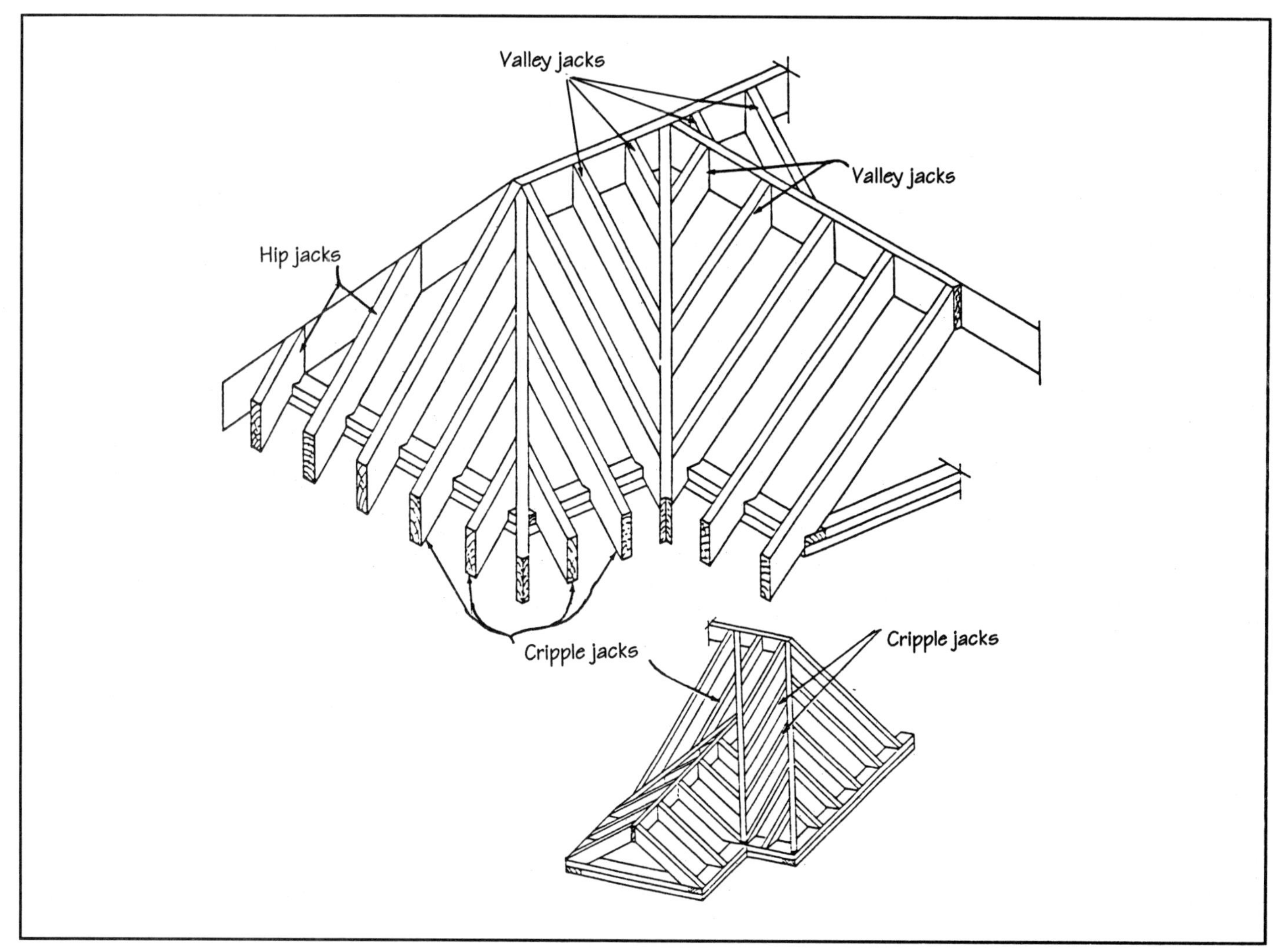

Roof framing is, for most carpenters, a mysterious joining of rafters that—with a little luck—fit the first time. For some, it is a nightmare of frustration. Most of us have no problem with common rafters or an occasional hip or valley rafter, but jacks often seem mind-boggling.

The key to understanding all rafters is to think of them in terms of horizontal and vertical coordinates (lengths of run and rise) rather than as the diagonal lengths they really are. They should also be visualized in plan and elevation views. If thought of in these ways, all rafters are quite easily understood.

Jack rafters are simply shortened commons that intersect with a hip or valley or both. Hip jacks run from the top plate to the hip; valley jacks from the ridge to the valley; and cripple jacks between hips and valleys. Since they are shortened versions of commons, it make sense to view them as types of commons.

To follow this discussion, you'll need some familiarity with the use of a framing square. I use a Stanley AR100 aluminum framing square. It has the most complete set of scales and tables available, and it is easy to read. Also, because it is aluminum, it will not rust. I also use "stair gauges," movable clamp-on-fixtures that allow the user to represent the units of rise

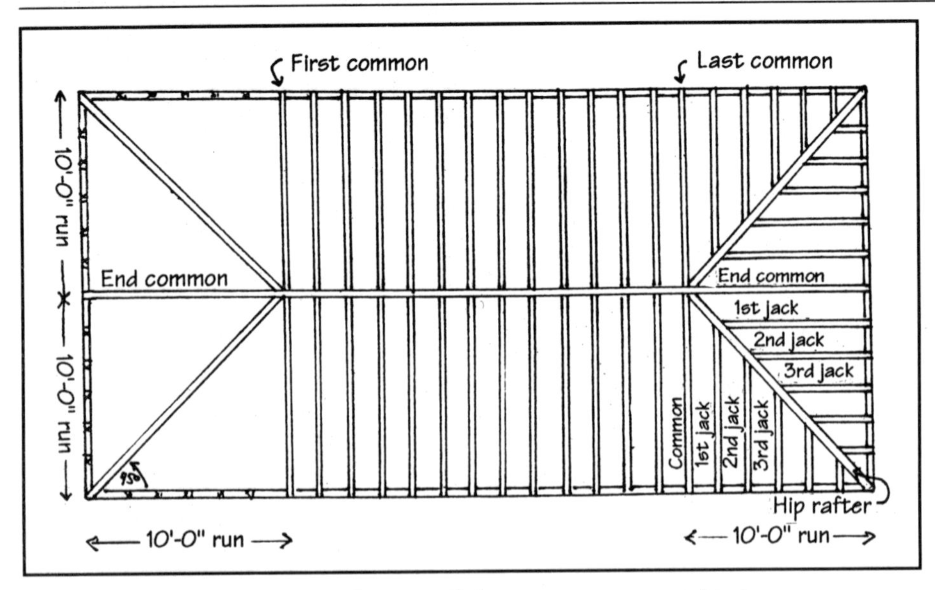

Figure 1. ***Roof plan:*** *Hips are always at 45 degrees to commons and jacks.*

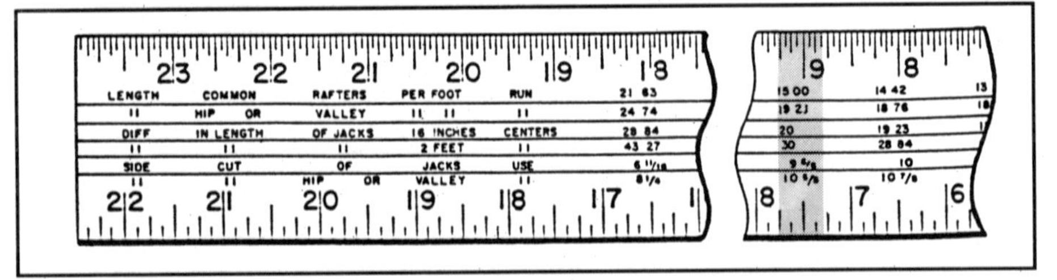

Figure 2. ***Framing square:*** *Shows that 20 inches is the "difference in length of jacks" for a 9/12 roof framed on 16-inch centers.*

and run on the tongue and blade (body) of the square. (I use Starrett #111 stair gauges because they offer greater accuracy than General's brass "stair buttons.") I also lay out my plumb and level lines with a sharp #2 or #3 pencil, not a carpenter's pencil.

Laying Out a Hip Roof

Before we get into laying out and cutting jack rafters, let's review how a hip roof is laid out. For the system illustrated, the span is 20 feet, so the run of the common rafters is 10 feet.

Starting at the outside of each corner, measure in 10 feet (the length of the run) along the top plates and mark these points. These marks show the center-line position of the first and last common rafters, and the two end commons.

Next, lay out the centerlines of each common rafter, starting with the first common and marking the top plates at the appropriate distance (every 16 or 24 inches) on-center. Following that, mark the commons on the opposite plate. Since each mark is for the center of the rafter you'll need to square a line 3/4 inch off each centerline for the edge of each rafter. These marks can be transferred from the plates to the ridge.

Now you're ready to mark the hip jacks. These should be measured from the first commons, last commons, and end commons—working your way into the four corners. Keep the same on-center intervals.

Make a Pattern

Now select a straight and defect-free piece of rafter stock to make a pattern. Calculate and lay out the length of a common rafter on the pattern piece—complete with crow's foot (birdsmouth), and rafter tail. Cut and install all the commons and the ridge. But don't use the pattern for a common rafter, since it will also be a pattern for the jacks.

Next, lay out one of the hip rafters, cut it and test it, then use it as a pattern for the rest. Be sure to adjust the depth of the crow's foot on the hip rafters so that the top edges of the hip lie down in the same plane as the common rafters.

Most of us can get to this point in the framing process without too much difficulty. But now the jacks must be cut. In Figure 1, you can see there are five jacks at 16 inches on-center. In the plan view, each is shorter by the same amount: 16 inches. That's because the hips are always at 45 degrees to the commons and jacks.

Laying Out the Jacks

The easiest way to figure the lengths of the jacks is to calculate and lay out their length on the common-rafter pattern. To do this, first find the "difference in lengths of jacks" on your framing square for the roof pitch and rafter spacing you are using. In this case (see Figure 2), for a 9/12 pitch at 16 inches on-center, the difference is 20 inches shorter than the previous one. The longest jack–measured on its centerline–will be 20 inches shorter than the common rafters.

The trickiest part here is finding the length of the first hip jack. Start by measuring along the top edge of your pattern piece, extending the hook of the tape past the end of the rafter to the theoretical center of the ridge. (This is easily done by nailing a piece of 3/4-inch scrap to the end of the rafter as in Figure 3a). Make a mark at 20 inches—the difference in length found in the rafter table—and square it across the top edge of the rafter.

Now you have to adjust for the thickness of the hip rafter. To do this, make a second mark 1 1/16 inches past the first—measured on the horizontal. (You'll do this by dropping a plumb line down the face of the rafter, and measuring over to a second plumb line. One and one-sixteenth inches is half the diagonal thickness of the hip rafter, based on nominal 2x lumber.) Now square this second plumb line across the top edge of the pattern rafter (see Figure 3b). Put an X where this line crosses the centerline of the rafter.

The next step is to mark the so-called side cuts. To do this, go back to the framing-square tables and find the number called side cuts and jacks under the appropriate pitch. Under 9 (for 9/12 pitch) we find the side cut is 9 5/8.

Adjust your framing-square stair gauges to 9 5/8 on the tongue (16-inch leg), and 12 inches on the blade (24-inch leg). Hold the square to the edge of the pattern and mark on the blade side a line through the X you just made. This diagonal line is your side cut (see Figure 3c).

Finally, extend a line down the side of the rafter from the side cut to make your plumb cut (see Figure 3d). This plumb cut now represents the length of the longest jack to its long mitered point. The second longest is 20 inches shorter, and so on.

If you were going to use this pattern as your first jack rafter, you would set your circular saw to 45 degrees and cut along the plumb line. But to use it as a pattern, you should instead cut it square along the plumb line, so it can be used to mark both left and right jack rafters.

A Pair of Jacks

All the same-length jacks should be cut together to ensure uniformity and accuracy. Begin with the longest and shorten the pattern by 20 inches for each successive jack. All pieces of each length should be in matching pairs, cut plumb at 45 degrees, and have the same crow's foot and rafter tail as the common rafters.

It's best to try the first pair of jacks of each length before cutting the rest. Jacks, like all rafters, should be installed by first nailing the crow's foot into the plate. But be certain to do them in pairs so that their upper ends line up at the hip directly opposite each other. This will automatically straighten the hip. The upper ends of the jacks should be nailed through the cheek into the hip.

Valley Jacks

Valley jacks are easier to lay out, because the cut is measured directly from the plumb cut at the crow's foot on your common pattern. There are no hips or ridges to adjust for.

A centerline is not necessary. Just mark the "difference in length," square

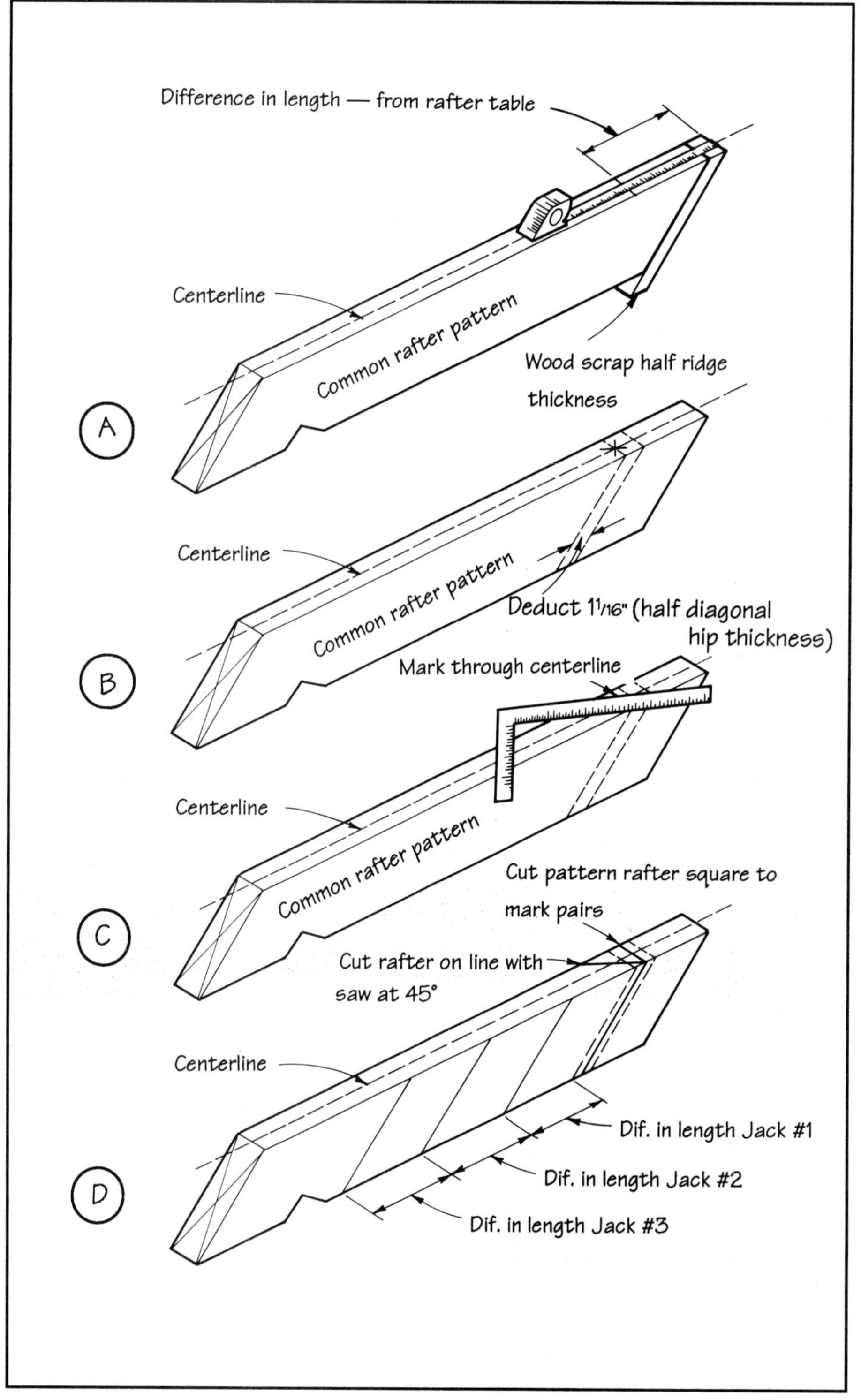

Figure 3. *Laying out hip jacks.* *Starting with the common-rafter pattern (A) mark difference-in-length-of-jacks. Next, subtract half the diagonal thickness of the ridge (B). Third, mark the appropriate sidecut angle (C), and cut on the plumb line (D) with the saw set at 45 degrees. (Or cut square on the plumb line to make a jack-rafter pattern to mark left and right jacks.)*

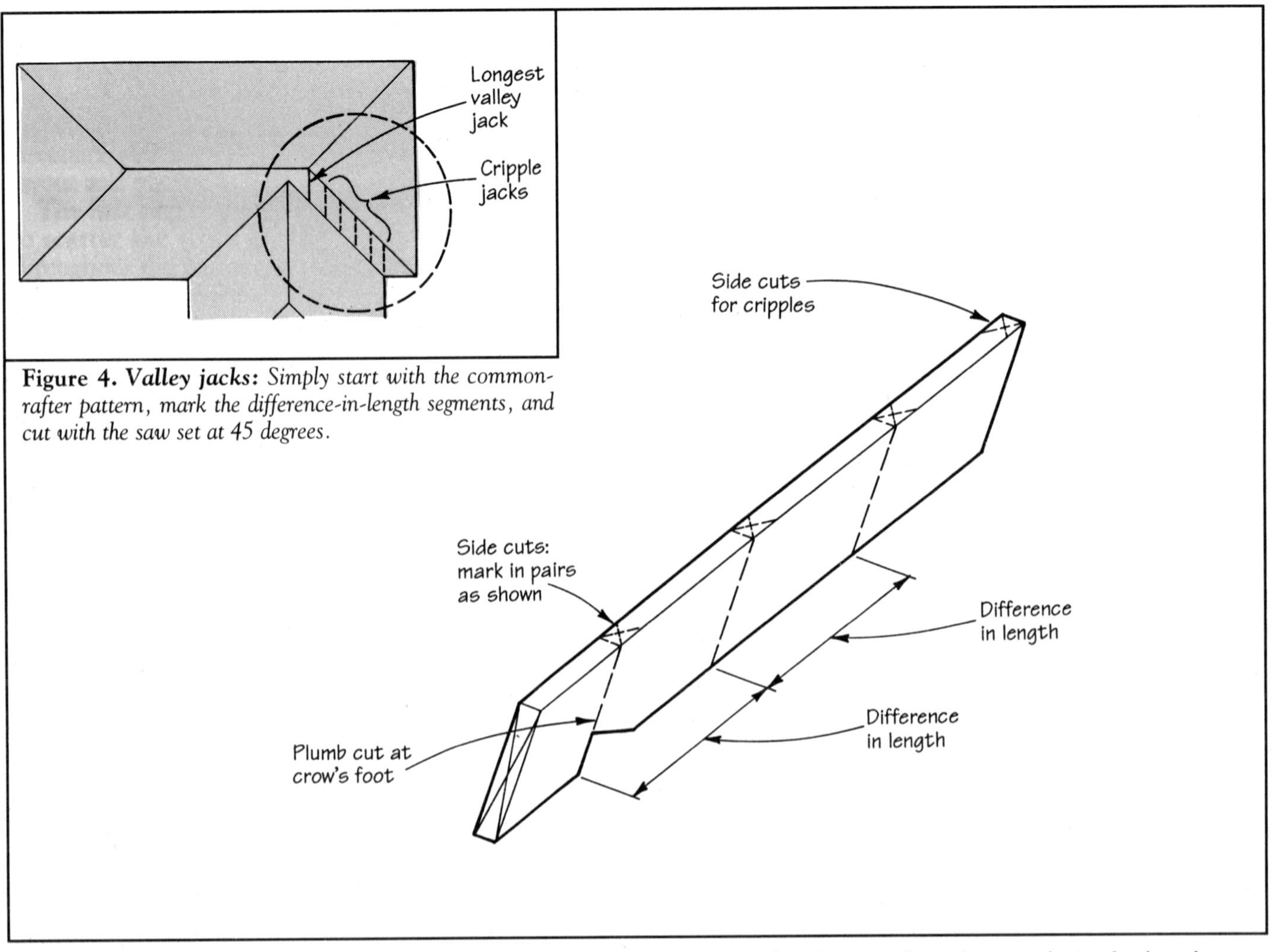

Figure 4. Valley jacks: *Simply start with the common-rafter pattern, mark the difference-in-length segments, and cut with the saw set at 45 degrees.*

Figure 5. Cripple jacks: *These are all the same length if roof pitches are equal. Use the last, longest valley jack as a guide. Its plumb cut becomes the long point of the cripple jack's sidecut.*

the mark across the top edge, then draw the plumb cut down from the face (see Figure 4). This line marks the long point of the miter. You can mark the side cut, as you did before, using the framing square and the stair gauges. But if you are cutting with a power saw, it's not really necessary to mark the side cut. Just set your saw to 45 degrees and cut along the plumb cut.

As with the hip jacks, the pattern should be cut square (on the plumb line) to mark lefts and rights. Then the next shorter valley jacks are successively marked and pattern-cut.

Cripple Jacks

Finally, on a complicated roof you might encounter cripple jacks, which connect a hip and a valley. Cripple jacks are laid out similarly to valley jacks. Start with the last, longest valley jack (see Figure 5) and use its plumb cut at the ridge as the long point of the side cut on the upper end of the cripple. If the intersecting roofs are the same pitch, cripples are the same length. Unequally pitched roofs are another subject.

Jacks are not difficult to cut and install when using this method. All calculations and layouts are done only once on a pattern, and the first pair for each length is checked for accurate fit. The only tricky part is finding the length of the first hip rafter. This takes a little thought, but it guarantees proper on-center spacing from the full-length commons to the last jack. You'll appreciate that when sheathing and insulating—and particularly if the plan calls for exposed rafter tails. ■

Robert Adam has been a carpenter and preservationist for over 25 years. He is department head for the Preservation Carpentry program at North Bennet Street School in Boston, Mass., and does consulting from his home in Shirley Center, Mass.

Framing a Gable Dormer

This technique leaves room for a vaulted ceiling and puts the load where it belongs

by Richard Cooley

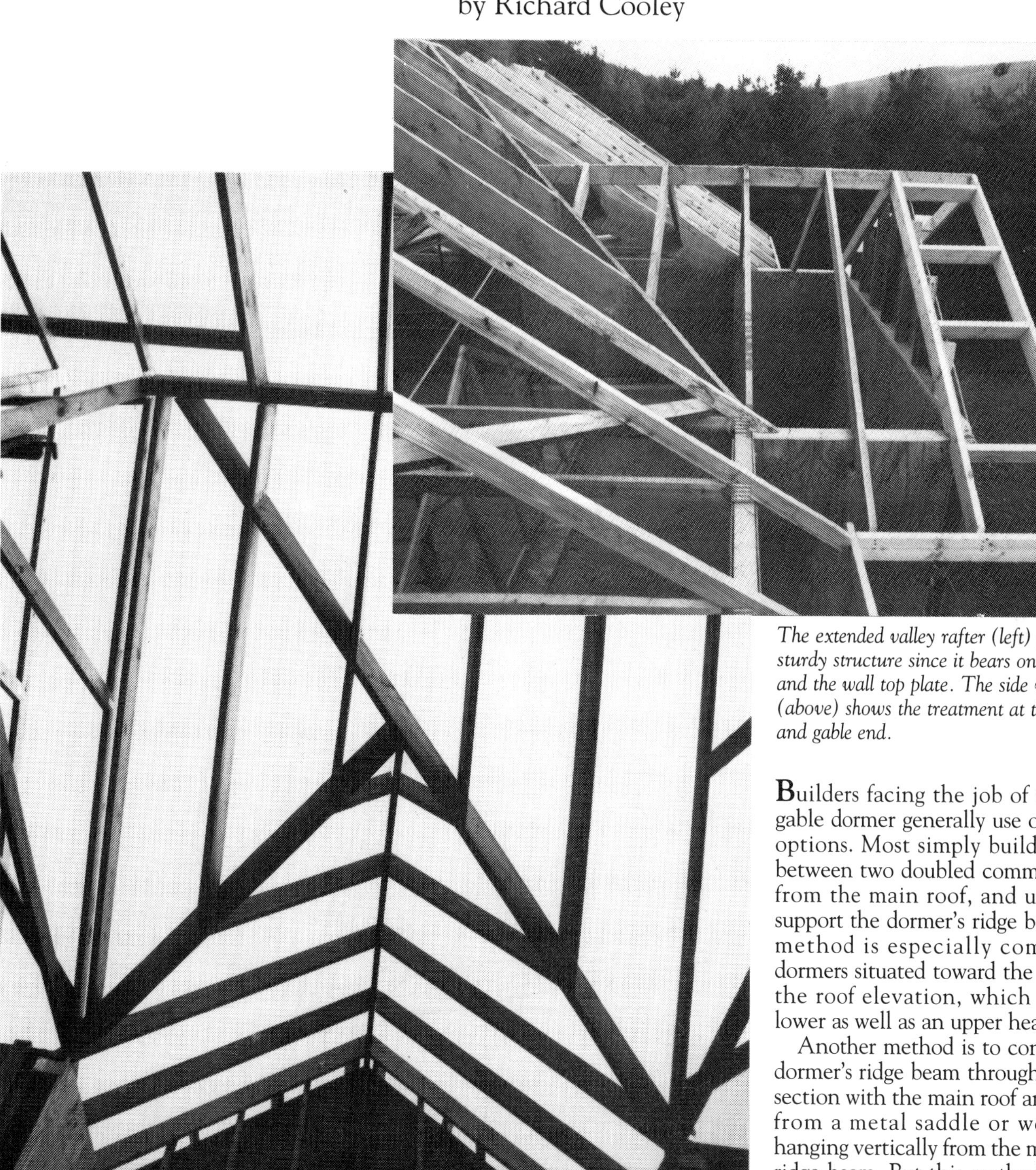

The extended valley rafter (left) provides a sturdy structure since it bears on the ridge and the wall top plate. The side view (above) shows the treatment at the eaves and gable end.

Builders facing the job of framing a gable dormer generally use one of two options. Most simply build a header between two doubled common rafters from the main roof, and use this to support the dormer's ridge beam. This method is especially common for dormers situated toward the middle of the roof elevation, which require a lower as well as an upper header.

Another method is to continue the dormer's ridge beam through its intersection with the main roof and hang it from a metal saddle or wood 2x2s hanging vertically from the main roof's ridge beam. But this method compromises strength and can't be used with vaulted ceilings.

To get around these limitations, and

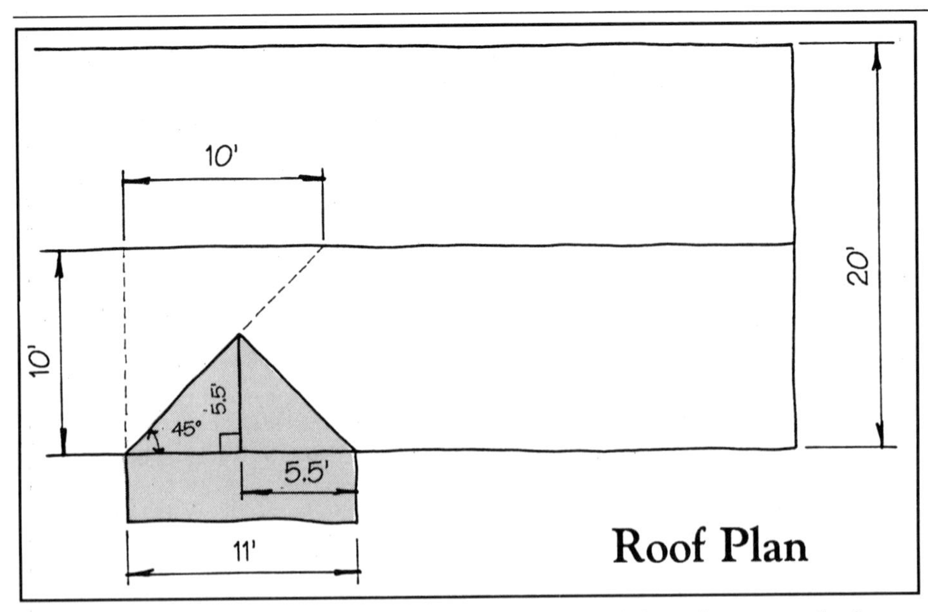

Figure 1. Roof plan for a gable dormer. *Because both roofs have the same pitch, the extended valley rafter forms a 45° angle in plan. Lay out the extended valley rafter as if the dormer were extending all the way to the ridge of the main roof.*

to put the load of the dormer where it belongs—on the main roof ridge and the house's exterior walls—I like to use a third method: I extend one of the dormer's valley rafters upward from the house's top plate until it intersects the main ridge beam. This works particularly well with vaulted ceilings. The method takes a bit of figuring and a little eighth-grade geometry, but the design versatility and structural rigidity it provides are worth it.

Mixing Trusses and Conventional Framing

For the job I'll describe here, I had a unique situation in that the dormer helped form a vaulted ceiling over the house's central hallway and stairwell, but standard 8-foot ceilings were used on either side of this. This let me simplify things by using trusses for the 8-foot ceiling areas, while using my

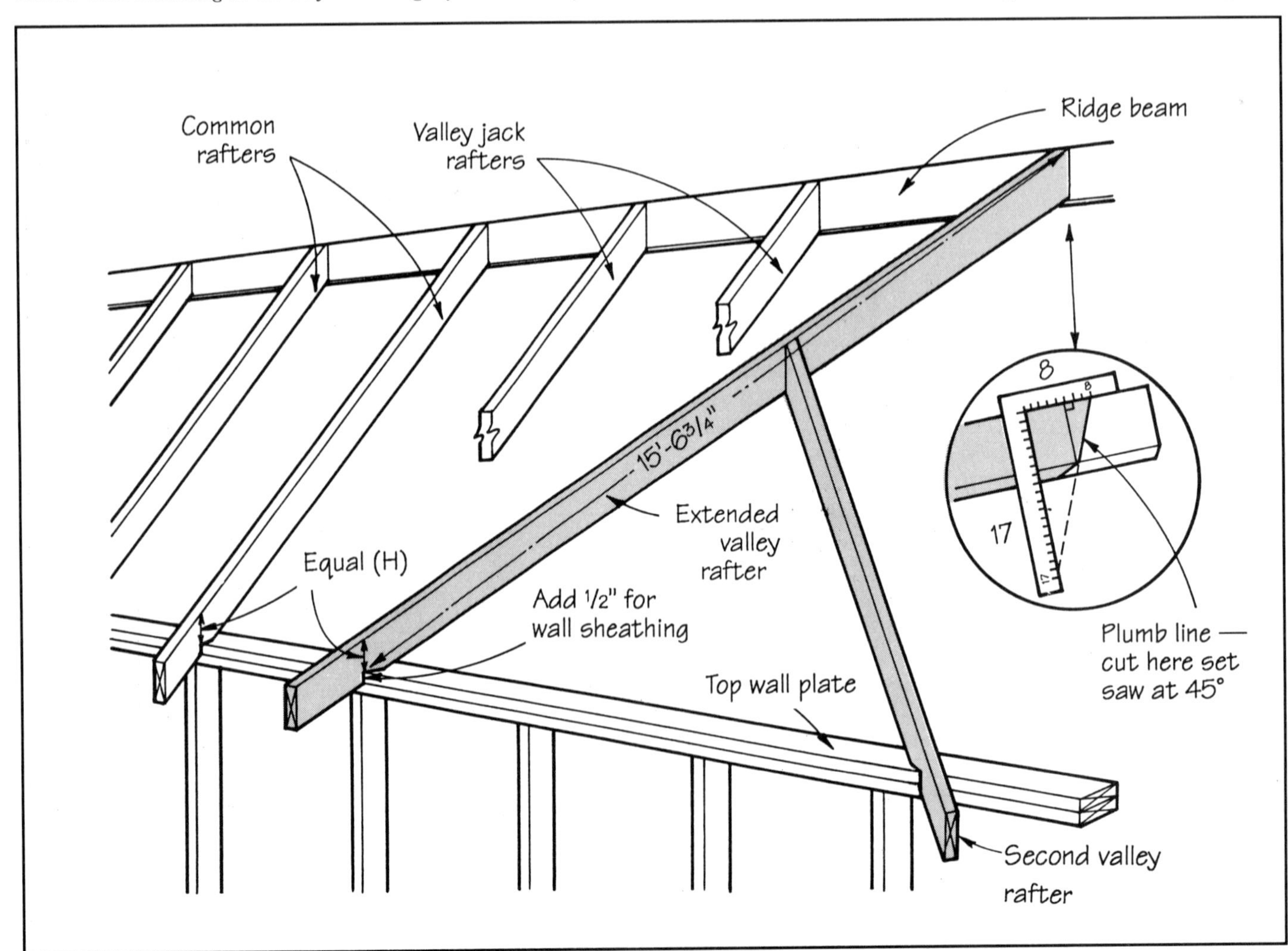

Figure 2. To measure the extended valley rafter with a tape: *Run your tape from the top long point where it meets the ridge to the outside edge of the top plate (at the top of the birdsmouth). Here, it is 15 feet 6 3/4 inches. For the top cut, mark the side of the rafter at 8/17 and cut with your saw blade set at 45°. The plumb at the birdsmouth is located so that the height above the cut (H) equals the corresponding height on the common rafters.*

extended-rafter approach over the vaulted area. The extended valley-rafter saved me some truss expense, for had I used the header method, I would have had to get specially engineered trusses to take the load of the header mid-truss.

Siting the Extended Rafter

The first step is siting the extended valley rafter (see Figure 1). Our dormer, which was to be 11 feet wide, was to have the same pitch, meaning the valley would form a 45° angle when viewed from above. Extending that angle to the main roof's ridge in a plan view, we find it meets the ridge 10 feet over from a line drawn up the roof from the eaves line of the dormer. This distance always equals the run of the roof because it is the leg of a 45° right triangle.

After marking the start and finish points of the extended rafter on the ridge and top plate, I found the rafter's length by measuring with a tape between these two points. It was 15 feet 6 3/4 inches from the ridge to the outside of the top plate of the exterior wall; this represents the length from the top long point of the rafter to the top of the birdsmouth (see Figure 2).

If you're uncomfortable measuring this length with a tape, you can find the rafter length by looking it up in a hip/valley rafter table or figuring it on your steel square. In either case, find the length for a valley rafter that makes the run and rise of the main roof rather than the dormer. Also, keep in mind that either of these methods will give you the theoretical length of the rafter—from the center of the ridge to the birdsmouth plumb cut, as measured along the rafter's top edge. For the actual length, you reduce the theoretical length by half the 45° width of the ridge—about 1 inch for 2x lumber.

We cut the rafter from a piece of 2x12 Douglas fir. The top cut is a compound cut. The run of a valley rafter is longer than the run of a common, while the rise is the same. So, on a standard valley with same-slope roofs, the valley rafter has 17 (16.97) inches of run for every 12 inches of run on a common ($17 = 12 \times \sqrt{2}$). Therefore on this 8/12 roof, the valley rafter needs an 8/17 slope. The angle from the plan view is 45°, so you simply make the cut along the plumb cut line with your saw set at 45°(see Figure 2).

Once you've made the top cut, you can measure along the rafter to find and mark the spot for the bottom cut, which is a birdsmouth plumb cut into the bottom edge of the rafter.

Pythagoras Makes His Play

With the extended valley rafter in, it's time to locate and cut the other dormer valley rafter. This is where the geometry comes in, namely the Pythagorean theorem ($A^2+B^2 = C^2$), which enables you to find the length of a right triangle's third side if you know the other two.

The length of the second valley rafter forms the hypotenuse (longest side) of the right triangle whose other legs are defined by the ridge (length A in Figure 3) and the first full-length dormer common rafter (length B in Figure 3).

Length A will be half the dormer's total span as long as the dormer's roof pitch is the same as the main roof's. In Figure 1, we find that dimension A is 5 feet 6 inches, or 5.5 feet.

We already know the dormer roof's run (5.5 feet) so we can derive dimension B (the length of the dormer's common rafters) from our roof framing charts. For an 8/12 roof, the length is 8 feet 3 inches, or 8.25 feet.

Whipping out the calculator, we figure $A^2 + B^2$:

$$(5.5 \times 5.5) + (8.25 \times 8.25) =$$
$$(30.25) + (68.06) = 98.32 \text{ feet.}$$

Pushing the square-root button on our calculator, we find that the square root of 98.32 is 9.92 feet, or 9 feet 10 7/8 inches. That will be the length of our second valley rafter.

We measure this length along the main valley rafter and mark the point of intersection. This is also the length of the dormer valley rafter we need to cut. The intersection of the extended valley rafter and the second valley rafter is 90° as shown in Figure 1. This is not a compound angle, but a simple cut laid out with a framing square for an 8/12 pitch. The bottom cut is the same as for the extended rafter.

Back to Banging Nails

This completes the valley skeleton of the gable dormer—everything else is routine fill-in, and our job as mathematicians is done.

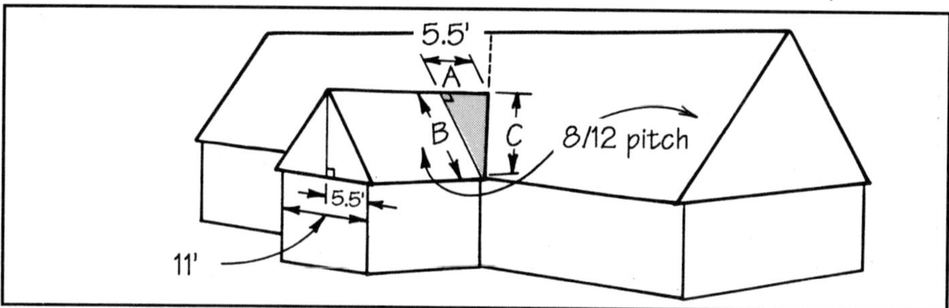

Figure 3. *The dormer valley length C is the hypotenuse of the shaded triangle. B equals the length of dormer common rafters. And A will equal the run of the dormer rafters if the dormer pitch equals the main roof pitch.*

The main roof jack-rafters will be compound cuts similar to the cut made at the top of the extended valley rafter—the saw will be set at 45°, and the line cut on the rafter will be for an 8/12 pitch. The dormer ridge will fasten into the crotch of the valley rafters, with dormer jack-rafters and common rafters filled in last.

As always, when measuring actual rafter lengths, pay close attention to whether you're measuring the long points or the short points of the angles you're preparing to cut. Also, be sure to allow for thicknesses of ridge beams and outside sheathing when converting theoretical lengths to actual cuts. And for any rafter that will help support a roof overhang, be sure to include the overhang length.

Last comes the fascia board and roof sheathing to complete the job. This leaves an open, vaulted ceiling inside for the drywallers—and the loads resting on the exterior wall and the ridge, right where they belong. ■

Richard Cooley is a builder in Schenectady, N.Y.

After being framed, sheathed, and trimmed on the deck, the tower roofs are lifted into place by a crane (right).

Framing Tower Roofs

The quickest way to construct a tower roof is to build it on the ground and lift it with a crane

by Will Holladay

I've spent most of my framing career in the high-end southern California building market, and I've often been called on to frame conical tower roofs. There are many ways to go about the job. Which framing method I choose depends mostly on the type of material—either beam stock or 2x—that's called for. In this article, I'll describe one method that I use with 2x lumber to form a conical roof with a vaulted interior ceiling.

Starting Out Round

The quickest way to construct any tower roof is to build it on the ground and lift it into place with a crane. But to make this work, the walls must be perfectly round and the top plates absolutely flat to receive the roof. Once built, the conical roof structure is intensely strong and rigid, and won't settle out to a bow in the wall plates. So be precise when scribing and cutting the plates for both the wall and the roof.

To make the top plate, I mark the circumference on sheets of plywood with a big compass (see Figure 1). I use three Stanley trammel points on a long piece of 1x4. Two of the trammels are fitted with pencils and spaced the wall thickness apart, so I can draw the exact width of the plates in one pass.

Most large radius curves can be cut with a circular saw. Working the saw this way, the blade will get hot and warp, so I stop for a few seconds every few feet to let the blade cool off.

I usually cut wall plates out of 1 1/8-inch plywood. If the wall sits on concrete, the bottom plate can be cut out of 2x12 redwood. The walls are built in curved sections and the plates are joined together with an overlapping circular layer of 3/4-inch plywood.

To plumb the tower walls, tack a 2x4 across the diameter on the top wall plates. Measure the radius from various points along the outside circumference to locate the exact center on the 2x4. Drive a nail straight down through the point and attach a plumb bob to the nail on the underside of the board. Rack the tower around until the bob is right over the center mark on the floor below and nail up temporary sway braces on the inside. Finally, sheathe the outside with two layers of 3/8-inch plywood.

Once the walls are up, I assemble a giant "donut" to build the roof on. For this, I find a flat floor space somewhere in the house that's large enough to scribe out the tower circumference. I use this large circle as a pattern to assemble my donut from the pre-cut top plates. Like the wall plates, I construct the donut out of a layer of 1 1/8-inch plywood overlapped by a layer of 3/4-inch plywood.

Once it is nailed together, move the donut to an area outside to assemble the rafters. The donut must be positioned so the crane can move it easily later. Place the donut on a six- to eight-sided rack that is blocked up level and is high enough off the ground for the drop in the overhang.

Rafters

Since the rafters radiate from the top point of the cone, the rafter lengths are all the same. The rafter run is the radius of the tower. With the run and the pitch of the roof you can figure the length of the rafters.

While most folks use rafter tables or the Pythagorean theorem, I find it's faster to use a line length (LL) ratio (see Figure 2). The LL ratio is the unit rafter length, or hypotenuse,

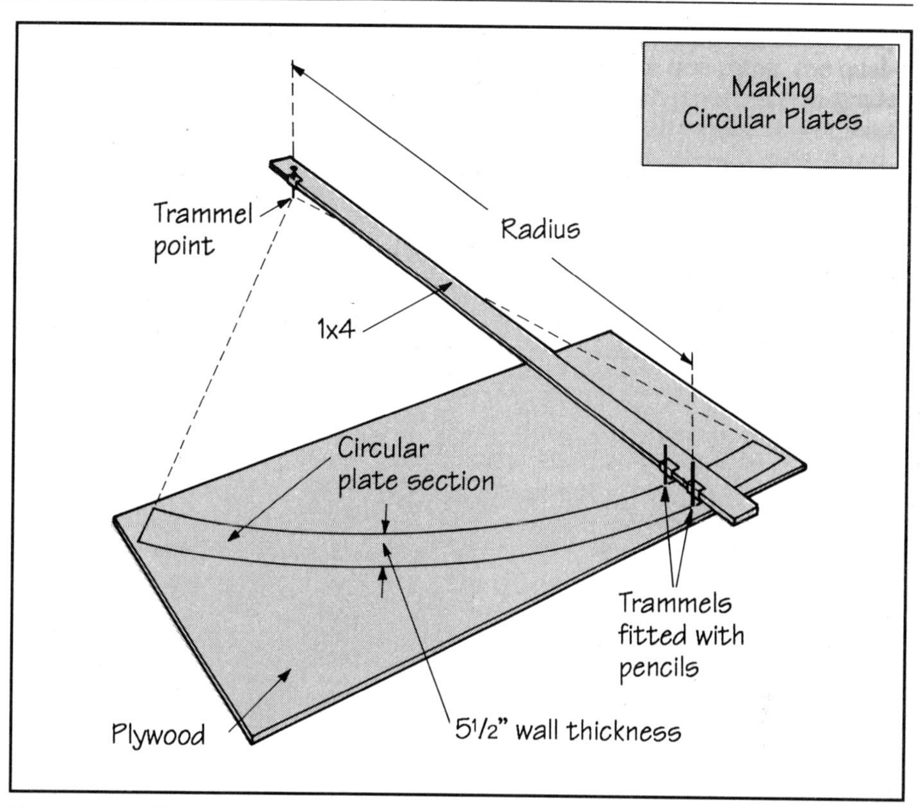

Figure 1. *To lay out circular plates, the author makes a big compass from three trammel points on a piece of 1x4.*

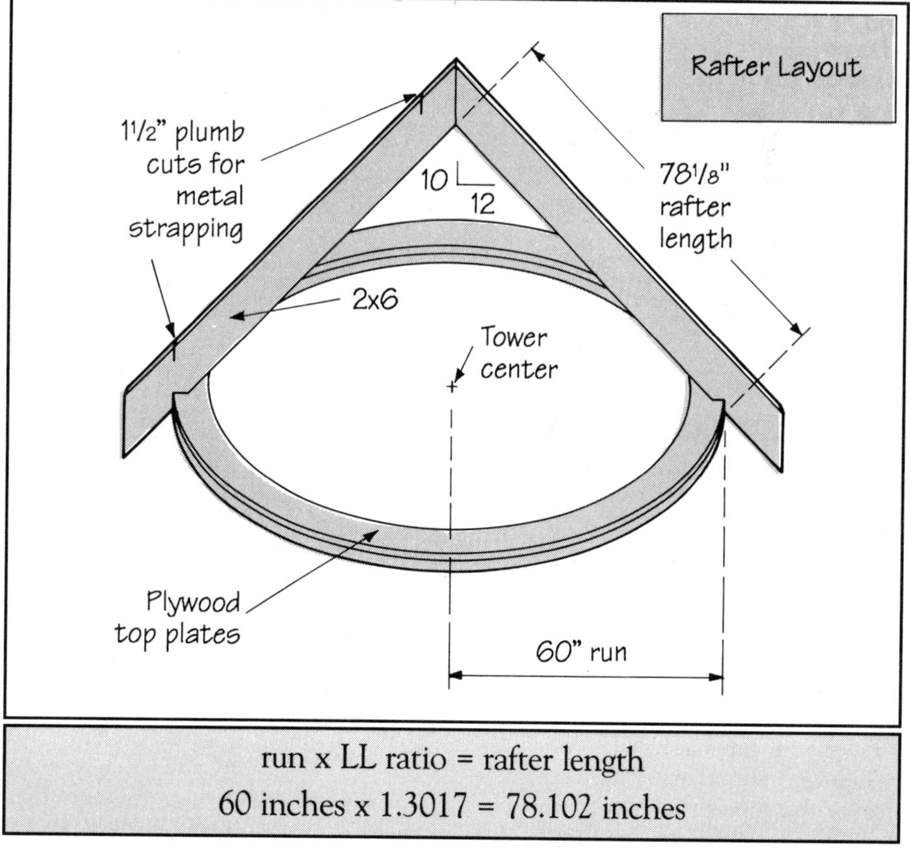

run x LL ratio = rafter length

60 inches x 1.3017 = 78.102 inches

Figure 2. *Tower rafters are cut as commons with a run equal to the outside radius of the plates. To figure the rafter lengths, the author uses a line length (LL) ratio.*

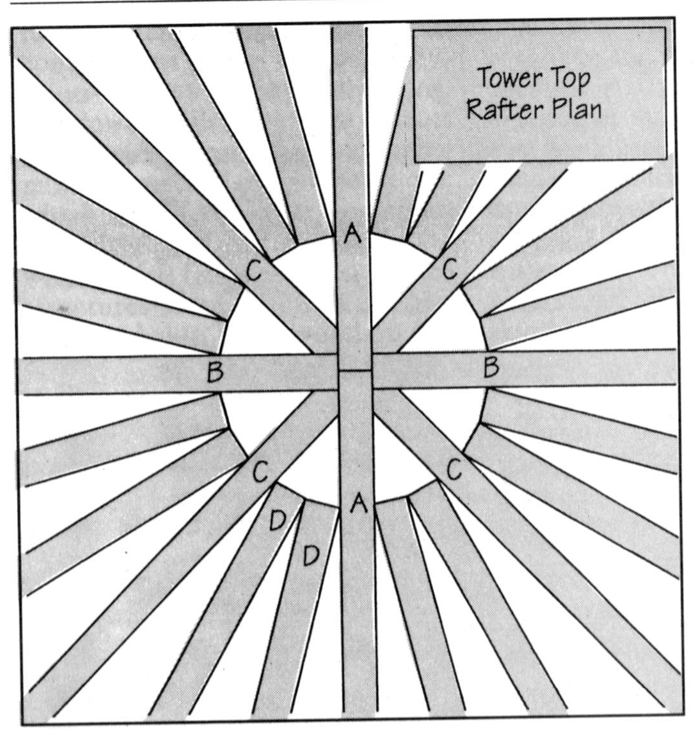

Figure 3. *With 2x lumber, the author uses this framing connection at the top. Cut the rafters labeled* **A** *to the calculated rafter length, and shorten rafters* **B** *and* **C** *by 3/4 inch and about 1 inch, respectively. Rafters marked* **D** *are paired and dropped in as a unit.*

Figure 4. *If the rafters (including the tails) are under 96 inches, the roof can be sheathed with pie-shaped pieces cut from a sheet of plywood. The author uses two layers of 3/8-inch plywood staggered.*

over the unit run, and is figured beforehand for each roof pitch. For a 10/12 pitch, as shown in the examples, the LL ratio is 1.3017.

Figure 3 shows the rafter connection at the top of the cone. This connection works for any multiple of 8 rafters. To find the actual number of rafters needed, I first figure the circumference of the circle using the equation:

$$circumference = 2\pi r$$

(If you slept through geometry class in school, remember π equals 3.14.) I then divide the circumference by 16 (the on-center spacing of the rafters on the plates). For example, for a roof with a 5-foot radius:

circumference = 2 x 3.14 x 60 inches
= 377 inches

377 inches ÷ 16 inches = 23.5

Next, choose the closest multiple of 8 (in this case 24), and divide this number back into the circumference to get the exact on-center spacing:

377 inches ÷ 24 = 15 3/4 inches

Lay out the plates using a small scrap of lumber as a spacing pattern to measure around the circumference.

The first two rafters (labeled A in Figure 3) are cut at the calculated rafter length, but the other rafters are shortened. The next two rafters, labeled B, will butt into the A rafters at right angles, so they each need to be shortened by 3/4 inch (half the thickness of the framing material). The rafters labeled C will be nailed into the corners, so they are shortened about 1 inch (half the 45-degree thickness of the material). These shortening distances are measured perpendicular to the plumb cut.

Since not all 24 rafters can squeeze together at the top of the cone, the rafters labeled D are shortened even further. The easiest way to find this shortening distance is to find the circumference on the cone at the point where all 24 rafters can fit in a circle.

In this case, 24 x 1 1/4 inches = 36 inches. The radius of this circumference is then found from the equation:

radius = circumference ÷ 2π

radius = 36 inches ÷ 6.28
= 5 3/4 inches

Remember that this is a radius, so it too is measured perpendicular to the plumb cut on the rafters, not along the line of the rafters.

These rafters (D) are paired and dropped in together after the other rafters are in place. I nail the two rafters together before installing them and spread their tails apart to the on-center spacing. I then put the two in as a unit, which is spiked through to the adjoining rafters with a 60d nail.

With 2x rafters, I use three toenails to secure the birdsmouths to the plates. If I'm using larger beams, I lag up through the plates into the seat cut of the birdsmouth.

Bracing

Without ties, the roof will want to flatten out under load like those paper parasols you find in exotic cocktails. To prevent this, I collar the rafters with a couple of bands of 1¹/₄-inch, 16-gauge galvanized metal strapping. The bands are let in after the rafters have been put up. I set the saw table to the angle of the plumb cut and cut across the tops of the rafters. One band is positioned on the exterior plate line, and the other just below the top (see Figure 2). This second band is there mostly to support the cone when it is picked up by the crane. To tension the strapping, I hook it with my hammer claw and pull it tight before nailing.

I also put in one or two temporary cross ties to further stabilize the cone when it's lifted by the crane.

Sheathing

The trick to sheathing the tower roof is to cut the plywood into curved pieces. Here again I use a big compass that's as long as the calculated rafter length plus the tails.

I use two layers of 3/8-inch AC plywood for sheathing, and stagger the vertical seams between the layers.

If the roof diameter is under 10 feet, I cut pie-shaped pieces out of a sheet of plywood since the rafters (including the tails) are under 96 inches (see Figure 4). Between 10- and 14-foot diameters, I'll even special-order 10- and 12-foot-long sheets of plywood, since using pie-shaped pieces saves so much time. But for larger diameter roofs, you have to use two pieces along the rafter run. In these cases, I stagger the end cuts.

Since the radius of the curve is tight up near the point of the cone, you have to kerf the back side of the plywood near the point, otherwise the face will crack. I set my saw to about a 3/16-inch depth and make a series of cuts along the radius lines, about an inch or two apart. On the last project I did, we nailed the sheathing off with roofing nails, which helped hold the pieces down where the radius was small.

The toughest part of the whole job is the fascia. On the last job, which had a 5-foot radius, we used 1x10 clear redwood. After splitting one piece, one of the carpenters soaked the boards in water to make the wood more pliable. He used galvanized screws to secure it to the rafter tails.

The style in southern California is to leave the rafter tails exposed and stucco up between them. In other parts of the country where soffits are preferred, you can cut 2x4s that tie to the rafter tails and return level into the side of the building. A 3/8-inch plywood soffit can then be scribed using the big compass with two trammel points spaced apart by the width of the overhang. This work would be done, of course, after the roof is installed.

The Crowning Touch

Lifting the roof on is the easy part. If the roof is made of heavy timbers, we hold it from at least four points, and use long, threaded eyebolts to lift from. But with a light, framed roof such as the one shown in the photos, two pickup points are sufficient.

For the pickup points, I drill through the sheathing on each side of the first two opposing rafters (marked A in Figure 3—these two rafters are supported by a temporary cross tie) and thread a rope loop through. These loops should be positioned above the midsection of the cone, so when the roof is picked up, it will be weighted towards the bottom.

On our last project, we made two loops out of some rock–climbing rope that one of the carpenters swore by. The rest of the crew took bets on whether or not the rope would break. Climbing rope is very elastic and it stretched a lot when the crane lifted up. But it held. ■

Will Holladay is a framer from Santa Barbara, Calif. He is the author of A Roof Cutter's Secrets to Framing the Custom House, *from which this article is adapted.*

Section 5. Remodelers' Specialties

Bob Terenzoni Courtesy: High Tech Dormer

Demolition With Care

Dismantle buildings carefully to save headaches later on

The Farrell Company is a father-and-son remodeling and restoration team that has been in business for thirty years. The company serves some of the pricier communities in the San Francisco Bay Area, and grosses $1.1 to $1.2 million a year. Ten to twenty percent of this figure is for demolition. Because demolition is such a significant part of remodeling, The Journal of Light Construction *asked son Steve Farrell how the company goes about this time-consuming, but vital, part of the business.*

Michael Rosenfield

Preparing For The Job

JLC: *How do you prepare customers for the disruption a major demolition brings?*

Farrell: During a preconstruction meeting, I educate clients about what will happen and I try to paint a worst-case scenario. I warn them that they are likely to panic a little when they see their house coming down.

On small projects, we're more likely to arrange for the client to remain in the house. If need be, they can survive with a bedroom and bathroom only, along with a hot plate and microwave. But we tell the homeowner that it's going to be dusty and unsafe, and we warn them about long periods of time without gas, water, and electricity.

On large projects, it's easiest if the occupants find another place to live for a few months.

JLC: *How much do you spell out in your contract, and how much do you communicate verbally?*

Farrell: We have a specification sheet in our contract that says what rooms will be demoed and what we are and are not responsible for. We give a room-by-room breakdown. For example, we say a kitchen or bath will be demoed down to the subfloor or stud walls.

One clause says that if we find conditions that are not up to code, or extensive wood decay or termite damage, we will repair at our cost, but as an extra. Similarly, I have a clause in my contract that spells out the procedure in case we find asbestos.

We don't have clauses that specifically deny responsibility for things going wrong, because clients don't want to see a lot of disclaimers in the contract. Instead, I try to limit our potential liability by limiting the scope of our work.

For example, I specify that breakables and valuables must be removed. If I let an item remain, I become responsible for it. Pets are also a concern. We'll build a shelter for pets, but we won't be responsible for taking care of them.

JLC: *What steps do you take to ensure building security?*

Farrell: That depends on the size of the job. If we're demolishing a room or section of a house, we just plywood the doorway or hallway entrance. For large demos we rent a fence with barbed wire across the top and one or two gates. A cyclone fence will run $800 for a year. That's minimal compared to our liability for building security or a backyard pool.

To protect valuables, we encourage clients to rent a large bin that will remain at the job site. They come in various sizes and configurations, such as half office, half storage. If we use part of the bin for tool storage or office space, we'll split the cost.

Of course, bins are ugly, so we ask the clients to make sure their neighbors don't mind. Also, in some towns they need permits.

JLC: *What types of barriers do you use to keep dust and debris out of the rest of the building?*

Farrell: We put up a double layer of 6-mil poly over openings, and we plywood over that.

Sometimes we have to go in and out of an opening, so we'll frame it into a temporary door. We'll use an old door or make one from 1$^1/_8$-inch plywood. Stapling scrap carpet around the stop helps keep down dust, but no matter how you block off an area, you can't entirely avoid the dust.

JLC: *How do you supervise the demo subs?*

Farrell: I always have at least two journeymen carpenters on large projects to point out structural parts of the house that we don't want removed or that could be dangerous if removed improperly. Also, we don't want to make work for ourselves, such as having to replace a gas line or a wall that should have remained in the first place.

A carpenter knows how to take things down in the reverse order that they went up in. The risk element goes way down with good carpenters on the job.

Bidding

JLC: *How do you avoid surprises when you bid a job?*

Farrell: During the bidding stage, my dad and I walk through the house and go over the job with a fine-toothed comb. We use both written and mental checklists. We look at the condition of the water heater, furnace, roof, and masonry. We identify plumbing chases, the main vent stack, antiquated wiring, and the way the joists run.

If I'm skeptical about something I see, I bring in the appropriate sub. For example, I may want to double-check that the sub figured on moving an electrical meter.

Asbestos is another big issue—especially with plaster and acoustic ceilings. On a pre-demo walkthrough, we'll note what we think is asbestos and get samples tested. If I find a problem, I stop the job and call an asbestos removal contractor. It's not worth fooling with and the fine for improper removal is substantial.

JLC: *Do you typically cut away existing finish materials to see things when estimating?*

Farrell: On preinspection we go into a crawlspace or attic, but we rarely open walls unless we see something that is really suspect.

For example, on one large demo we spotted a badly cracked head jamb over a double door. A tremendous load fell on top of the doors. So we opened up the wall and found that the header had completely snapped in half. We immediately shored up the area, then fixed it.

JLC: *How do you charge for demo?*

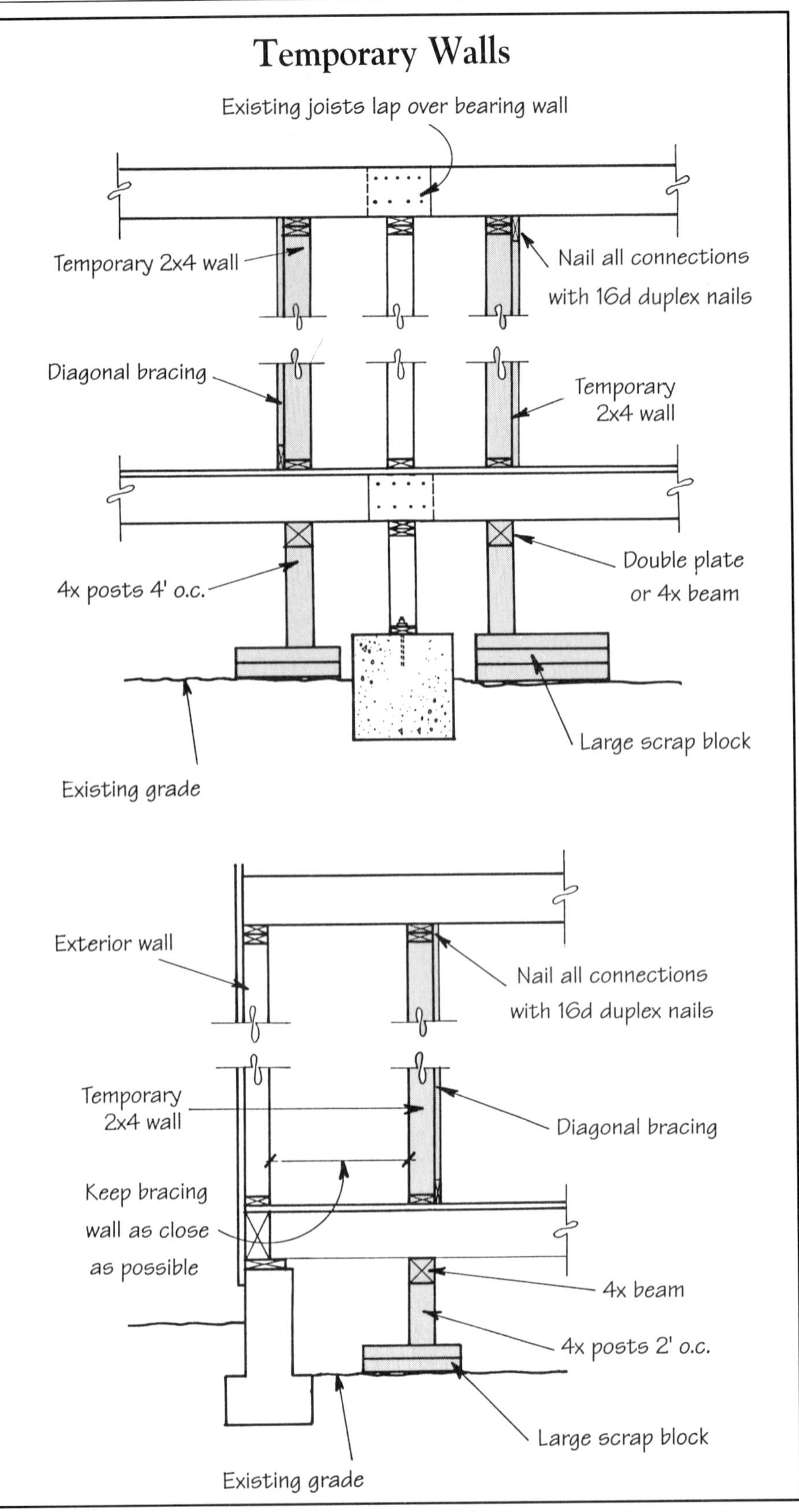

When removing an interior bearing wall (top), Farrell builds temporary bracing walls on both sides, fully supported from the ground up. Large glulam scraps at grade level serve as footings for 4x4 posts, 4 feet on-center, which support a 4x4 header. The header distributes the weight from the temporary 2x4 wall above. When removing an exterior wall (above), Farrell builds a single bracing wall on the inside, as close to the exterior wall as possible.

Farrell: There are a lot of variables in demo work. Are we responsible for moving everything out of the house? For security? Are we demoing a second story, where the first floor will be exposed to the weather? Do we have to move debris through a clean part of the house?

We haven't found any magic formula. We walk through the job in our minds, figuring about how long it will take us, based on past experience. Then we add a third to a half of this figure—although this is negotiable if the client balks.

This markup is to cover the unexpected—for example, if we find 12-inch-thick concrete instead of the usual 4-inch.

Structural Evaluation

JLC: *How do you determine what is load bearing and what isn't?*

Farrell: Once you've removed the drywall, you can see the relationship of the wall framing from one floor to the next.

As any good builder knows, if joists split over a wall, that wall is load bearing. With framed and truss roofs, the load-bearing walls are generally beneath the eaves, though with trusses there might be a bearing wall under a web-chord intersection. The gable ends of a house with a gable roof are not usually load bearing, nor are walls that run parallel to the joist. Of course, there are always exceptions.

JLC: *Any other critical points to look for?*

Farrell: Yes, point loads and shear walls.

Point loads occur beneath posts or columns. Sometimes posts run straight up to the peak of the roof and support the ridge pole. Other times, they'll be embedded in a wall.

Some walls function as shear walls, an important issue in seismic and high-wind areas. These walls may have let-in braces, steel straps, or plywood. You can remove the let-in braces from shear walls during the demolition phase, but you have to replace them with metal straps or 3/8- to 1/2-inch plywood.

Make sure all load-bearing walls and point loads are supported from the foundation up. And always demo bearing walls and posts from the top of the structure down, unless you don't have a choice.

JLC: *Do you ever call in an engineer or architect?*

Farrell: When we discover hold-downs, ties, or major bearing walls, code officials make us call a structural engineer or the architect who drew up the plans. We also have to contact an engineer if we discover structural flaws in the building.

This has been a big headache in old homes which were not built to plan. For example, we were doing a demo recently on a house with a second-story addition. The city had signed the building permit for the previous work, but we found a serious structural flaw.

The whole ridge, with a tile roof, sat on a center bearing wall, but there was no foundation or central girder under this wall—just a grid of 4x6 girders over a tight crawlspace. The 1 1/8-inch subfloor sloped to the middle of the house. I called the architect and told him there was no way to get a big beam in there. After discussing alternatives, we decided to place reinforced concrete pads, jack up the bearing wall, and support it on posts.

Mechanical Work

JLC: *What kinds of electrical and plumbing problems can you anticipate?*

Farrell: In demolition you always expect the unexpected. The most common plumbing problem we find is that waste lines do not slope adequately downhill. Also, electrical connections made in the wall rather than in a junction box are quite common.

When we discover problems like these, we share the discovery with the homeowners, give them our evaluation, and our cost to fix it. Our contract explains that anything in the walls or in the direct area of the remodel will have to be brought up to code, at their expense.

JLC: *Do you use mechanical subs during demo?*

Farrell: I always ask the subs how they want us to handle demo. Many times they can walk us through the demo, and we'll do the mechanical work ourselves. We'll shut off the water and remove the plumbing right to the floor level. Or we can shut off the electrical panel and remove wiring. If the crew tears out too much, however, it makes the sub's job harder.

In particular, I like the electrician to be on the job while electrical is being removed so he can keep what he wants intact and can label wires.

JLC: *How do you arrange for power on the site?*

Farrell: If the owner is not living in the house, the electrician kills the panel and sets up plugs outside the house–one 220-volt line and four 20-amp, 110-volt lines.

If the owner is living in the house, we keep the area where they're living hot and make sure the demo crew knows where the wiring is hot.

Shoring and Bracing

JLC: *What type of shoring or temporary bracing do you typically use when removing load-bearing walls?*

Farrell: Before we begin removing any load-bearing walls, we make sure we have the following items on hand:

- Plenty of large blocks for use as pads. We use 8x, 6x, and 4x materials 1 to 3 feet long. We save all our glulam scraps for use in shoring and bracing.
- Plenty of 2x4s and 4x4s from 8 to 20 feet long.
- Duplex nails, both 8d and 16d.
- Hydraulic and screw jacks. We use the 20-ton variety.
- Large beams of any lengths, such as 4x12s, 6x12s, 8x12s, and any glulams we have, to spread loads over large areas.

Safety is the number one concern in shoring or bracing any wall or floor. This is one time when overkill is the way to go. Whenever there's a question, we always add more shoring and bracing than the minimum that would be required.

JLC: *Any helpful tips or techniques for installing bracing?*

Farrell: When removing a bearing wall or beam, we start bracing at the lowest point, which is ground level, and brace up through the area to be removed. This can go up to the second or third story as long as braces line up vertically with the braces below.

With an interior bearing wall, we brace on both sides of the wall. We start on the ground with block pads placed 4 feet apart. Glulam scraps, which come in a variety of thicknesses, work well here. On top of these scraps, we place 4x4 posts which support a 4x4 header. The header helps distribute the load from the bracing wall above. The posts should never be more than 4 feet on-center.

For the bracing walls, we use 2x4s, 16 inches on-center. Pre-cut studs don't work; we cut each stud to length for a tight fit. We also use a double top plate or a 4x4 header. All connections, from top plate to ceiling, bottom plate to floor, etc., should be secured with 16d duplex nails. If the bracing wall started to move, or someone accidentally fell into it, you'd want the security of the nails to hold it in place.

You also need diagonal bracing, such as a 2x4 nailed at a 45-degree angle, on walls taller than 8 feet. The diagonal member should be nailed with two 16d duplex nails at the top and bottom plate. Once the bracing is in place, the bearing wall can be taken out, and you can install a beam or repair the wall.

JLC: *Does this differ for exterior walls?*

Farrell: The approach is similar on an exterior wall, except that bracing is done from one side. In this case, we put posts closer together, at 2 feet on-center, in the crawlspace or basement. Try to keep the cantilever of the existing floor joists to a minimum by placing the bracing wall as close to the exterior wall as possible.

If we're demoing a bearing wall that has a lot of braces or shear panels, we'll shear our bracing wall as well. Also, when we remove longer walls, or the whole side of a house, we go to more trouble to provide shear panels.

To create shear panels, we nail vertical 4x8 sheets of 1/2-inch plywood to the top and bottom plates. If we're removing a wall the length of a building, we put plywood at each end of our temporary wall to tie it to the outside walls. If we're removing a 40-foot wall, we'll panel every other 4-foot section in addition to the end sections.

JLC: *What if the saw binds when you're cutting through a stud during demolition?*

Farrell: If I cut studs—for example, the trimmer and king stud—and they bind, I'll put a post or brace next to the area I'm cutting. Also, if I'm cutting nails with a Sawzall and the saw binds, I shim up the area with a temporary wall.

JLC: *What steps can you take to minimize cracking plaster, binding doors and windows, etc., on the rest of the structure?*

Farrell: If the wall finishes are to remain, you don't want any movement in the existing structure. So you want your temporary supports to be very tight without being stressed.

If we're taking out a bearing wall in the center of a house, we measure the height of the bearing wall and frame ours the same. We don't jack unless the wall has a sag. Then we straighten with jacks.

Jacking 1/2 inch per day is a pretty safe level. Then we let the building sit overnight to absorb the movement. Drywall is more flexible than plaster; if you jack plaster it will take a beating.

Even without plaster, too much jacking can cause problems. Last summer we worked on a job where a portion of the foundation had sunk 2 inches. We went under and poured pads, but when we had jacked the walls up 1 1/2 inches, the plates were starting to compress. There was simply too much weight to get it completely level. We didn't want to crack the roof or walls, so we stopped short of perfection.

JLC: *When you replace a large section of wall with a beam, how do you raise the new beam into place?*

Farrell: We use screw jacks and hydraulic jacks when we have to raise large beams into place. The adjustability is a plus because it keeps posts tight against the beams.

We have more confidence in screw jacks, but we sometimes have to use hydraulic jacks when we're in a tight space and don't have room to turn the handle on the screw jack.

Storage and Removal

JLC: *Where do you start when you take down a multi-story building?*

Farrell: We always work from the top story down. Removing material from a second story or roofline, we use a large Sonotube. We place 2x4 braces on the sides and extend these to the ground. Or we use plywood chutes into a truck. Sometimes we'll back a scissors-lift truck up to the house.

JLC: *Any special techniques for carrying debris out of the building?*

Farrell: When we can't pitch debris out a window or off a roof, we use scoops, like a large snow shovel, and wheelbarrows. If we can't get to the room with a wheelbarrow, we throw debris in a large, heavy tarp, get a guy on each end, and roll it up. Then we'll carry it out to the truck.

The most economical way to remove the material is to heave it right into a truck. The demo crew we use owns two large trucks, and as soon as the truck is full they haul it away.

JLC: *How about storing debris on site?*

Farrell: If we need to, we'll make small piles throughout the job site and haul them off as soon as we have a truckload.

However, you never want a large pile sitting around. On one job, a fire started that way and torched a house we had just about finished. The fire marshall concluded it was someone tossing in a cigarette. Now I pick up piles every week.

Safety

JLC: *What kinds of safety tips do remodelers need to remember?*

Farrell: There are a lot of ways people can get injured on a remodeling job, so we pay a lot of attention to safety.

We always fix a broken subfloor with plywood or diagonal sheathing. When we're doing foundations and have exposed rebar sticking up in forms or in block cavities, we use blocks of wood, or plastic balls available from the rebar supplier, to cover the tip of each piece of rebar. If we have a long row, we'll build an L-shaped box of 1 1/8-inch plywood or 2x material, and wire the box over the rebar. A fatal injury from someone falling on a piece of rebar is easy to prevent.

We always make sure our temporary braces are nailed and tight. Temporary walls can move.

We never work under someone else. Even with a hard hat, you can get a nail or debris in your eye, or a hammer could fall from a loop and hit you.

We never do anything out of balance. For example, we'd never lean out a window and swing a sledgehammer. We also never work under an item we're trying to demo. If we're trying to lift or remove a large beam, we don't stand under it. We stand to the side and prop it with sawhorses, ladders, or braces. ■

100

Sill Repair & Jacking

Set your jacks on solid footings, shore as you go, and don't skimp on common sense

by Roger Desautels

In old-house remodeling projects, owners are usually far more concerned with the new living space than with the structural integrity of the building. Foundation problems and rotted sills, however, are common in older buildings and should be put in good repair before renovations begin upstairs. It's the contractor's responsibility to protect the homeowner's investment and his own reputation by examining the foundation and sills and making any necessary repairs.

Inspecting the Sills

To inspect the sills, start by walking through the interior of the house, taking note of any humps or sags in the floor within 3 feet of the exterior wall. Also, look for cracks where interior partitions join exterior walls. These cracks can indicate settlement of the exterior wall, which is often caused by advanced sill rot.

Next, jump on the floor near the exterior wall to feel if the floor is solid or springy. Flexing next to an outside wall may indicate that the joists are no longer secured to the sill because of rot in the beam pockets. Looking at the house from the outside, sight along the clapboards, looking for gradual sags. Look for peeling paint, mold, or dampness, indicating moisture problems in the lower section of the wall. Pay particular attention to areas of poor drainage, dense foundation plantings, extreme shade, stagnant air, or where rain gutters are missing.

Also note any concrete or stone steps that trap moisture against the sill, and note poorly maintained house features such as leaky thresholds or windowsills that allow water to reach the sill from above.

Any of the above areas that are suspect should be inspected more closely by probing with an icepick (an icepick is much better than a screwdriver or an awl because of its smaller diameter and sharp point). Although it is sometimes necessary to remove a skirtboard or clapboard, probing with the icepick from the outside where the masonry meets the wood is usually sufficient. It is also important to go into the basement or crawlspace to probe. Sometimes sill rot is caused by interior dampness or plumbing leaks, so every section of sill should be checked, no matter how difficult or unpleasant the access.

Once you've determined the amount of decay, you can plan the method of repair. The factors to consider are: the type of framing system, foundation type, access to the repair area, the owner's budget, and the type of jacking equipment available. Although every repair job is different, the same basic principles apply to all. These are:

- Do not allow the house to settle as a result of the work.
- Replace the sill with the longest new pieces possible.
- Do not skimp on splices and fasteners.
- Remove the cause of the rot in order to protect your new work.
- Be very conscious of safety.

Repair worker Dana Myrick surveys the damage to the sill and band joist on an almost-new home built over a damp crawlspace.

Post-and-Beam Sill Repair

Start the job by removing the bottom 2 feet of the siding to expose the frame. Then remove any insulation, dirt, and obviously rotted wood. Be sure to check for electrical wires and plumbing, and if necessary, make arrangements with other tradesmen to have them removed.

Replace the most rotted section of the sill first. Often this area has settled enough so that the studs exert very little weight on it and the sill can be removed without any jacking. If there is still obvious weight here, it may be necessary to set one or more jacks under the sill on either side of the area to be repaired (see "Jacking Tips," page 104). When the weight is relieved in the repair area, remove some of the stone or block foundation material under the sill to provide working room. (If the foundation is poured concrete, you'll have to use a smaller dimension sill, or cut short the studs a little, to create working room.) At the same time be sure to shore up any floor joists or cross beams that are supported by the sill section being removed to prevent them from settling.

To remove the old sill, cut it into short sections with a gas or electric chainsaw and pry these sections away from the studs. You may find studs or floorboards nailed to the sill. These are most easily cut with a metal-cutting blade in a reciprocating saw. Where cross beams or posts tie into the sill, try to save any non-rotted tenons for tying in the new sill.

Once you've removed the sill section, clear away all debris and check to make sure there are no nails sticking down from the floor boards. At this time, repair any floor joist ends that have rotted from proximity to the rotted sill. How you repair a floor joist end depends on the degree of rot. A small amount of rot on the bottom of an old half-round joist can simply be cut and chiseled away—leaving a slightly smaller joist end.

If the rot is more extensive, a pressure-treated 2x4 ledger lagged to the sill may be sufficient to catch the joist. If there's extensive decay, complete joist replacement or sistering may be necessary. If you sister, make sure to lap and lag sufficiently to make a rigid splice (see Figure 1).

Also, if you are only replacing a short section of sill, now is the time to cut half-laps in the remaining sill on either side (see Figure 2).

Next, cut the new sill timber to fit

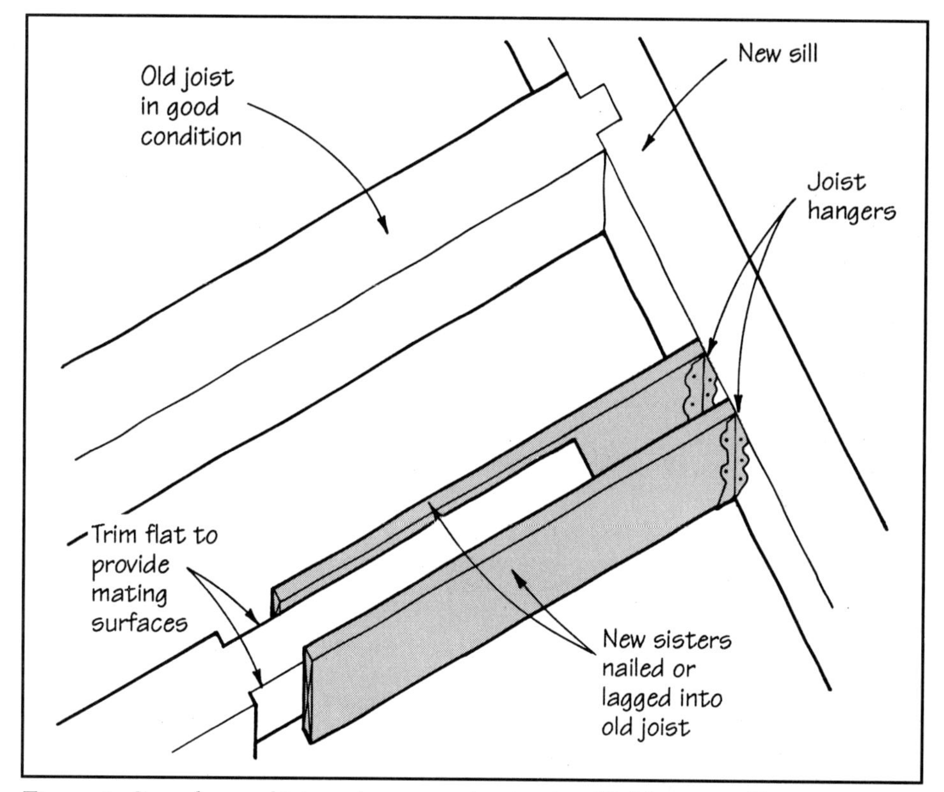

Figure 1. *Severely rotted joist ends may require sistering. With less rot, adding a pressure-treated ledger to the sill may be sufficient.*

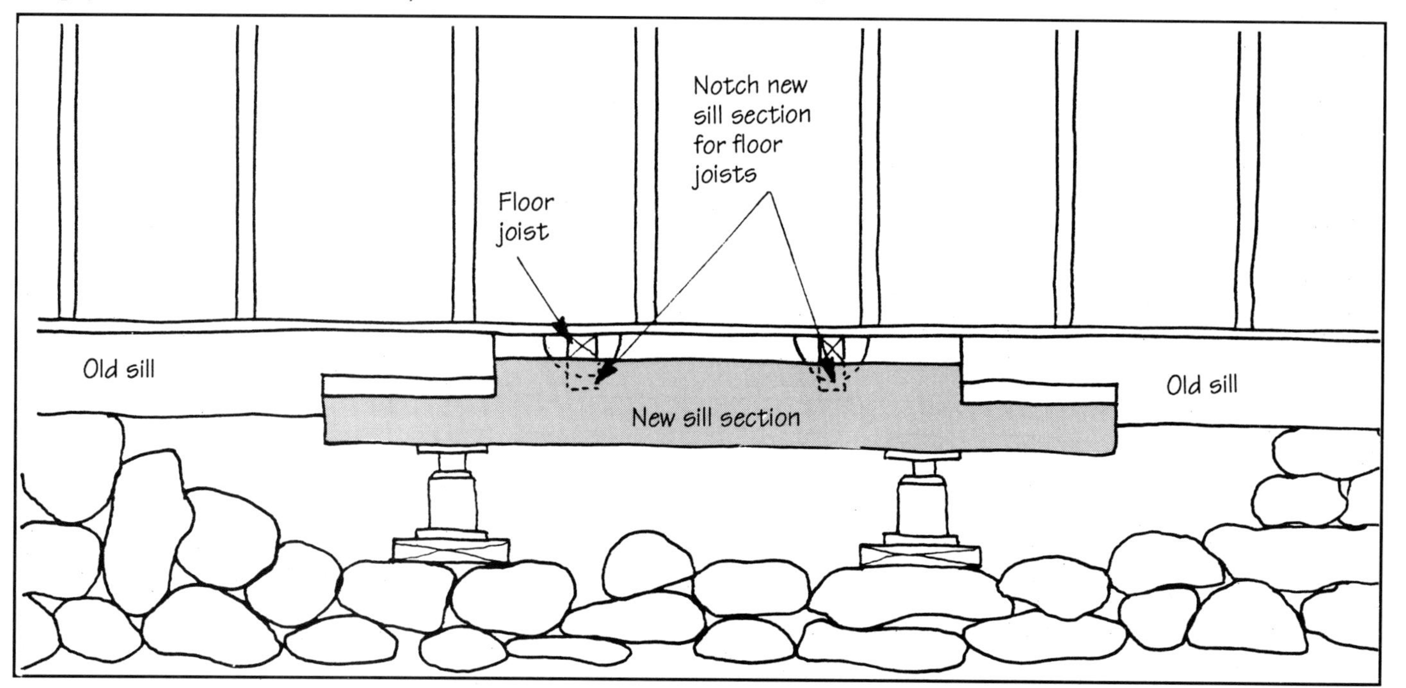

Figure 2. *When replacing a section of sill, cut matching half-laps in the old and new sections. Force the new section into place with crowbars, sledges, and jacks.*

Jacking Tips

I use three kinds of jacks in sill repair: screw, hydraulic, and mechanical (see Figure A). A fourth type, the adjustable floor post, can also be used when working in full basements. The traditional screw jack is very dependable although slow to use in tight places. Keeping it well oiled will make its operation much easier. Hydraulic jacks in 12-ton and 20-ton capacities are easier to use and, in most cases, they will provide more lifting power than screw jacks. In most situations, screw and hydraulic jacks can be used interchangeably. Mechanical jacks were originally used by railroads and are commonly called railroad jacks. They have three main advantages over the other jacks. They will provide more vertical lift in one set-up and, because of their toe-lift feature, they are inherently more stable in most situations. The toe-lift also allows the jacks to be set up in a smaller working space.

When jacking the sill of a house,

▲ **Figure A.** *Shown left to right are a railroad jack, a screw jack, and a hydraulic jack. A fourth kind—the adjustable floor post—is shown in Figure 3.*

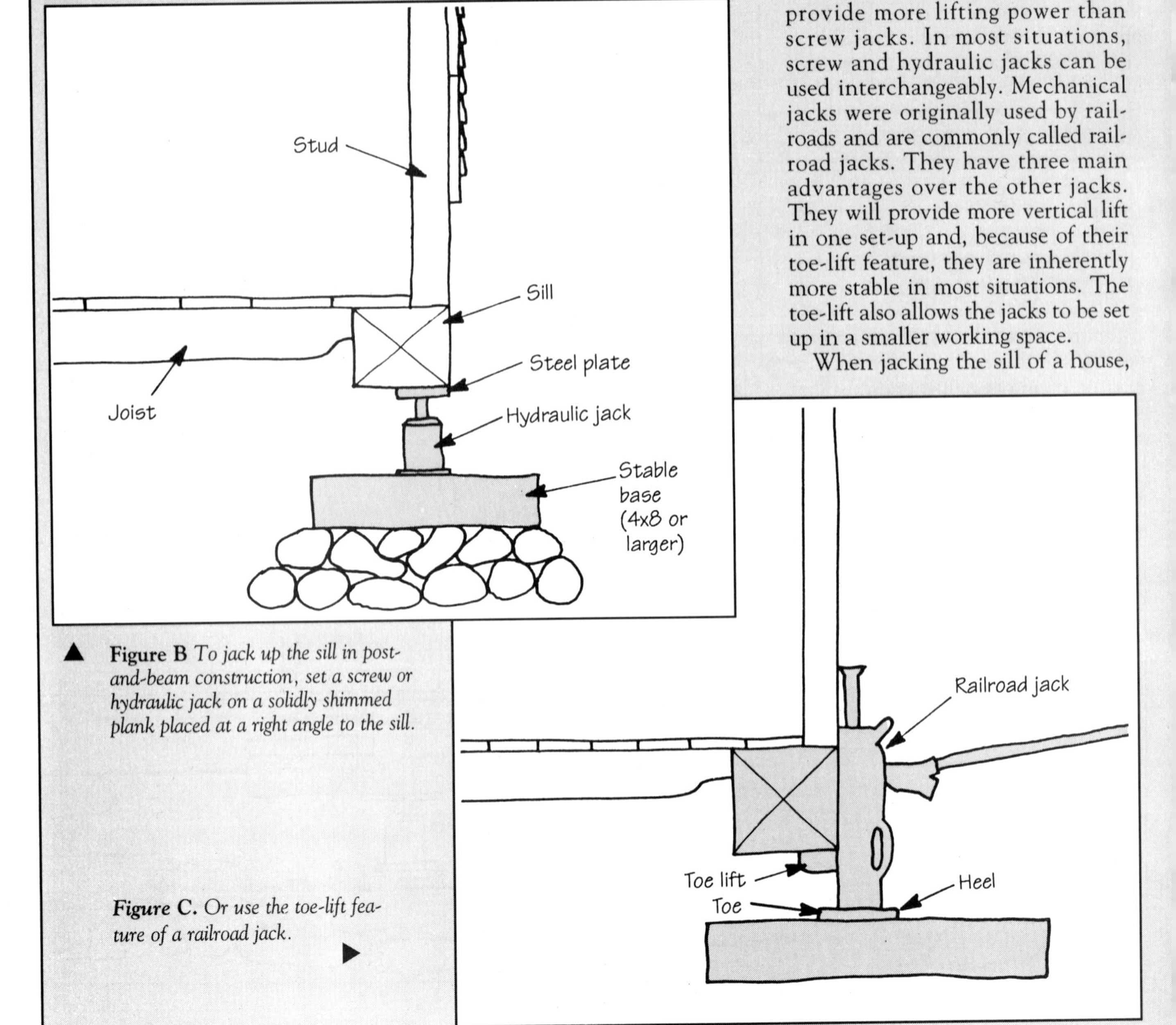

▲ **Figure B** *To jack up the sill in post-and-beam construction, set a screw or hydraulic jack on a solidly shimmed plank placed at a right angle to the sill.*

Figure C. *Or use the toe-lift feature of a railroad jack.* ▶

the most important step is to build a good level base for the jack to rest on. If you don't, the jack is likely to tilt, causing it to stress the house and become dangerous. With screw or hydraulic jacks, you are most likely going to place the jack between the foundation and the sill. This requires roughly leveling the foundation wall at the jack point and setting a well-shimmed plank over the stone at right angles to the sill. Place the jack so it is centered on the outer edge of the sill, where the studs exert the most weight (see Figure B). This placement is particularly important in cases where the house is sagging.

The railroad jack can also be used for lifting directly on the sill (Figure C). By using the toe-lift feature, less foundation needs to be removed and the jack is quite stable due to a low center of gravity.

After establishing a good stable base, slide the railroad jack tightly against the sill. Often, the upper portion of the jack will hit remaining upper sheathing boards, and it may be necessary to shim the toe of the jack to provide clearance. This placement slightly off-level is not a problem as long as you use the jack in the lower half of its operating range. When the jack is centered on its base, tightly against the sill and clearing the sheathing boards, place a steel angle plate between the toe-lift and the sill in order to spread the pressure over a wider area. You are now ready to jack.

Usually if you need to jack in one place, you should be jacking in other places at the same time. This is to keep from applying all the pressure in one spot, which can be hard on the building. When your jacks are set to lift, apply pressure at the lowest jack point first and then follow with the other jack points, always avoiding too much pressure on one spot. You may find that you need to set an extra jack or that you have one too many. Experimenting, noticing how the building acts, and changing your jacking positions are best done early in the process—before you have engaged the full load. Always check to make sure that your jack bases are not settling, and always keep the sill blocked with wood shims as you go up so that the building won't come down even if you have a jack failure.

The railroad jack is very versatile and can safely be used to jack a building from points other than the sill. You can suspend the wall of a post-and-beam building from above by gaining access to the upper plate and pushing on it at an angle with a push-pole set up on the toe-lift of the jack (Figure D). The timber plate must be in good condition, the push-pole should be a straight 6x6 with no serious defects, and the jack base should be dug into the earth at a slight angle on firm soil. Cut the 6x6 to the exact length needed, use a steel plate on the toe-lift, and adjust the angle of the jack so that it is parallel to the pole. You adjust the angle of the jack by shimming the heel or toe of the jack with a good wood shim. Apply jack pressure and readjust your jack base if it shows differential settlement. When you jack the wall this way you have the opportunity to replace your sill in fairly long lengths, although you will probably use other jacks to adjust the sill to level once it is in place.

Some people use a push-pole on top of a hydraulic jack. However, this practice is dangerous and should only be done on a very good jack base and when the push-pole is heavily braced to the building. I personally have never used this technique and do not recommend it.

Jacking can be a safe procedure or a dangerous one. It is your attention to the details that makes the difference. Always keep an eye on your jack base to note settlement. If your jack starts to tip, stop and correct the problem immediately—it won't fix itself! And, most importantly, always keep your sill blocked and shimmed tight. Never allow more than 3/4 inch space between your blocking and the sill. — *R.D.*

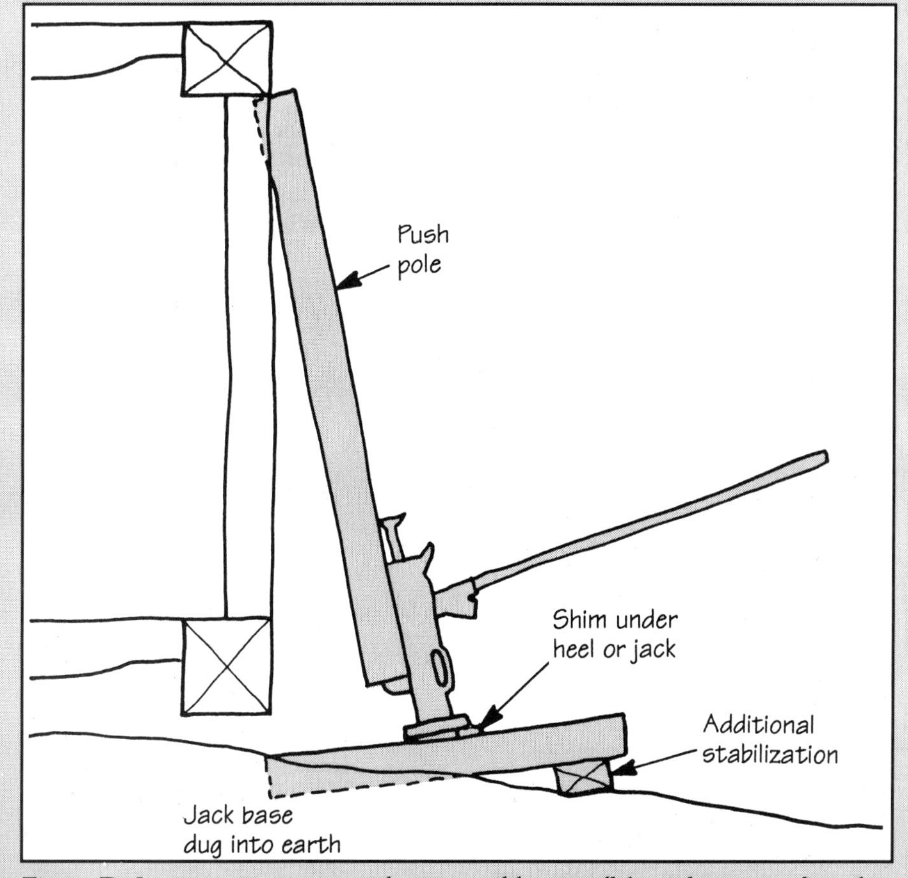

Figure D. *In some cases you can jack a post-and-beam wall from above to work on the sill below. Use a railroad jack with a 6x6 push pole to lift the wall from its top plate.*

the hole you've just prepared. Notch the new sill section, as needed, to receive tenons or the ends of floor joists. It helps installation to cut these notches slightly oversized and to chamfer the corners. On some post-and-beam frames, both vertical and horizontal tenons may have been saved when the old sill was cut out. If these are close together, however, you may have to cut off one of the tenons to allow the sill to slip into place.

The replacement piece is now installed with the aid of crowbars, sledgehammers, battering rams, and jacks. Once the sill is in place, align the studs with its outer edge, set jacks underneath the sill, and lift it to raise the studs and remove the sag from the building. If the sag is severe you may need to put a considerable amount of lift on your new sill section. This operation will most likely relieve the weight on adjacent sill sections, which can be helpful if you are replacing these sections as well. If so, you are now ready to repeat the process just described. If you are not going to replace any more sill in the adjacent area, don't worry about lack of weight in this section; it will settle in a few days time or can be shimmed when you repair the foundation.

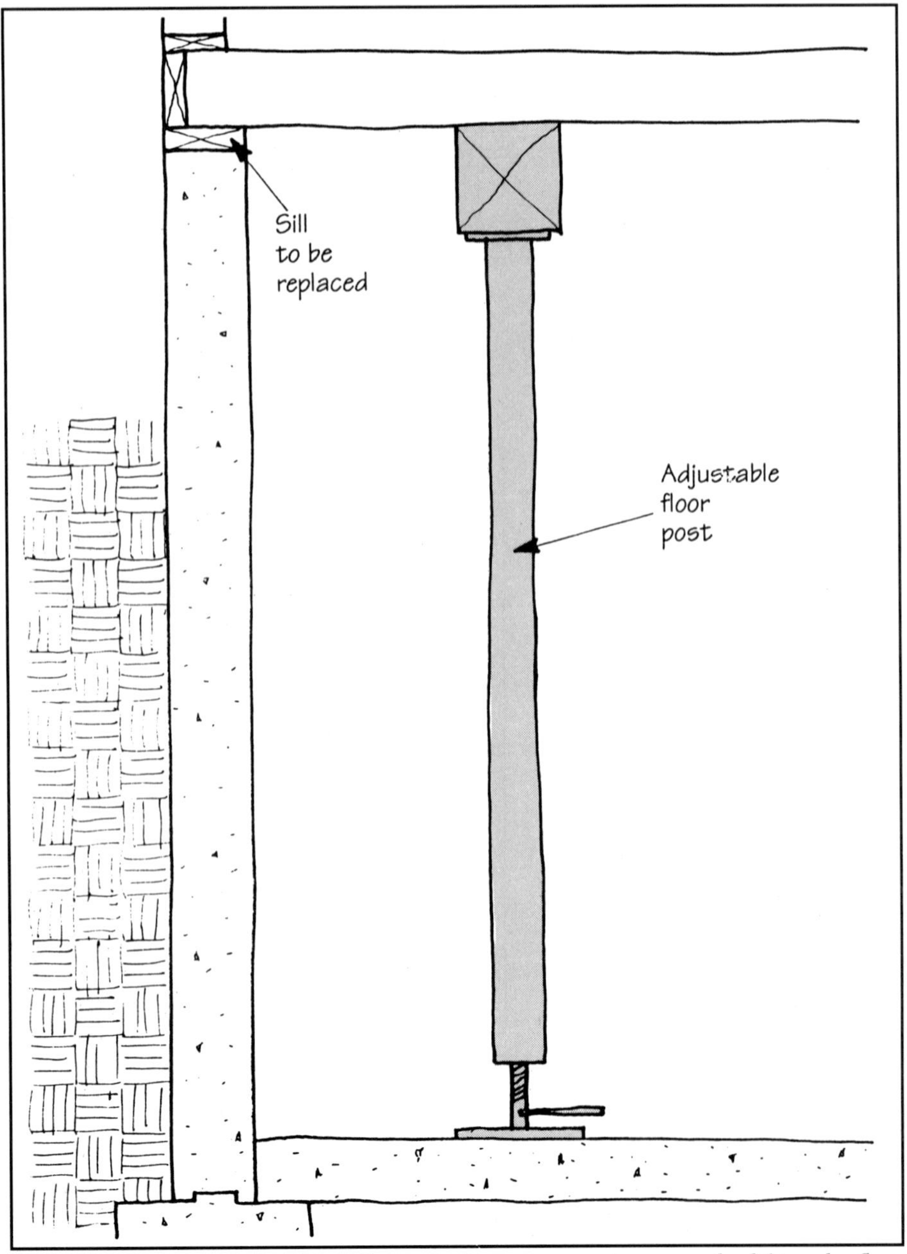

Figure 3. *To repair or replace sills in platform construction, start by lifting the floor joists at points just inside the foundation wall. Use an adjustable floor post, as shown, or railroad jacks with push poles.*

The next step is to fasten the new replacement section to the remaining good sill. You can use white oak pegs, drilled and driven in at an angle. But more commonly, bolts, threaded rods, or large spikes driven in at different angles do the job. After securing the new sill, shim the studs if necessary, and renail them. Also, any mortise-and-tenon joints in the repair area should be pinned in whatever way possible.

Any beams (either vertical or horizontal) that have had their tenons removed should be fastened by custom-made angle plates, lagged or bolted into intersecting beams. If the job involved replacing the sill at a corner of a building, a simple half lap is the easiest joint to use there. But sometimes more complex joints are desirable at the corner—for example, if you want to preserve the tenon at the bottom of the corner post.

Plywood can be used as replacement sheathing and serve a second purpose as a well-nailed gusset on the sill splices. Once the sill is fastened into place, it's time to repair the top of the foundation wall that was previously removed.

Platform-Construction Sill Repair

Platform construction requires a different form of sill replacement. In many respects, it is easier to work on than post-and-beam. First, remove the bottom 2 feet of the siding, exposing the framing. If the sheathing is plywood, cut it off in a straight line about 10 inches above the top of the foundation (higher if the band joist is also being replaced). This will make a good point for renailing the sheathing back later because there will be good nailing into the band joist.

Also, remove the nuts from the anchor bolts in and adjacent to the repair area. Then lift the building slightly by jacking up the floor joists at a point just inside the foundation wall (see Figure 3) using railroad jacks or floor jacks (adjustable floor posts). As the floor comes up, you can cut the toenails fastening the joists to the sill with a metal-cutting blade in a reciprocating saw. Do not jack any more than is necessary to loosen the sill—in order to minimize the stress on the rest of the building. After the sill is loosened, cut slots with the reciprocating saw to allow the sill to slide out from under the joist. Put the new sill in much the same way the old one came out: by cutting slots for the sill to slide around the anchor bolts. The anchor bolt nuts are then put on with oversized washers, generously lubricated, and tightened down securely. Then lower the building, fasten the joists with hurricane clips, and replace the exterior siding.

Choosing the Materials

Originally the buildings of post-and-beam construction were framed with long lengths of 8x8 or larger timbers. The species of wood used were white oak or slow-growth "punkin" pine; both of these were naturally very rot resistant. Later, hemlock and spruce were used although these woods did not tolerate moisture as well. For repair work of the highest quality, white oak is still the first choice, although it is difficult to work with, and therefore more expensive to use. Slow-growth pine is no longer available. Hemlock is available and good structural wood, but is quite heavy to handle when green from the sawmill. Spruce is readily available in long lengths, has a low moisture content when still green and therefore is relatively light to handle. For special projects, southern yellow pine that has been pressure-treated with preservatives can be ordered, although it is expensive and hard to handle due to its weight.

Any of the woods mentioned can be additionally protected against rot with the application of chemical preservatives. Because the sills are mostly hidden and people do not come into contact with them, use as strong a preservative as is available (some old timers swear by a liberal application of motor oil).

On major sill repair jobs where there is plenty of jacking equipment available, there are advantages to using long lengths of timber. Using long lengths minimizes the number of splice joints required and helps save time during the leveling process. Also, in most cases this gives the building better structural integrity.

Summing Up

Structural repair work is not something you learn overnight. But even

A half-lap joint lagged and bolted into place securely ties the new post-and-beam sill section to the old (above). The foundation window cutout provided a convenient place to set jacks without disturbing the foundation.

Metal Fasteners

Although post-and-beam frames were originally built with only wood joints, we must usually rely on steel to simplify our repair techniques. Galvanized or stainless-steel nails are necessary when fastening pressure-treated wood. Steel bolts or threaded rods are very useful in bolting together structural half-lapped or scarf joints.

When a tenon has been cut off for access reasons or rot, a heavy steel corner plate can be used to perform the same structural job. The steel plate should be slightly narrower than the beam it is being fastened to and be at least 1/4-inch thick. The length of each leg of the corner plate should be determined by the load it has to carry, the condition of the wood it is fastened to, and the available space. Drill 1/2-inch holes in each leg in a pattern that will allow room to use wrenches, and will avoid minor rotted areas or splits in the beam. The plates can be fastened to the beams with 1/2-inch through bolts, although 1/2-inch lag bolts in long lengths will do an adequate job. If you use lag bolts, you should drill a 1/2-inch pilot hole the length of the threaded area to ensure maximum holding power. Be sure to oil the lag before you put it in to ease its installation and provide extra protection against rusting. I drive in lags with a socket driver attached to an electric drill.

—R.D.

This ten-year-old house in central Vermont suffered from negative grading, standing water in the crawlspace, a wooded site, and bathroom vents exhausted into the attic. Plus, the owners insisted on running humidifiers all winter long.

with years of experience, you'll have to experiment some in placing your jacks. Make your best guess as to where to place them; then adjust as necessary.

Whatever your level of experience, keep safety concerns paramount. Two key points are (1) Always work off a stable base. If you try to cheat, it will always work against you; and (2) Always block and shim as you jack. Never leave more than about 3/4-inch between your blocking and the sill.

Finally, remember that after straightening a building, secondary stresses will be relieved for a year or two afterward. Tell your clients to expect some plaster cracking over time. A building that took a hundred years to sag isn't going to straighten out in a day. ■

Roger Desautels is a third-generation structural repair contractor and building-renovation consultant, based in Irasburg, Vt.

Raising the Roof

Sometimes it pays to build up—and take the roof with you

by Henri de Marne

Raising or moving an existing roof is often more economical than building a new one from scratch. I've worked on two roof raisings. One was to create a dormer on a Cape. On that job, we hinged most of one side of the roof, then lifted and held it with a crane while the carpenters raised the wall they had built in the attic the day before.

The other job involved an unusual situation where part of a trussed roof was removed to add a second story to part of a house.

The piece of roof happened to be exactly the size of a carport planned for the opposite end of the house. It was carefully picked up by a crane and moved, while two men with a rope swung it around 180° so the gable would face correctly.

Setting Up

In the case shown here, we raised the roof of a one-story house to add a second story. Typically, a new roof is built over the old, which is later removed piece by piece once the new one is weatherproofed. But you can raise an existing roof in a fraction of the time—and with considerable savings in materials. The rafters, sheathing, and roof covering are all reused with minor alterations or repairs. Consider how veteran building mover Royce Lanphear of Waitsfield, Vermont, dealt with the situation.

The house was 40 feet long and had a few complications: an ell on the back, a two-car garage at one end, a stone fireplace at the other, and a brick chimney in the middle. Moreover, the roof was built with standard rafters; trusses would have made the job a lot easier.

The first thing Lanphear did was to erect steel staging on one gable end at the level of the attic floor. To reach the working platform, he placed against it a set of site-built stairs with a shallow pitch and wide treads. Lanphear says this speeds up the job, since workers can walk up to the staging carrying full loads rather than needing one hand to hold onto a ladder.

A "doorway" was then cut into the gable and covered with a plastic curtain. The curtain was held at the top by a cleat and lowered at the end of the day to seal the opening against the weather.

Building Trusses

Next, a plywood deck was laid over the existing attic floor joists to provide a working platform. Another set of joists was set on top of the plywood deck and nailed to the ends of the rafters. These 2x6s became the bottom chords of new roof trusses when 2x4s were nailed between them and the existing rafters in the familiar W pattern. A 2x6 was nailed to the gable-end studs to brace them.

It was now safe to free the roof rafters from the rest of the structure. Workers crawled to the rafter seats and pulled or cut the nails holding the rafters to the floor joists and plates. In some instances it was easier to do this from the soffits outside, which had been opened to free the outlookers from the walls. The gable was cut just below the new 2x6 end joists.

Next, the asphalt shingles were removed where the roof had to be cut. Since Lanphear planned to lift the roof in two 20-foot sections (he thought the 40-foot roof was too big to lift in one piece), the shingles were removed at the midpoint of the roof as well as at the joints with the garage and the ell.

Midway across the main roof, workers cut the sheathing next to a set of rafters, removed the flashing at the chimneys, and cut four holes in the roof at the eaves about 4 feet in from each end of each roof section to be lifted. At these points, 6x6 blocks approximately 4 feet long were braced in place at the eaves between the top and bottom chords so the cables could be tied around them.

The new second-story walls, which had been framed and sheathed on the ground in manageable sections, were hoisted and stacked on the staging platform. The first roof section was now ready to be lifted by a 45-ton hydraulic crane.

Robin Foster

Using a custom lifting frame, house mover Royce Lanphear picked up the first half of the roof with four cables tied around the lengths of 6x6 braced in place near the eaves. The 20-foot section was stored on the ground while the pre-framed walls were raised.

The Big Lift

Royce Lanphear has designed and built a lifting frame. It is made of two I-beam sections that are spread apart and held together by four pieces of galvanized pipe. The crane picked up this frame by cables attached to its four corners. These, in turn, held four cables tied around the 6x6s placed at the lifting points at the eaves.

In addition, two guide ropes were attached to the front of the lifting frame so the roof section could be guided or positioned as needed whether on the way up or down.

The main problem the chief carpenter on the job, Fred Gilbert, had to contend with was the central chimney; the slightest bump from the roof could have sent it tumbling down to the basement. Talk about precise coordination between the crane operator, those on the guy ropes, and Gilbert on the roof directing traffic!

Once it had been lifted off the house, the roof section was deposited on the ground next to the house, and the carpenters quickly erected and braced the sections of walls that had been stacked on the staging platform. The crane then lifted the roof section back up and placed it over the new walls, where it was fastened with nails and metal anchors. The procedure was repeated with the second roof section.

Cleaning Up

Once in place, the two roof sections were secured together by doubling up the rafters at the joint between them. The free edge of plywood was nailed into the doubled rafter, the eight lifting holes were patched, and new shingles woven back in. All this was done in one day, so the finished lower level was never exposed to a change in weather.

Windows were installed, repairs made to the soffit and gables, and new siding was put on. The chimneys were raised and reflashed.

The existing 2x8 first-floor ceiling joists were adequate to support the new finished second story; only a few were doubled up as needed under partitions. The plywood decking, used as a working platform, was firmly glued and nailed as the new subfloor. Through all this action up top, the ceiling below remained attached and undamaged.

The finishing work was ready to proceed once the mechanical systems were roughed in and the insulation was finished.

With the first half of the roof replaced, the second half was lifted (left) and stored on the ground, while the pre-framed walls were raised. Next, Lanphear's crew replaced the second half of the roof (above left), locked it to the first, and sealed the shell to the weather (above)—all in a day's work.

Risky Business

As tempting as it may be to tackle this type of job, a remodeling contractor should not lose sight of the enormous liability involved and the danger to workers. Lifting a full dormer requires precise planning and bracing to prevent the remaining half of the roof from caving in or the hinged section from collapsing. But lifting an entire roof, even in sections, requires experience and skill acquired over years.

While any skilled contractor can prepare a building for roof lifting—and complete the work that follows—the actual lifting should be left to an expert. Wisdom recommends that this phase be subcontracted to a building mover.

Adding a second story this way can give you a competitive edge, and be very profitable, too. Also, without a doubt, it can be a tremendous public-relations coup. Announce it in advance and make sure there is a crowd is watching. Of course, you'd better make sure that everything goes well. ■

Henri de Marne is a building consultant in Waitsfield, Vt.

Everything You Always Wanted to Know About Removing Collar Ties

Can you add attic headroom and still comply with the codes? Maybe.

by Harris Hyman

With the downturn in new construction, we're getting into a lot of remodeling jobs. One type of project that comes up again and again is the conversion of the attic space into a real room. Inevitably, the space is full of collar ties. Should we raise them a couple of feet, the way the owner wants? Or cut them away altogether? We may alter them and give our client a big confident smile but later we begin to worry.

What Do Collar Ties Do, Anyway?

Collar ties are peculiar items. Their job is pretty simple but their loading can be deceptively large. Basically, they keep the rafters from spreading. Just that. Unlike trusses, collar ties do not stiffen rafters or help carry loads.

Roof loads are transferred to the wall structure through the rafters, *in the direction of the rafter*. At the eaves, rafters push both down and outward. The collar ties help handle the outward forces; the wall studs handle the vertical loads.

However, collar ties do get in the way when you want to use the attic space. And everyone knows somebody who has just sawn them out and done a pretty nice-looking conversion. The idea of doing the job this way doesn't feel quite right, though. Should we repeat this ourselves?

I'll look at two common situations where you might want to move or remove collar ties to get more headroom in an attic. In the first case, the rafters come right down to the attic floor; the second case has the rafters resting on top of half-walls.

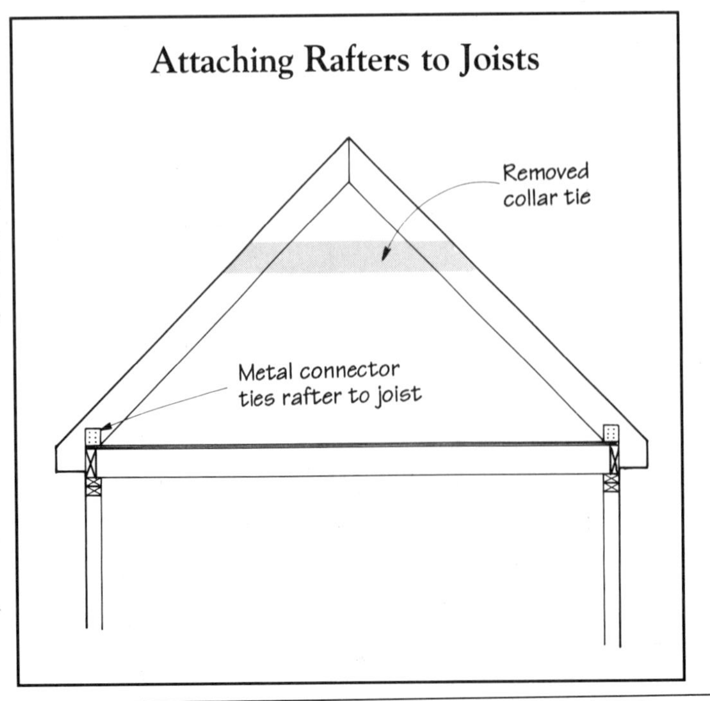

Figure 1. *In a typical gable roof, collar ties can be removed with no real effect, as long as the rafter ends are securely fastened to the floor joists. The author recommends metal connectors for restraining roof thrust.*

The Easy Situations

Rafter-to-joist connections are critical. If the rafters end at the attic floor (see Figure 1), they can be fastened in and the collar ties removed *with no real effect*. As an example, on a 24-foot span with 12:12 rafters 24 inches on-center, the necessary restraining force at the bottom of the rafter is 600 pounds under a 50-psf code load. Ties are available to handle this. The best fastenings are metal strap ties, such as those Teco makes. There are a variety of ties available to fit various situations, and they are described in the catalogs, along with their load capacities. There is a good collection of catalogs in the Sweet's collection, which your neighborhood architect will probably let you scan. You should always check the catalog specs for the strength of the fastener you want to use. The counter person down at the lumberyard may be real friendly and helpful, but has no liability for your projects.

Do not depend on toenailing, which, even when carefully done, is exceptionally weak. Toenailing should only be used to hold sticks in place until a firmer fastening method can be applied.

If the rafters are face-nailed to the

Code Vs. Reality, or Why Most Things Don't Fall Down

A reasonable technical analysis of collar ties suggests that they should *never* be removed, at least not in snow country. This "reasonable analysis" is based on building-code roof loads of 35 to 55 pounds per square foot. However, for the most part, in the typical house where the attic is to be converted to a room, actual loads are rarely above 10 pounds per square foot.

Just writing this makes me cringe a little, as I picture my liability insurance agent's face turning to chalk when she reads it, and as I try to frame answers to the letters from responsible engineers. Still, let's look at the physical world (not to be confused with the "real world," where lawyers exist).

In reality, attic conversions are almost always done on steep roofs, 9:12 or steeper. In most cases, a shallower roof is simply unsuitable for adding living space. And a little observation shows that snow usually doesn't pile up for long on a 10:12 or 12:12 roof. The lower 3 or 4 feet might accumulate an inch or two of ice. This looks frightening, but an inch of ice weighs only about 5 pounds per square foot. Added to this external load is the typical roof structure and shingles, in the neighborhood of 5 or 6 pounds per square foot. A design load of 15 or 20 psf is actually reasonably conservative.

Now, I am not enough of a fool to stamp designs to this criterion, but the information is offered to help explain why some have gotten away with cutting out the collar ties. It's like a stop sign on a road with no traffic: I can say the road is empty, but it's irresponsible to advocate that you ignore the stop sign. *I strongly encourage you not to neglect the building codes.*

As an example, consider a 24-foot span under a 12:12 roof, with the rafters 24 inches on-center and the collar ties 7 feet above the top plate (see illustration). Engineers make a common assumption for analysis: The bottom ends of the rafters are held up by the walls, but are unsupported against spreading–as if they were resting on short, platform-framed half-walls. So under a code load of 50-psf, an engineer's calculation finds a tension in each of these collar ties of 1,440 pounds.

Don't get tricked into thinking that half the tension can be carried by each end—a small but unforgivable violation of Newton's First Law. The total tension in the collar tie must be held on both ends. The tension must also be supported in the middle of the tie with wood. Construction softwoods have a tension design stress of about 300 psi, so a 2x8, which has 5.3 square inches of cross-section area, will handle the 1,440 pounds.

If that sounds like a lot of force, it is. And the really difficult problem becomes how to fasten the ends of the ties to the rafters to withstand such tension. You could do it with a pair of 2 1/2-inch split-ring connectors, or a dozen 3-inch screws, if you can find space for them! Bolts are unsuitable because there just isn't room for three 1-inch bolts with washers; although smaller bolts would never break, the larger bolts are required to spread the load and avoid crushing the wood fibers.

In reality, of course, we expect a roof load more like 15 psf. In that case the tension is reduced to 432 pounds, which can be supported by five 12-penny nails. And we may also find some support provided where the rafters tie in at the eaves, which helps to explain why a couple of nails banged into each tie seem to have held.

This issue comes about because of the gap between physical reality and the codes. Perhaps the right way to approach it is to advocate change in the code, but our working world is not that of the code-making official. We have to expand the attic space, and tomorrow's revised building code will not get today's job done.

One piece of advice: If the only thing involved is a little money, play it conservatively.

— H.H.

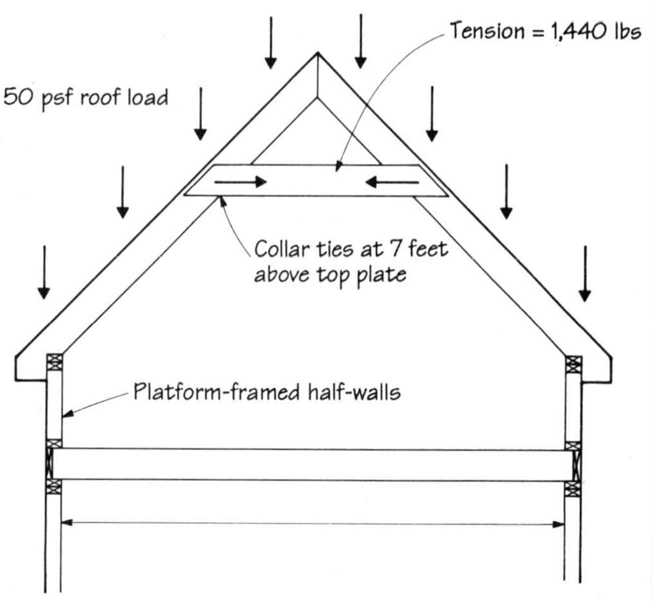

The higher the collar tie, the greater is the tension. In the example above, using a design load of 50 psf, the tension in the collar tie is 1,440 pounds. The problem for the builder is not the strength of the wood, but finding a practical fastener to withstand that kind of force.

joist ends, a common practice in attic spaces, you may want to beef up the connection before removing the collar ties. A workable rule of thumb for nail strength says that 10-penny, 12-penny, or 16-penny nails in 2x spruce hold about 100 pounds per nail in shear, or 60 pounds against withdrawal. Nails can be clustered about a dozen diameters apart—about 1 3/4 inches for 12-penny nails—without splitting construction softwoods.

Kneewalls an option. If it's not possible to make a strong enough connection between joists and rafters, another option is to build a kneewall. A short wall a few feet inside of the eaves line shifts some of the load to the deck (see Figure 2). The triangle formed by the floor, the rafters, and the kneewall acts as a buttress. This contains the outward thrust, and reduces it some by reducing the actual span of the rafters. The rafter loads are carried to the floor, which becomes a massive tie.

Kneewalls are often good architectural solutions, since they trap off some messy, unusable space at the bottom corner of the roof pitch and allow the space to be filled with insulation. If the wall is a little higher, say 5 feet, and farther in from the eaves, it creates some storage space.

Kneewalls must be built carefully, and securely fastened to both the rafters and to the floor joists. The plates must be nailed directly to the studs, not toenailed. For adequate strength and stability, the upper plate of the kneewall must be nailed directly to the rafters, and the lower plate must be nailed directly to the floor joists. No special sheathing is needed on kneewalls; ordinary drywall is generally strong enough.

With kneewalls, vertical loads are transferred to the ceiling joists, but the kneewalls are usually close enough to the eaves that the joists are not overstressed.

Half-Walls

A messier situation exists when the rafters rest on a short wall extending above the deck (Figure 3). If the building is old enough or new enough to have been balloon-framed, the wall can handle the push of the rafters if it isn't too high—about 20 inches under code loads for 2x6 studs 16 inches on-center.

Where the rafters rest on platform-framed half-walls, however, *there is no transverse strength at all.* In this case, the collar ties are resisting the outward thrust of the roof, and should not be removed. But can you move them higher?

Figure 4 shows a hypothetical situation: a 24-foot cape with 6-foot platform-framed half-walls, a 12:12 roof with rafters 24 inches on-center, and collar ties at 7 feet off the attic floor. The owner wants to raise the ties to 10 feet! Can you do it?

You definitely crank up the tension when you do this. Under the 50-psf load specified in some model codes, the original tension in the tie is 655 pounds. Raising the tie to 10 feet pushes the tension up about 40%, to 900 pounds. That's 10 nails, brother! Even if that looks okay, then we find another problem: bending stress in the rafter. Code loading demands either a

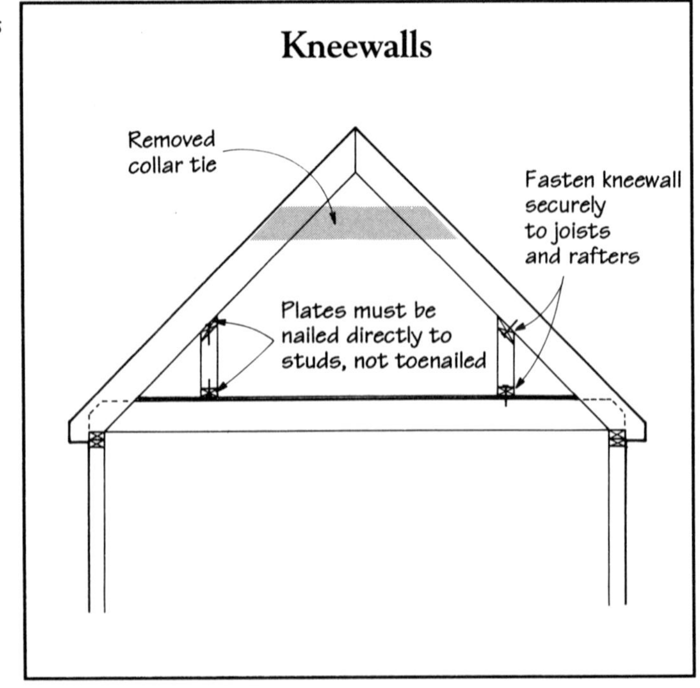

Figure 2. *Kneewalls close to the eaves can serve to make a strong rafter-to-joist connection. The kneewall must be carefully built and securely fastened to joists and rafters.*

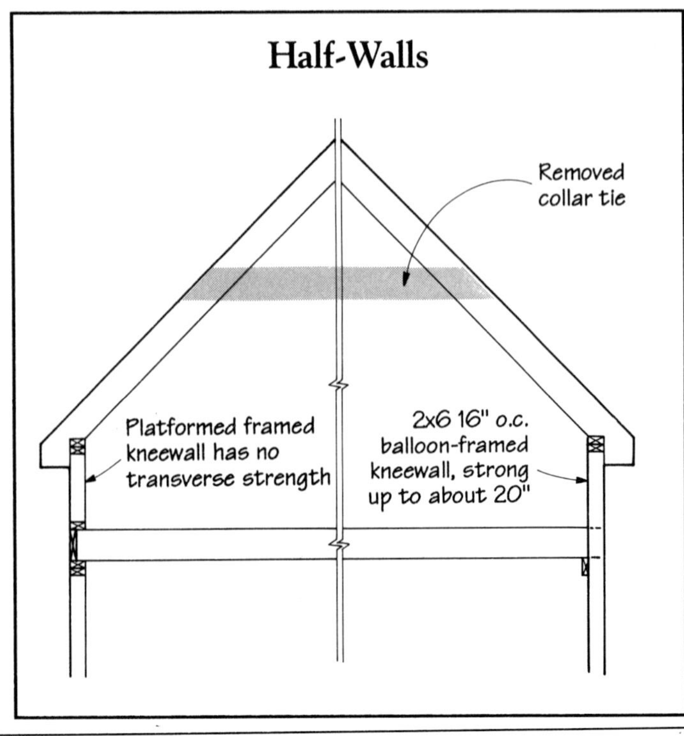

Figure 3. *With story-and-a-half structures, you must be extremely careful about removing collar ties. Platform-framed half-walls (like the left half of the drawing) have no strength to resist the thrust of the rafters, so the collar ties must remain in place. With balloon-framed 2x6 half-walls framed 16 inches on-center (at right), collar ties can be removed if the half-wall is less than 20 inches high.*

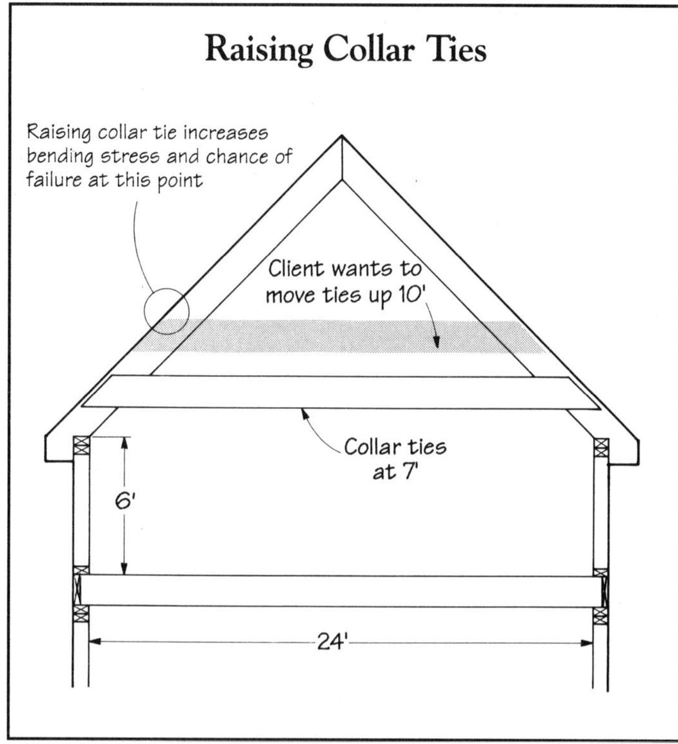

Figure 4. *Raising collar ties to create more headroom can raise the tension in the tie so high that the tie cannot easily be fastened to the rafter. The raised tie also adds bending stress to the rafter.*

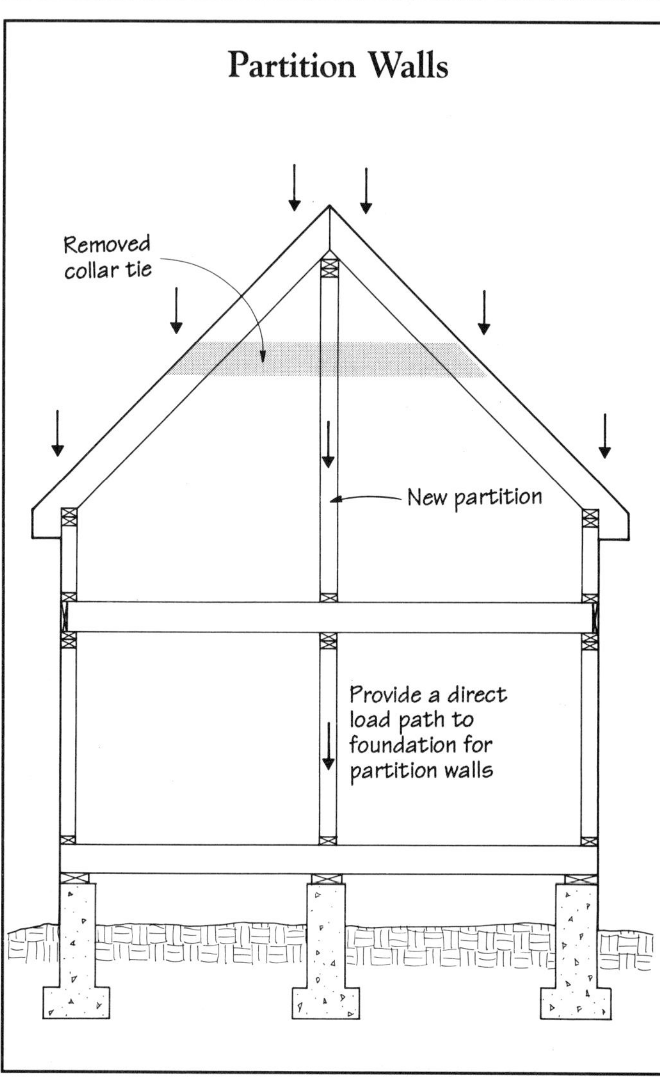

Figure 5. *Load-bearing partitions can remove the roof's outward thrust, but you must be sure you have a direct load path to the foundation.*

double 2x10 or an engineered material (glulam, LVL, Parallam, etc.) for the rafters. Of course, the rafters are already in place, otherwise there wouldn't be a renovation job.

But, you might say, the house has been there for 80 years and the ties have only 3 nails in the ends and there's been no trouble. This is because real roof loads and code roof loads are often not the same (see "Code Vs. Reality," page 113). Without suggesting a violation of code, I'd say raise the ties as little as possible.

Removing alternate ties. And what about taking out every other tie? Doing this *doubles* the tension loads in the remaining ties, since you have half the supports. "But my cousin Jack did it on the Farthing job and it worked good!" Ayuh, code vs. physics again. My advice is to be prudent and not to remove any ties unless you understand the loading in the particular structure. With platform-framed halfwalls, chances are the structure is already marginal and should not be made worse.

Solutions

So how do you get some space in the attic? Although you can't fight the collar tie problem directly and still conform to code, there are two general ways to circumvent it. The outward thrust can be removed, or it can be contained. The more effective method is to remove the thrust by holding up the rafters.

When the rafters are supported by other structure, they do not lean on each other and there is no outward push at the bottom ends. Containment of the thrust is achieved by holding the lower ends in by some method other than collar ties. (To get a feel for this, rest your elbows on the desk, clasp your hands, and push down on your clasped hands with your chin. You can contain the thrust with your triceps, but it's easier to support your chin with a structure—a stack of books.)

Partition walls. One solution for providing structure is to install partition walls that support the rafters (Figure 5). This takes practically no engineering and does a very effective job. Structurally speaking, the absolute

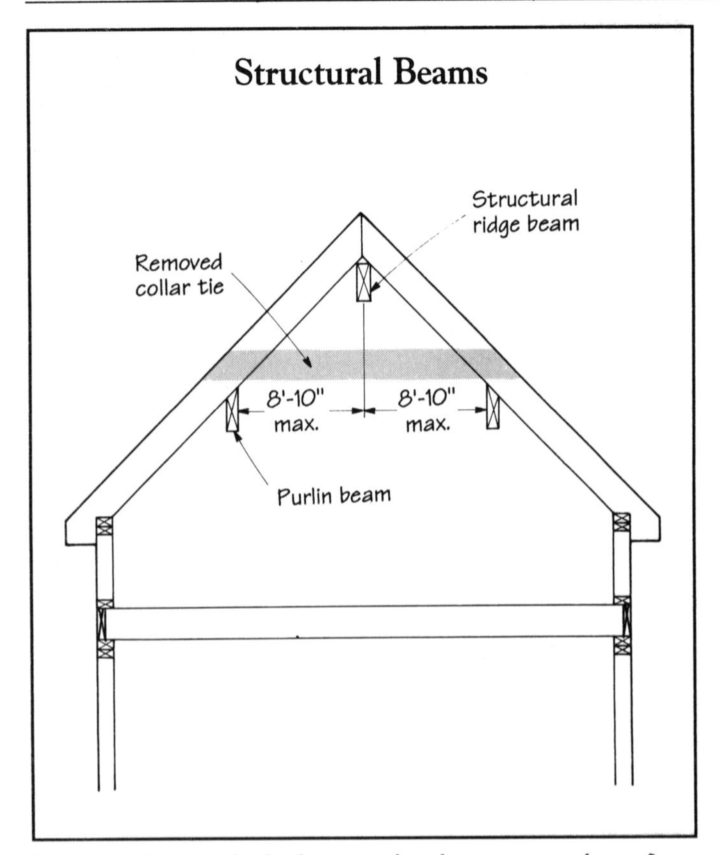

Figure 6. *Structural ridge beams and purlins can carry the roof's load to the gable ends, which must be framed adequately to carry the loads to the foundation.*

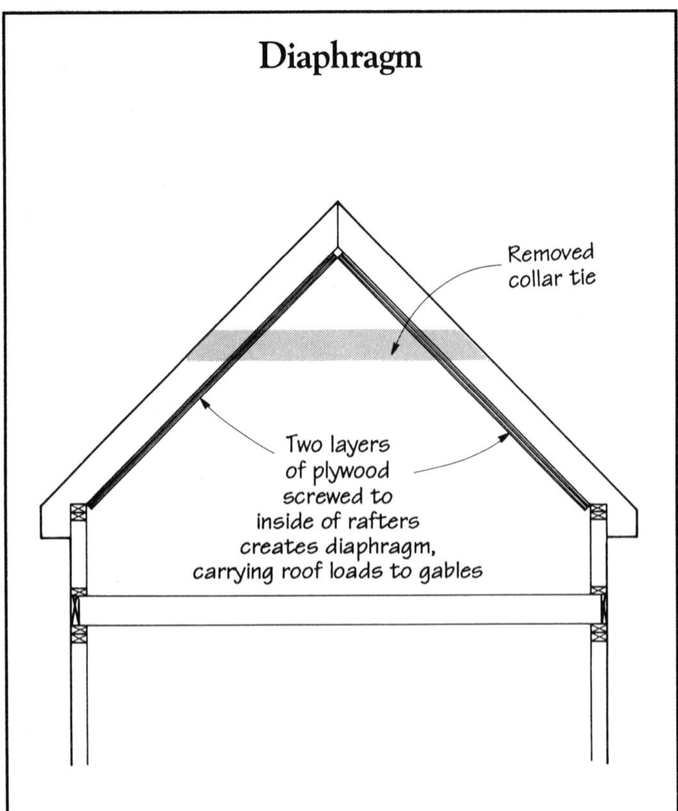

Figure 7. *An unusual solution for carrying roof loads is to skin the inside of the roof with two layers of plywood, creating a diaphragm. The entire assembly acts like a beam, carrying the roof loads to the gable ends.*

best place for a partition wall is under the ridge, but transverse walls (perpendicular to the ridge) also work, functioning as massive ties. The roof rests on the partition walls and the lower ends do not push outward.

The builder should be careful with partition walls, so that the loads in the walls are carried to the ground without overstressing the floor joists. This is relatively easy to accomplish in a house, where lower-story partitions provide support for the new attic partitions.

However, partition walls are not always good architectural solutions, since getting some nice space under the roof is why we got into this mess in the first place.

Structural beams. A more difficult solution for removing thrust uses ridgewise support beams or purlins (Figure 6). Since the rafters hang from the beams, the low ends do not push outward. To install these beams, you often have to open up the gable ends. Beams also require posts in the gable end to carry the loads to the ground.

Structural beams can be installed anywhere between the ridge and the midpoint of the rafter span. They should usually not be more than 8 to 10 feet apart. They can be either engineered lumber or steel; ordinary wood is generally not strong enough to span from gable to gable. A 4x16 glulam or LVL ridge beam will hold up a roof measuring 24 feet eaves-to-eaves and 20 feet long with a 50 psf load. If two purlins and a ridge beam are used, 2x16s or 4x12s can be used.

If you choose to use steel for the ridge beam, a W12x14 beam or 12x4x5/16-inch tube will hold up the 24x20-foot roof described above. For purlins, 8x4x3/16-inch tubes will do. Rectangular tubes cost more and are heavier than W-sections (I-beams). But tubes work much better for ridge beams since it is easy to Ramset a facing board onto the side of the tube, and attach joist hangers to the facing board.

Diaphragms. And finally there are the exotic solutions, such as the folded plate or diaphragm (Figure 7). In an attic situation, this involves covering the underside of the rafters with an extremely strong integral plate. A workable plate consists of two layers of 1/2-inch BC plywood with the joints staggered. The plywood must be screwed to the bottom of the rafters with 1 3/8-inch and 1 3/4-inch drywall screws, about 4 inches on-center.

This system carries the roof load directly to the gables, with the plane of the roof acting as a beam. The flow of the load to the gables also stresses another component that must be watched carefully: the gables themselves. Often, the gables need to be strengthened with the same sort of careful double-sheathing as required on the pitches.

Using the inner skin of plywood demands some careful eaves venting to get rid of moisture, since both the inner skin and the exterior shingles act as vapor barriers. A diaphragm is neat and workable and can be installed directly inside the attic space.

Horizontal box beams. Another exotic is the horizontal box beam or steel beam at the eaves (Figure 8). This type of beam provides a transverse resistance to the outward thrust of the rafters. These beams must be carefully built and installed. Like diaphragms, they also transfer loads to the gables, which must be built accordingly. On our 24x20-foot roof, an engineered wood restraining beam should be 6x10 or 4x12, and a steel tube should be $6x4x^1/_4$ inch or $4x4x^3/_8$ inch.

Folded plates and eaves beams are rarely used, and there are no generally available guidelines for their design. Also, with these techniques, the load flow is not obvious and most builders do not have a feel for where to apply meticulous care and where to cut corners.

Better call you neighborhood engineer on these. ■

Harris Hyman is a civil engineer in Portland, Ore.

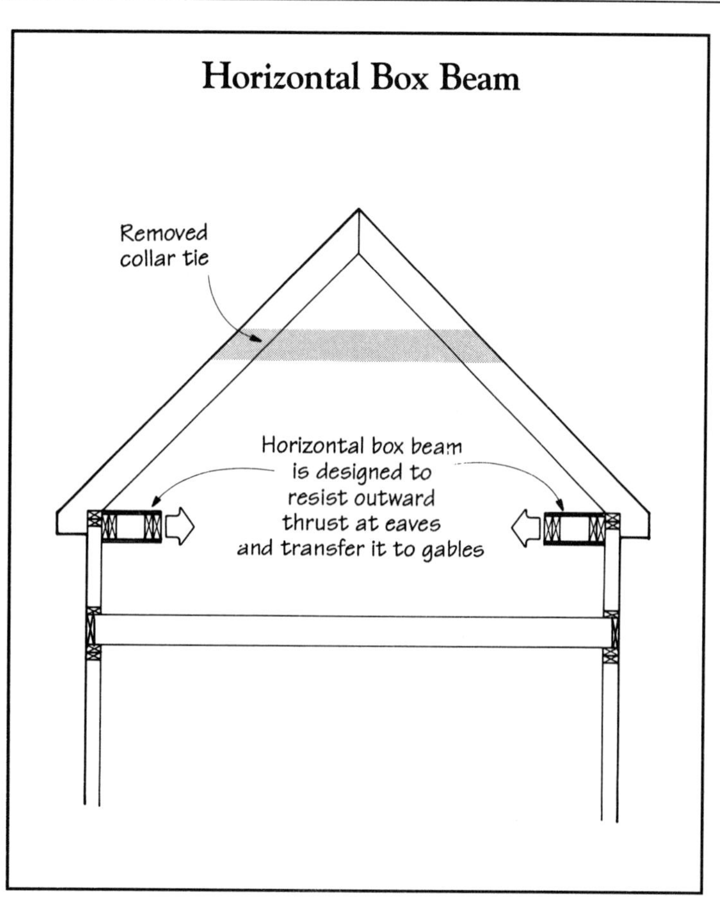

Figure 8. *A horizontal box beam can be designed to resist the outward thrust of a roof at the eaves. The loads are carried to the gables, which must be designed to carry the loads to the foundation.*

Adding A Bay Window

Build attractive, economical bays with fast layout techniques and simple detailing

By Clayton DeKorne

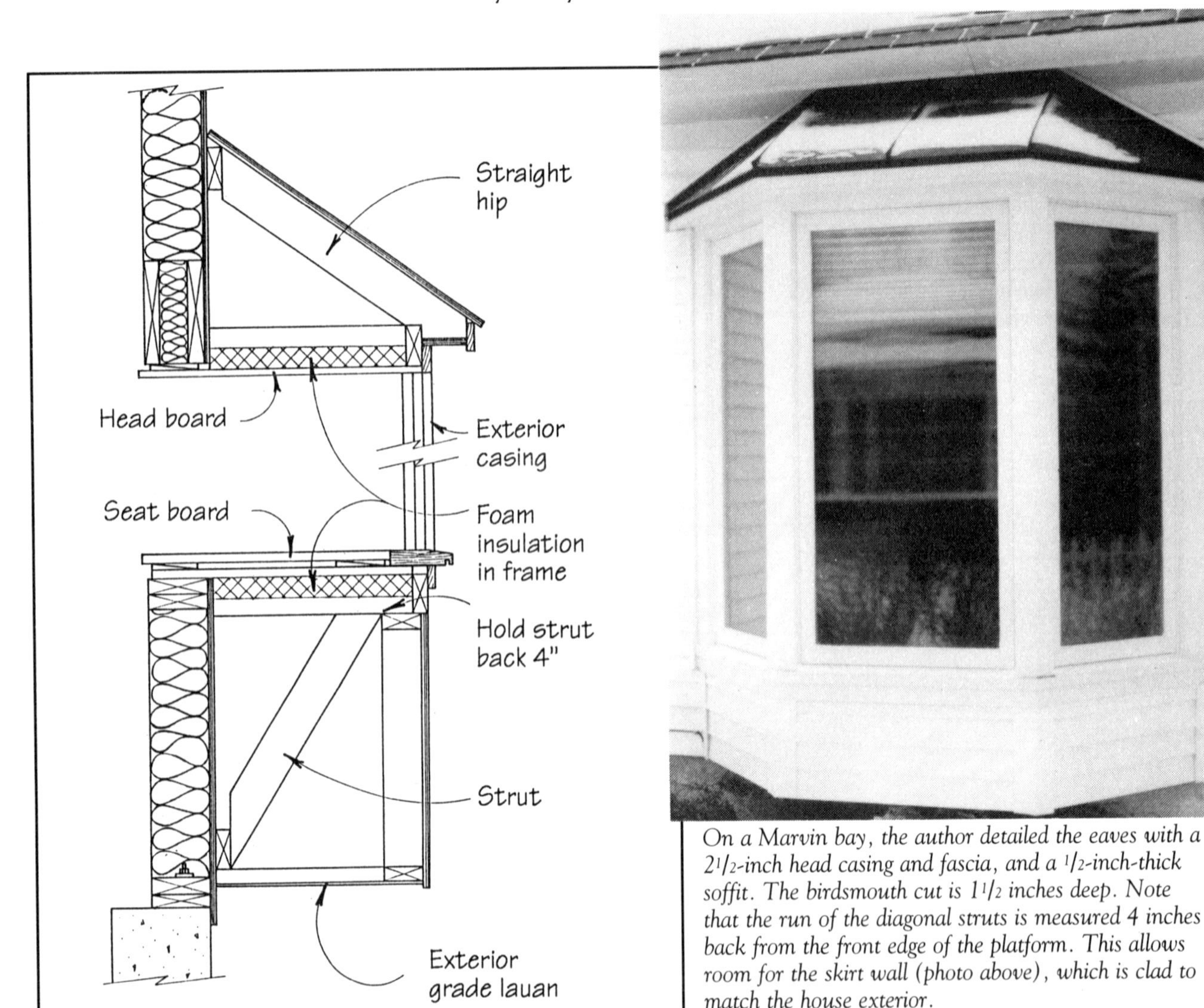

On a Marvin bay, the author detailed the eaves with a 2 1/2-inch head casing and fascia, and a 1/2-inch-thick soffit. The birdsmouth cut is 1 1/2 inches deep. Note that the run of the diagonal struts is measured 4 inches back from the front edge of the platform. This allows room for the skirt wall (photo above), which is clad to match the house exterior.

I blew the estimate on the first angled bay I installed, in part because the unconventional roof layout took me by surprise. The plans and sections in the window catalog offered few clues on the roof and base details. After several installations, however, I figured out the roof cuts, and worked out a few tricks to make these window installations more economical.

Choosing a Style

There's an enormous range of design options in manufactured bay windows. But to keep a bay in line with a tight budget, it helps if you insist on certain design details: First, stick with factory installed head and seat boards (the panels that cap the top and bottom of the windows). These rule out the possibility of a walk-in bay, but save the complications of cantilevered floor joists or angled foundations. Second, stick with 45-degree bays to simplify the roof cuts. Also keep in mind that, while pre-fab metal roofs and decorative brackets simplify the installation, it's often cheaper to frame a hip roof and base yourself.

There are two roof styles I've used.

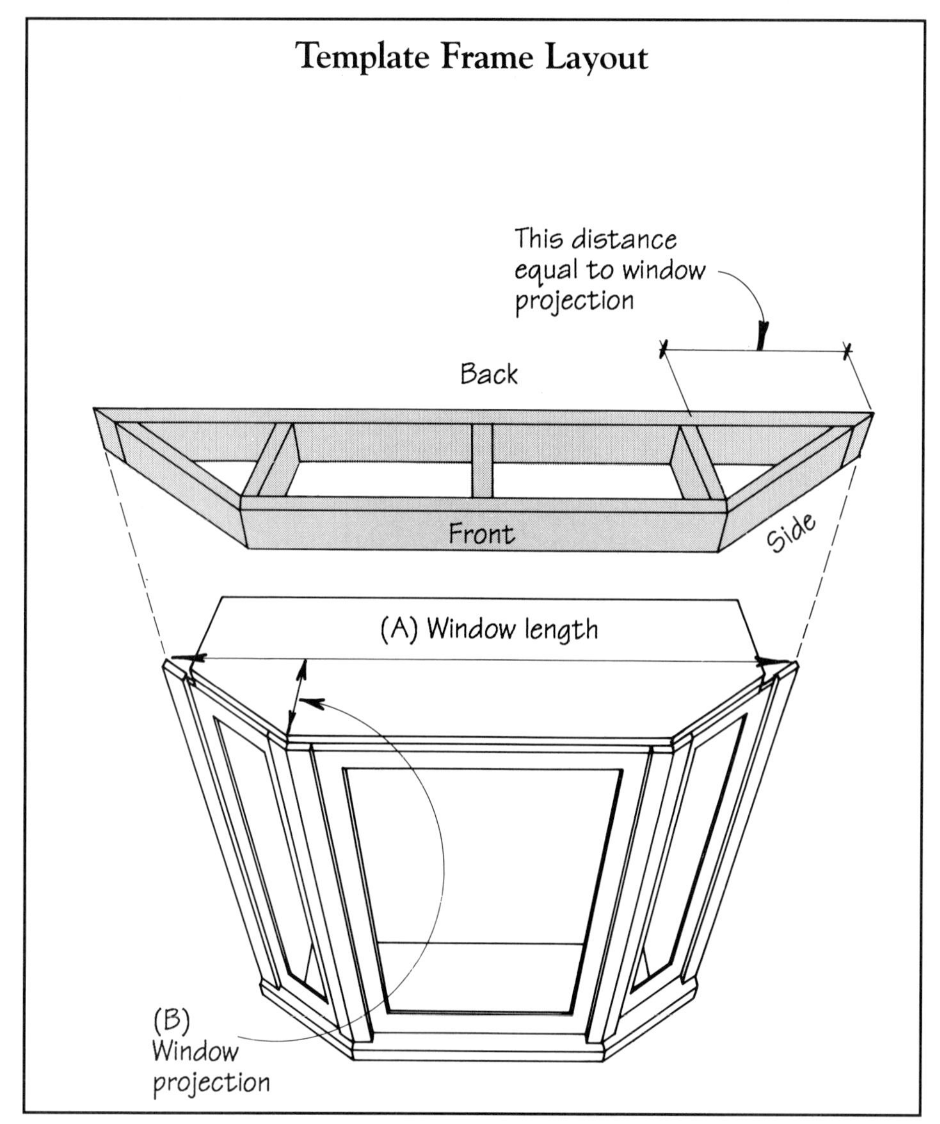

Figure 1. *The dimensions for the template frame (shown in gray) are taken off the top of the bay window. The back of the frame is equal to the window length (A) measured from the inside of the exterior casing, minus 1/4 inch to allow for play when setting the window on the platform. To establish the length of the front, the back is shortened by twice the distance of the window projection (B). The sides are then measured and cut with 45-degree angles at both ends.*

One, which I call a "straight hip," has hips that run perpendicular to the face of the bay. The other, which I call a "bastard hip," has hips that angle towards the center at 22 1/2 degrees. The straight hip roof is simplest, and is the method I'll describe in this article. It's easy to explain to your lead carpenter because all the rafters are treated as commons. This creates unequal overhangs, but it's fast and the pitch can easily be matched to the main house.

Below the windows, I prefer a vertical base that extends from below the window to the skirt board. This gives the impression of a walk-in bay that cantilevers off the top of the foundation (see lead photo). It's faster to build than an angled base that slopes back toward the house. If your client wants the sloping skirt, you can always build an upside-down version of the straight hip roof and cut the rafters without tails.

Building the Base

Before installing the bay into a rough opening, you have to build a platform to support the window. How you detail this deck depends on the kind of window you install. I've used Marvin (P.O. Box 100, Hwy. #1, Warroad, MN 56763; 800/346-5044; in Minn., 800/552-1167) windows, and like them because the bottom of the sill is flat. This allows you to place the window on a level platform without blocking beneath the sill to compensate for the angle. A lot of other manufactured bays are built like this, but not all.

If you do use Marvin, and build the deck as I describe, you'll have to increase the height of the window rough opening given in the catalog by 3/4 inch.

Use a template. To execute all the framing quickly, I rely on a template of the window footprint. Actually, I build two frames out of 2x4s to the dimensions of the window footprint. One serves as the platform (called the "platform frame"), and the other as the "top plates" for the roof framing. But before this top frame, which I call my "template frame," gets put into the roof, it will serve as an indispensable layout tool.

Since all my work depends on building the template frame accurately, I take time to measure the footprint dimensions carefully. This can be a little tricky because of the angles. But because this is a 45-degree bay, I can establish the angles and dimensions quickly, as shown in Figure 1.

I cut the length of the frames 1/4 inch short so there's a little play if the window doesn't get centered perfectly in the rough opening. At this time, I also cut two extra 2x4s equal to the length of the front of the frame. One of these I will use as a ledger for diagonal supports in the base, and the other as a "ridge" on the roof.

The platform frame is nailed to the house wall flush with the bottom of the wall opening. The entire surface is then decked with 3/4 inch plywood. You can save a lot of time by using the template frame for drawing the cut lines for the plywood decking, and adding the width of the framing sill. I cut to the inside of these lines to ensure that the plywood deck won't run past the framing. The ends of this deck are clipped back about 2 inches so the plywood will fit through the window rough opening (see Figure 2).

For now, I nail the deck to the window opening in a single line, leaving it unfastened to the platform frame.

Diagonal support. Under the platform I nail in diagonal struts that carry the load back to the house wall. The length and angles of the struts can be figured just like rafters from the rise and run. The run I use is equal to the width of the platform minus 4 inches to leave enough room for a skirt wall.

Even though I've calculated the diagonal distance, I still rely on a level when nailing in the struts since a twist in the wall can throw things off.

Once the struts are in, you can nail the deck to the platform frame and install the window. Before lifting the window into place, I mark the centerline of the rough opening on the inside above and below the window. Working off the centerline saves time positioning the window. I also use a 2-inch block to check for an even overhang of the sill on all sides.

Vertical skirt. The skirt can be built out of place as a single, angled wall. I start with top and bottom plates cut at 22½ degrees at the bay angles and 45 degrees at the house wall. To save time, I measure and assemble the plates on my template frame before toenailing the corners together.

I cover the bottom of the skirt with exterior-grade lauan, so I have to add a couple of flat 2x4 nailers for support across the bottom. Again, I use the template frame to lay out the lauan.

Before sheathing the wall, I install rigid foam insulation into the rectangle and triangles of the platform frames. Once again, I use my second frame as a template to mark the foam shapes and get a tight fit.

At this point I cut enough foam to insulate the top of the window as well, but I don't install the top frame until I've cut the rafters. Even though a bay window roof is small enough that you can tape off the rafter lengths by hand, I do the calculations to save time. By keeping the top frame on the ground for taping off the runs, I can cut the rafters on the ground and just climb up once to set them.

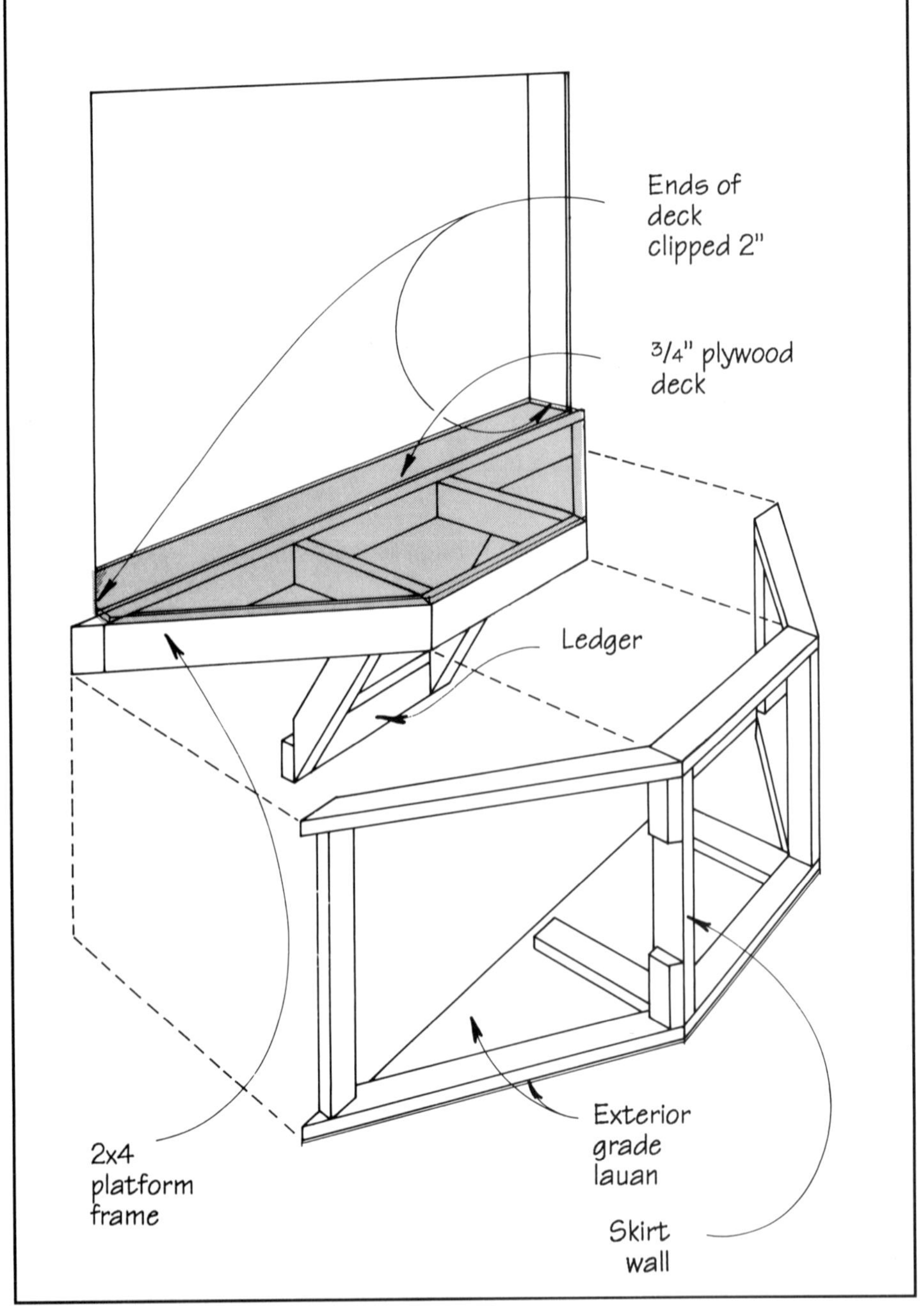

Platform & Skirt

Figure 2. *The 45-degree corners of the deck (shown in gray) are clipped so the plywood will fit through the window rough opening.*

Raising the Roof

I divide a straight hip roof into three sections: First, the "wall rafters" are cut as commons and then ripped at the angle of the roof pitch along the top edge. Second is the mid section, which is essentially a simple shed roof between the two straight hips. In plan the straight hips look like king commons, but they are beveled along their outside edges to carry the sheathing on the sides of the roof. Third, the side rafters get a double cheek cut and look like hips. But they are cut as commons with a shorter run.

Wall rafters and straight hips. If you've built your 2x4 frame carefully, you'll save time laying out the roof. The straight hips will be centered over the outside corners of the bay. So on the template frame the outside of the rectangle will mark the centerline of the straight hip. And the width of the template frame will equal the run of the wall rafters and straight hips, which will be equal to each other because of the 45-degree bay.

I usually match the pitch of the bay to the pitch of the main roof. Since I know the run, I can find the rise by cross multiplying to find equal proportions. With the rise and run, I can then use the Pythagorean theorem to calculate the rafter lengths.

I lay out at least four identical rafters—two wall rafters, which get nailed against the house wall, and two straight hips, plus any commons I need. The depth of the birdsmouth on all these rafters is determined by the height of the head casing on the window.

The straight hips are shortened $1^1/_2$ inches (measured perpendicular to the plumb cut) to allow for a "ridge" —one of the 2x4s I cut earlier—nailed flat against the house wall.

The wall rafters stay at the full rafter length, and are ripped along the top edge at the same angle as the plumb cut. When these rafters are eventually nailed in place, the long point of this bevel will stick up above the "ridge," as shown in Figure 3. (Alternatively, you can use a 2x6 ridge that is ripped at the pitch angle along the top edge.) The wall rafters also get a 45-degree fascia cut. I don't bother trying to angle the plumb cut on the birdsmouth, though.

The top edges of the straight hips are also beveled, but only to the centerline. The bevel angle is established by the pitch of the side rafters. I usually wait to make this rip until the rest of the roof is cut and I have nailed the wall rafters on. Then I can hold the hips in place and mark the point where the bevel on the wall rafter falls on the hip. This point and the centerline of the hip mark the correct angle.

Side rafters. The side rafters are cut like common rafters with a short run. Working off my template frame, I mark the midpoint of one of the angled sides. I then measure the rafter run between this midpoint and the opposite inside corner on the template frame (see Figure 3). If you measure to the inside of the triangle, rather than measuring to the line on the wall sheathing where the true point of intersection should be, you will automatically shorten the rafter.

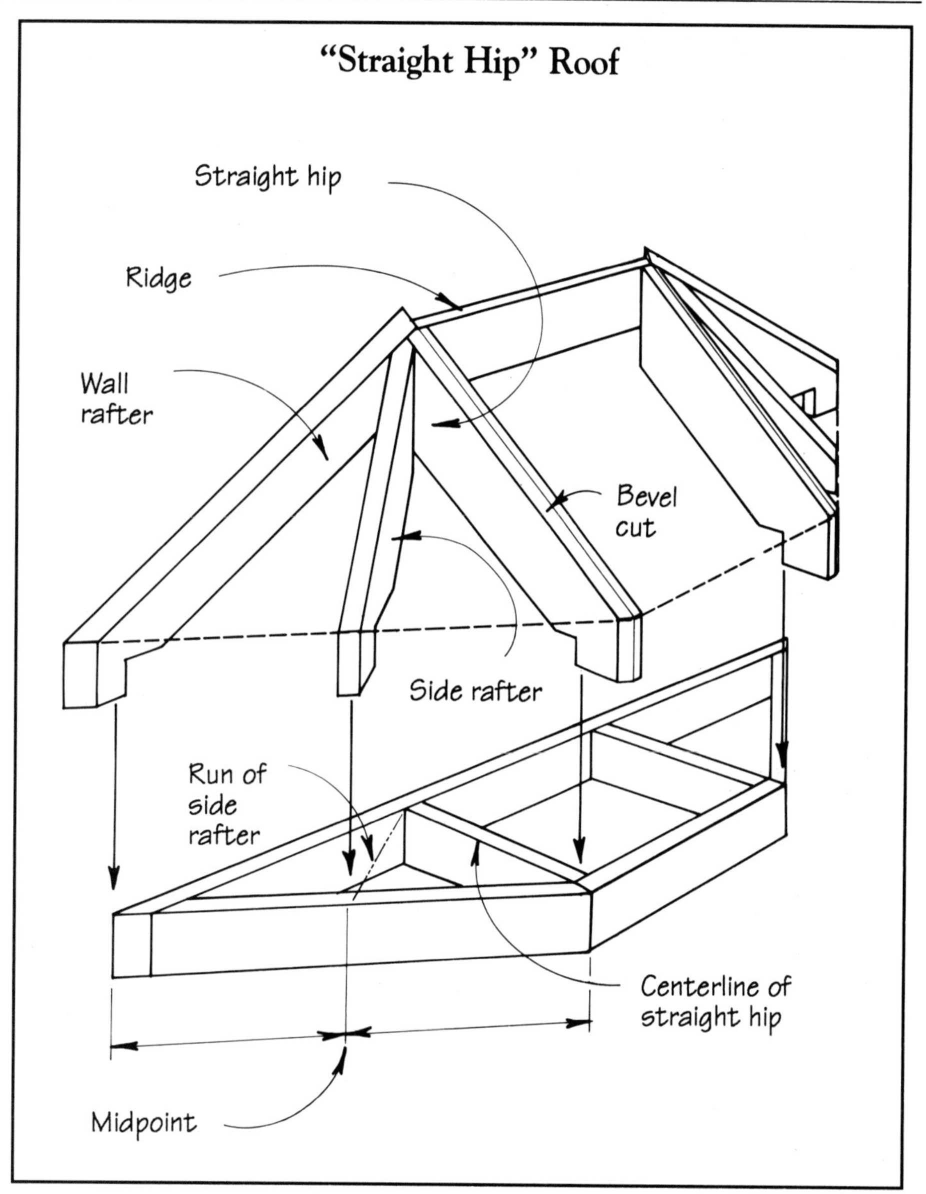

Figure 3. *The run of the side rafters can be measured square from the midpoint of the side of the template frame. The wall rafters are beveled, and the long point sticks up above the "ridge." All framing members are 2x4.*

After the side rafter is laid out, cut the top plumb cut on the side rafters at 45 degrees (like a cheek cut), and then clip $^3/_4$ inch off the point, cutting 90 degrees to the face of the cheek. When setting the rafters, this "sharpened" point is nailed into the corner where the wall rafter and straight hip intersect.

When laying out the rafter tail, maintain the same height above the plates, and use the same birdsmouth depth and fascia cut as you have on the other commons. The level cut, and consequently the roof overhang, will be shorter because the pitch of this side rafter is steeper.

The difference in the level cuts is only a problem when it comes to running the fascia. If you miter the fascia at $22^1/_2$ degrees, however, you won't notice a disparity. This offsets the fascia corner slightly from the overhanging roof sheathing, but the inconsistency will be imperceptible from the ground. ■

Clayton DeKorne is an associate editor with The Journal of Light Construction.

When You Have To Cut a Truss

Residential remodelers can safely modify a single truss if they understand the loads and closely follow basic rules

by Clayton DeKorne

The warning from truss manufacturers is very clear: You are risking structural failure if you cut into a truss. The warning shouldn't be taken lightly. However, on some remodeling jobs, trusses have to be modified.

When you must cut into a truss, the only path offered by the Truss Plate Institute is to "consult an engineer." That's fine when you need to modify a number of trusses for a large addition. But for small projects when you're only dealing with a single truss—to fit a skylight, chimney or hvac duct—it isn't likely you'll have the time or the budget to go that route.

One alternative to going to the full expense of hiring a consulting engineer is to go to a truss manufacturer. Most large truss fabricators have a full-time design staff that can advise on repairs. The Wood Truss Council of America (401 N. Michigan Ave., Chicago, IL 60601; 312/644-6610) can put you in touch with a truss fabricator in your area that offers design assistance. Some fabricators will extend the help as a professional courtesy, even if they did not build the original truss.

Another alternative is to learn how trusses support loads (see "How Do Trusses Work?"), follow some basic rules of thumb, and then liberally overbuild. This does mean taking on some liability, something you should give some hard thought to.

Some Basic Rules

This article examines two possible ways to restructure a roof when a single truss has been cut. The first case, from a truss designer, relies on new bearing walls to support the cut ends of the truss. The second, from a builder, shows how the cut ends can be headed off to the adjacent trusses. Both examples involve similarly sized, W-type trusses which are commonly used for residential buildings. Other types of trusses must be looked at separately. But regardless of the type of

How Do Trusses Work?

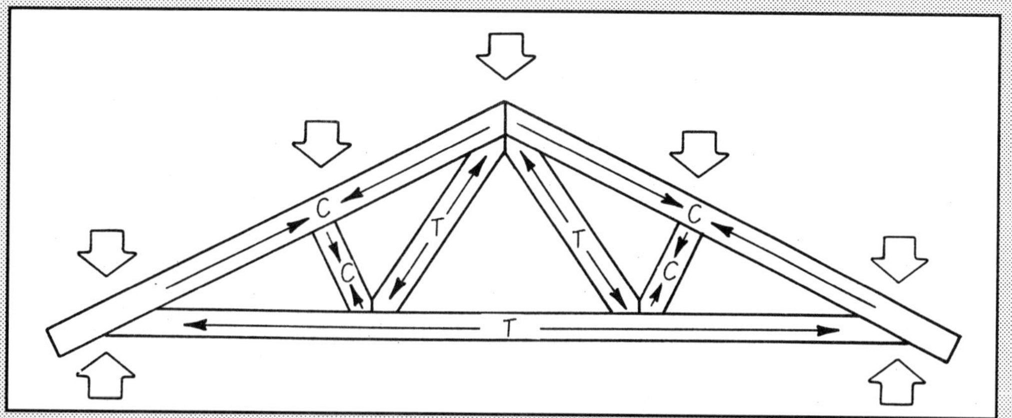

Truss forces: a balancing act. *Under load, the top chords are in compression and the bottom chord is in tension. The web members lend support to both the top and bottom chords. The webs leaning into the peak are pulled in tension while those leaning away from the peak are in compression.*

If you've ever watched a roof truss bow and flex at the end of a crane cable, you know this collection of spindly members is playing by a different set of rules than the rafters it has replaced on many homes. A look at the general principles of how a truss works will help differentiate these engineered components from their site-assembled cousins.

Rafters are sized like simple beams. That means they rely primarily on the strength of their wood fibers to resist bending and to transfer loads to the supporting walls. Rafters are sized large enough to do their job without additional support along their length.

Trusses, on the other hand, rely on their connections for strength. Trusses use the combined strength of many short, small-dimension members, each of which have relatively low fiber strength.

Collectively, the truss members have a great deal of strength because they are joined into a series of rigid triangles. A triangle is a naturally stable structure: the stresses acting along the length of any one side produce counter stresses along the adjacent sides. The whole configuration is thus held in a balance of tension and compression. Cutting even one member of a truss will offset this balance.

One way to visualize the forces at work in each member of a truss is to imagine that the top chords are connected with a hinge. Under load, the top chords are in compression. These stresses are working to open the hinge. The bottom chord is held in tension to keep the roof from spreading. The webs lend support to both the top and bottom chords. The web members that are leaning into the peak are pulled in tension (see illustration). The web members that are leaning away from the peak are in compression, and they bear the load placed along the length of the top chords.

It's All in the Connections

Without those little metal plates to hold the truss in a series of rigid triangles, the members would pull apart. The metal plates have a big job to do since the truss configuration amplifies the amount of tension along the bottom chord and in some webs. This puts great tensile stress on the connections. To complicate matters, tension joints are difficult to construct in wood. While wood is very strong in tension parallel to the grain, nails tend to pull out easily when tension is exerted along their length. And when the tension is exerted across a fastener, the small section of wood directly behind it tends to shear out. A bigger nail is even more likely to fail because it is more likely to split the wood. For this reason, steel truss plates have numerous small prongs to distribute the load over a greater area of wood.

According to Dave Matychowiak, of Wood Structures, in Biddeford, Maine, the strength of a 4x4 steel truss plate is rated at about 1,000 psi in tension. This is equivalent to over 40 10d nails loaded in shear. To get this many nails into a truss connection, the nails have to be distributed far enough apart to keep the wood from splitting and to get away from the end grain. Therefore, plywood gussets are sized according to the number and spacing of the nails needed.

Stressed Out

Just as stress-rated lumber is more predictable than standard visually graded material, so metal plates are viewed as more predictable than nails banged in on site. Trusses often use both metal plates and stress-rated lumber, so they are considered to be very predictable.

As a result, members are sized with very small allowances. In fact, it's not uncommon for the top chord of a truss to be designed so that, when fully loaded, it's using 97% of the wood strength of that member. Trusses are, quite literally, stressed out. This doesn't mean that trusses are underdesigned. Rather, they are designed just right.

In practice, this means that if a truss is cut and headed off, the adjacent trusses carrying the added load must be reinforced. Similarly, if you are going to add load to a truss roof by any means, whether adding tile, plastering the drywall ceilings, or tying in an addition roof, a "repair" would be necessary. But for these larger projects that involve more than one truss, you should (you guessed it) consult an engineer.

—C.D.

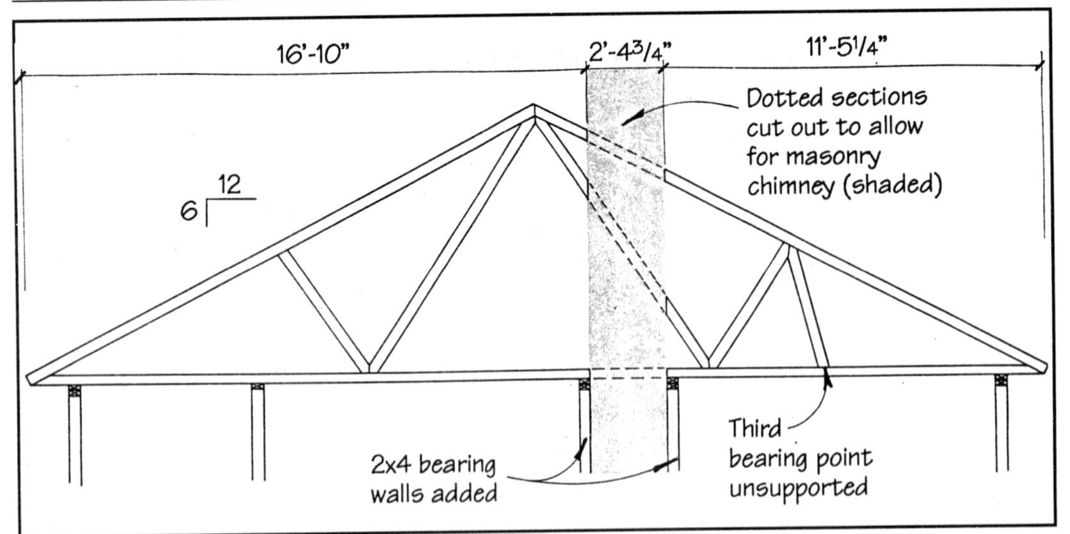

Figure 1. *To make room for a chimney, this W-type truss was cut in half. The contractor built bearing walls to support the cut ends, but the truss still needed rebuilding. To make matters worse, the truss, which was designed to have three bearing points, had originally been installed backwards: the third bearing point on the right should have been placed over the partition on the left.*

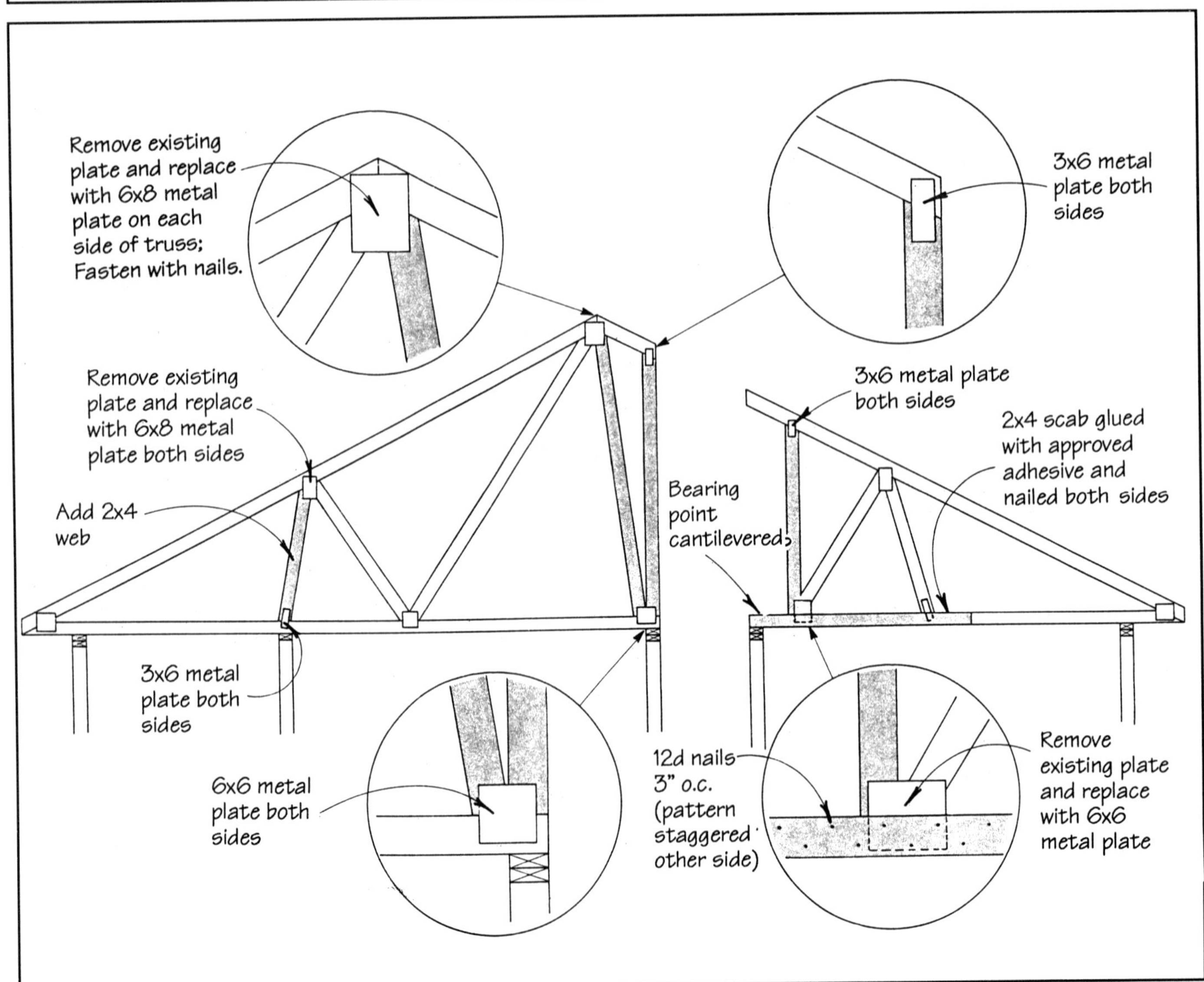

Figure 2. *To patch up the wounded truss, the engineer called for new 2x4s to tie together the cut ends of the top and bottom chords on both sides of the chimney. New web members were also added to form complete triangles inside the truss (all new work is shaded). The new configuration let each section function as a smaller, but complete, truss. Note the "scab" on the right-hand section to strengthen the cantilevered bearing point.*

truss involved, keep these basic rules in mind:

1. Trusses must be considered as a whole. Whether a single chord or an entire section is cut, the top and bottom chords (the outer members) must be tied back together so that the remaining sections form rigid triangles.

2. In addition, the webs (the members inside the truss) should all "triangulate," that is, form complete triangles within each section of the truss.

3. If trusses carry greater loads after the modification, they must be reinforced. This applies particularly to trusses that support headers from a cut truss. (The second case study gives an example of this.)

4. New connections made between truss members must be as strong as the metal plates they are replacing.

The application of these general guidelines can be seen in the following examples.

A Chimney Chase

David Matychowiak is the head of design at Wood Structures, in Biddeford, Maine. In addition to designing trusses, Matychowiak regularly advises on the repair of trusses. He described to me a job he was called to after a local code officer issued a stop work order. When Matychowiak arrived at the job site, he found a 30-foot W-truss that had been cut to make room for a chimney. Figure 1 shows the truss as he found it. The chimney had been built and bearing walls erected to close in the chimney and support the cut ends of the truss, but nothing had been done to restructure the truss itself.

Cutting this truss completely interrupted its transfer of forces. According to Matychowiak, the bearing walls were a step in the right direction, but nothing had been done to tie the top and bottom chords together or retriangulate the web members. The top and bottom chords on either side of the chimney now functioned as undersized rafters and ceiling joists and would probably have sagged over time.

To further complicate matters, the truss had originally been installed backwards. It was designed as a "tri-bearing truss," with three bearing points—two on the outside walls and a third on an off-center interior wall. Note in Figure 1 that the interior bearing wall is on the left-hand side of the truss while the bearing point is on the right-hand side.

To repair this truss, Matychowiak suggested removing the remaining pieces of the cut web member on both sides of the chimney. He treated each section of the truss separately, essentially creating two separate smaller trusses (see Figure 2).

On the left-hand section, Matychowiak called for removing the old plate at the peak, and adding a new web extending to the bottom chord to retriangulate this section of the truss. A new 6x8 NailRite plate, nailed off with N11 nails (10d common nails shortened to $10^{1}/_{4}$ inches) replaced the plate at the peak. NailRite plates are made by Hydro-Air, the company that supplies the truss plates to Wood Structures. Kant-Sag (U.S. Steel), Simpson, and Teco also make similarly sized steel mending plates that are widely available at lumberyards.

Plywood gussets can also be used and are often preferred because plywood is more readily at hand. Matychowiak emphasizes that plywood gussets should be sized to equal the strength of the steel plates they replace. "Too often the strength of a metal plate is underestimated," he says. For example, he notes that a 4x4-inch steel nailing plate is equivalent to a 2x2-foot plywood gusset nailed every 3 inches on-center along the truss members with 43 10d nails. The plywood may have to be even larger than this depending on how much nailing surface it covers on the truss.

Also on the left-hand section of the truss, Matychowiak recommended adding a vertical web member to tie the cut ends of the truss together. He used a 6x6 plate to tie both new webs to the bottom chord and a 3x6 plate to secure the joint at the top chord. Once secured, the new webs created a small "truncated truss" that worked as a self-supporting right triangle, much like a shed roof truss.

To correct the bearing problem originally caused by installing the truss backwards, Matychowiak replaced the missing web on the far left-hand side. This web formed a new bearing point over the load-supporting partition wall. Note that this new member joins the top and bottom chords at an angle to form a complete triangle in that section of the webbing.

The right-hand section of the truss was treated much the same way. Here a new vertical web was placed at the joint of the old cut web after the original plate was removed. Matychowiak called for scabbing an additional 2x4 along the bottom chord. This chord needed reinforcement because the bearing point would be cantilevered about 12 inches to the new wall. This created what Matychowiak calls a "bottom-chord tail bearing."

Matychowiak used the scab in this way because he wanted to minimize the time the builders spent crawling around in the webbing of the truss roof—kicking up cellulose and jabbing their scalps on roofing nails. An alternative, but more complicated, approach would have involved placing the vertical member directly over the wall and then replacing the next web member to form a complete triangle with the webbing.

The scab was installed after the web members were in place, and was nailed every 3 inches in a zig-zag pattern as shown in Figure 2. This pattern was reversed on the other side so that each 3-inch section of the scab had two nails in it. In addition to this nailing, Matychowiak recommended securing the scab with a construction adhesive that conformed to the American Plywood Association (APA) Performance Standard AFG-01. (The APA publishes a list of specific name brands that conform to this standard. Single copies of the list—publication V-450, "Adhesives for APA Glued Floor Systems"—are free from the American Plywood Association, P.O. Box 11700, Tacoma, WA 98411; 206/565-6600.)

A Skylight Well

In a simple bathroom remodel, David Pell of Hinesburg, Vt., had to remove a section of a truss in order to center a skylight over a bathroom vanity.

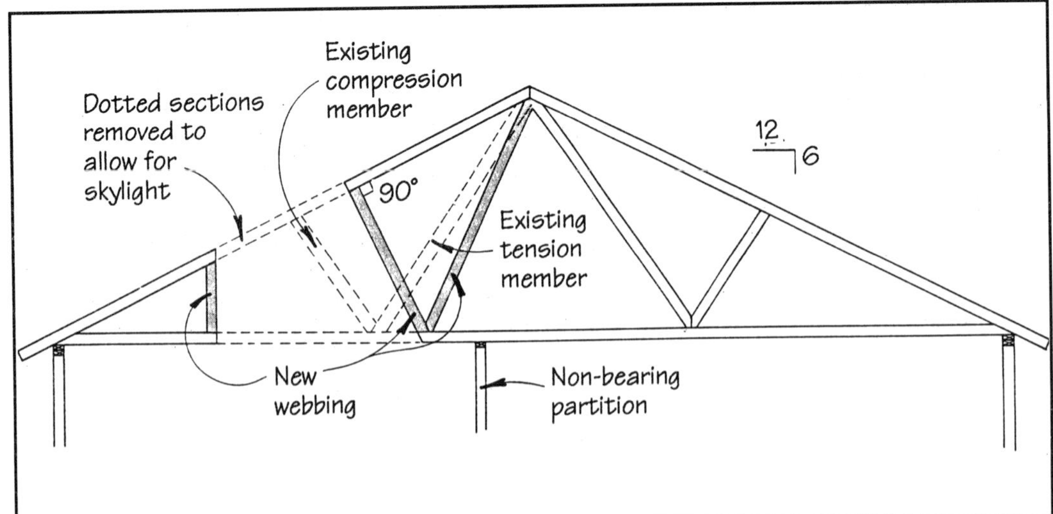

Figure 3. *To retrofit a skylight in a truss roof, a builder removed a section of one truss and headed off the cut ends to the adjacent trusses. Note how the back wall of the skylight well flares to form a right angle with the slope of the roof (new work is shaded).*

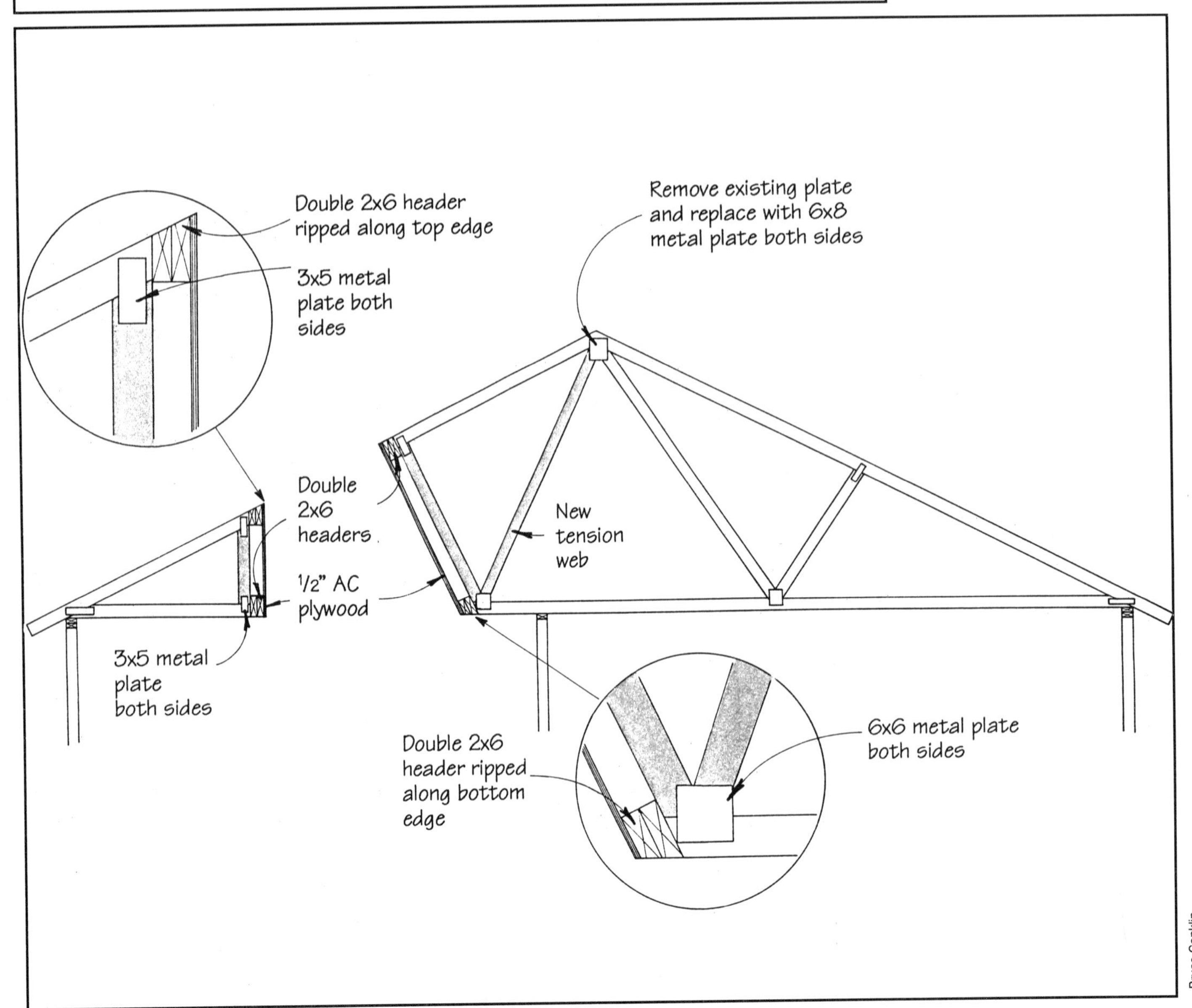

Figure 4. *The builder cut the top and bottom chords 3 inches back from the rough openings to allow for double 2x6 headers. To restructure the cut truss, he installed a new web behind each wall of the skylight well, and replaced one tension web at the peak to complete the triangle in that section of the truss. Half-inch AC plywood was installed around the inside of the light well to tie the structure together.*

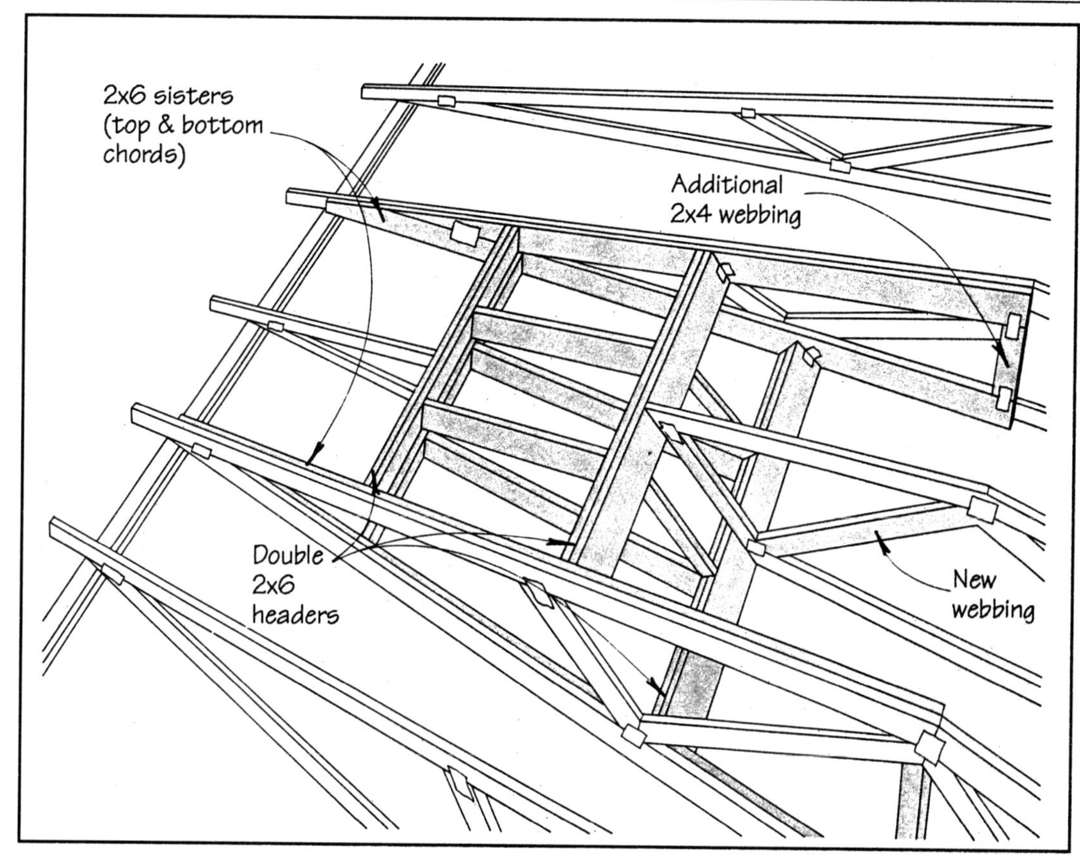

Figure 5. The load on the cut truss is transferred to the adjacent trusses along double 2x6 headers. The headers tie in to 14-foot 2x6 "sisters" nailed along the top and bottom chords of the two adjacent trusses. Additional 2x4 webs were added to tie together the top and bottom sisters on each side. (New work is shaded.)

The bearing on the right-hand side of the truss might have been transferred to the nearby interior partition using a scab, as in the previous example. However, the partition was not load-bearing.

Furthermore, the bathroom was on the second floor, so providing bearing would have required posting all the way down to the basement slab. Pell thought it would be simpler to head off the cut truss chords to the adjacent trusses. Here, as in Matychowiak's example, the remaining parts of the cut truss would be made to function as individual smaller trusses. However, instead of bearing on walls, the ends of each truss section would be supported by the adjacent trusses.

To position the skylight over the vanity, Pell needed to cut the top and bottom 2x4 chords of one 26-foot W-type truss. He designed the light well to flare open along the top wall so that it made a 90-degree angle with the slope of the roof (see Figure 3). This meant that one web in compression would be removed and a second one in tension would be cut.

Before cutting through the truss, Pell's carpenters installed new webbing. Eighteen inches uphill from the existing compression member, they installed a new 2x4 parallel to the old one (see Figure 4). On the downhill side, a new web was installed plumb. In addition, they nailed a new web from the end of the bottom chord to the peak to triangulate the remaining truss section.

Pell cut the new webs long enough to overlap the top and bottom chords and face-screwed the connections from both directions. He used steel corner clips to further reinforce the connection.

After installing the new webs, Pell then cut the chord 3 inches back from the rough openings on the top and bottom of the light well. This allowed him to install doubled 2x6 headers across each cut end of the truss. These tied in to 2x6 "sisters" that ran along the top and bottom chords of the two adjacent trusses (see Figure 5). The sisters were glued and nailed every 3 inches on-center along the top and bottom chords and joined together over the wall plates with Simpson 4x5-inch steel plates. Near the peak of the truss, additional webbing was added to support the reinforced chords.

To further strengthen the opening, Pell sheathed the inside faces of the light well with 1/2-inch plywood. This extra step required careful cutting and extra time, but the plywood tied the framework together securely. With a little sanding and three coats of oil-based paint, the plywood finished up nicely and no drywall was needed. It also provided a sturdy nail base for hanging plants.

David Matychowiak looked at a diagram of Pell's repair. He suggested that the new web connections could have been stronger and "more predictable" if the webs were placed in the plane of the truss and secured with flat nailing plates or plywood gussets. Figures 2 and 3 show the joints made correctly with steel mending plates. However, this was not a fatal flaw in Matychowiak's judgement, since the loads are relatively small, only one truss was cut, and Pell's overall approach was correct. ■

Clayton DeKorne is an associate editor for The Journal of Light Construction.

Section 6. **Troubleshooting**

Stephen Cridland Courtesy: American Plywood Association

Field Guide to Common Framing Errors

A building's frame is only as strong as its weakest link. Here are a dozen weak ones to avoid.

by David Utterback

Over the past 17 years, first as a builder and then as a representative of the Western Wood Products Association, I have traveled extensively, talking to builders and code officials to see how framing is done throughout the country. While I've found regional differences, I've also found a few serious framing problems that tend to crop up everywhere, again and again.

All of these problems are covered by the model building codes. A given problem might occur because the builder doesn't know better, or because framers are paying more attention to other construction needs. Either way, these framing defects not only cause trouble with code officials, but cause problems big and small down the line.

Here are some of the most common framing errors I come across, along with code-approved, structurally sound solutions.

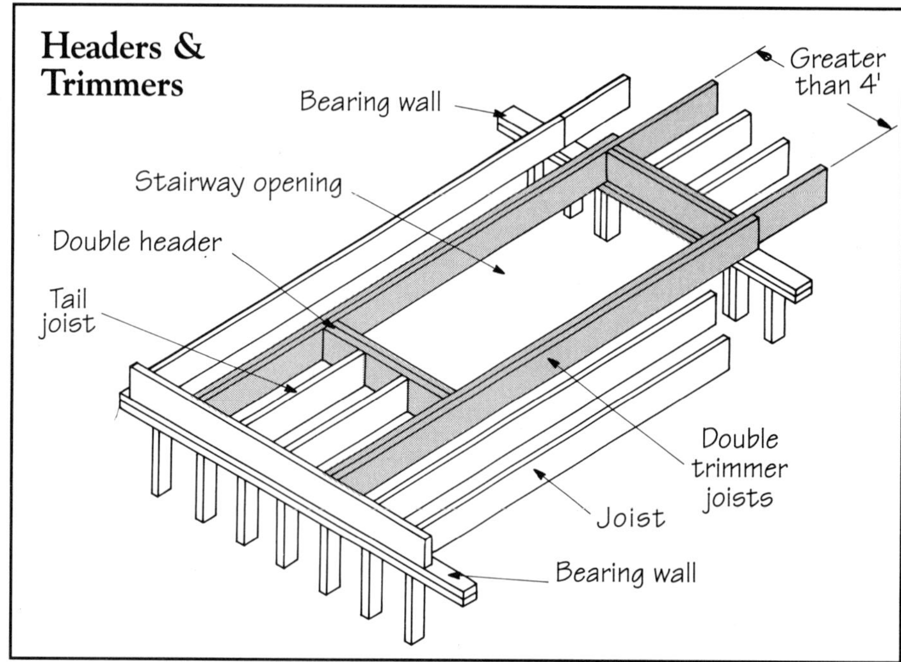

Figure 1. *Head off interrupted joists to transfer their loads to adjacent joists. If the header spans more than 4 feet, it must be doubled and the loads transferred to double trimmer joists.*

Framing Openings Cut in Floors

A common problem occurs with floors when subs cut through joists to make room for plumbing runs, hvac ductwork, or other mechanical elements. The loads these cut joists supported must be properly transferred to other joists. You can do this using header joists, end-nailed across the cut ends of the interrupted joists, to carry loads to the adjacent trimmer joists. Where the header has to span a space less than 4 feet wide, a single header end-nailed to the trimmer joists will do.

Things get more complicated if the header must span more than 4 feet, as in Figure 1. If that's the case, both header and trimmer joists should be doubled. The doubled trimmer joists must be nailed together properly (with spaced pairs of 16-penny nails every 16 inches) so that they act as beams. The header joists must be appropriately anchored to the trimmers. End nails will do for header spans up to 6 feet; beyond that, use hangers. Any tail joists over 12 feet should also be hangered.

When you're framing the floor, check the blueprints to see where any such openings might go, and header off any joists that might be in the way in advance. It's much easier than trying to work from underneath the subfloor later.

Holes and Notches

Whenever you cut a hole or notch in a joist, that joist is weakened. You (and your subs) should avoid this whenever possible. And when you absolutely have to cut or notch, you should know the rules for doing it in the least destructive manner.

Guide for Cutting, Notching, and Boring Joists

Joist Size	Maximum Hole	Maximum Notch Depth	Maximum End Notch
2x4	None	None	None
2x6	1 1/2	7/8	1 3/8
2x8	2 3/8	1 1/4	1 7/8
2x10	3	1 1/2	2 3/8
2x12	3 3/4	1 7/8	2 7/8

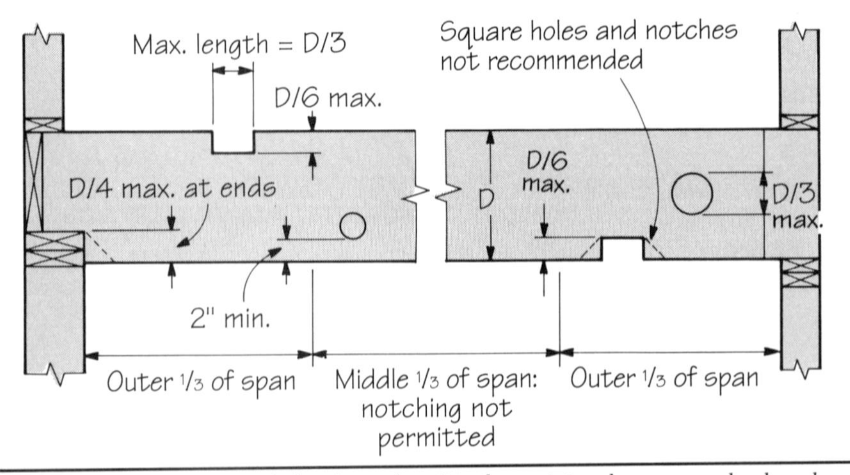

Figure 2. *In joists, never cut holes closer than 2 inches to joist edges, nor make them larger than one-third the depth of the joist. Don't notch the span's middle third where the bending forces are greatest. No notch should be deeper than one-sixth the depth of the joist, nor one-quarter the depth if the notch is at the end of the joist. Limit the length of notches to one-third of the joist's depth. Use actual, not nominal dimensions.*

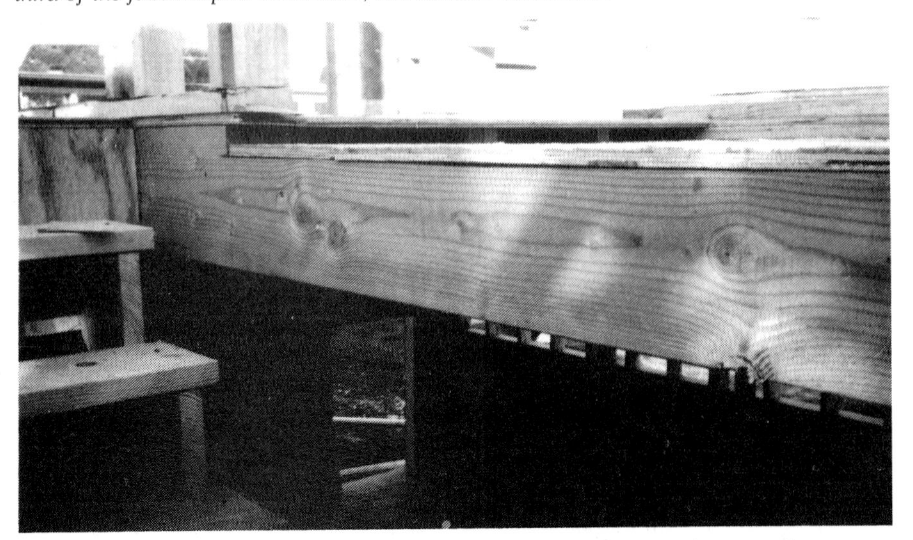

Figure 3. *Ripping long notches in floor joists, such as to make room for grouted entry floors, weakens the joists unacceptably and violates all codes. Instead, you need to use smaller dimension milled lumber of a higher grade, or set the joists closer together. If necessary, you can fur at the ends to bring non-grouted areas up to level.*

Figure 2 shows proper guidelines for cutting holes and notches. Straying from these guidelines weakens the joists and risks a red tag from the building official. Trying to fix such problems can be very costly, since it usually involves redoing the plumbing and electrical work along with replacing or doubling the joists.

When Notch Becomes Rip

Occasionally, what might be thought of as a notch turns into a rip, such as when floor joists at the entry of a home are ripped down to allow underlayment for a tile floor (see Figure 3). Unfortunately, ripping wide dimension lumber lowers the grade of the material, and is unacceptable under all building codes. You should frame these areas with narrow joists of a higher grade or stronger species, making sure they can carry the load.

Bearing Walls on Cantilevers

How far can a conventionally framed cantilever extend and still support a bearing wall?

Most of the confusion about how far a cantilever can extend beyond its support stems from an old rule of thumb used by builders and code officials alike: the rule of "one-to-three." This states that a joist should extend back inside the building at least three times the length of the cantilevered section—if the cantilevered section hangs 2 feet out, the joists should extend at least 6 feet in.

This rule works fine for nonbearing situations. But it does *not* apply to a cantilever that supports a bearing wall. In this situation, the maximum distance that joists can be cantilevered without engineering them is a distance equal to the depth of the joists, as in Figure 4. So if you are using 2x10 floor joists, the maximum cantilever for those joists supporting a bearing wall is 9 1/4 inches. Beyond this distance, shear becomes a serious factor, as does the bending moment at the support. This combination could eventually cause splitting of the cantilevered joists. The only way to work around this problem is to have it engineered.

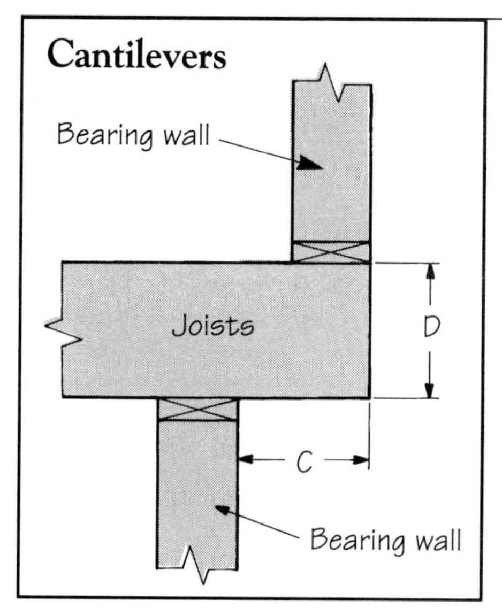

Figure 4. *When a cantilever supports a bearing wall, the distance it extends beyond its support (C) should not exceed the depth of the joist (D).*

Broken Load Paths

A similar alignment problem relates to maintaining vertical load paths. All loads start at the roof and transfer vertically through the building to the foundation. If they aren't transferred properly, you can end up with cracking of interior finishes or sagging framing. Many cracking problems written off to "settling" are actually due to what might be called broken load paths—paths that end up putting loads on areas not meant to carry them. This is one of the most common framing errors I see, and one to which many building inspectors pay close attention.

Misplaced struts. One example I see over and over again is a rafter-supporting strut carried down to a nonbearing partition below. Occasionally I even see these struts resting on "strong backs"—2x bracing run across the top of the ceiling joists to help brace them. This is a sure way to create cracking in the walls and ceilings below.

Of course, rafters do need to be supported when their lengths exceed their recommended clear spans. But the struts should carry down to bearing partitions, as shown in Figure 5. The struts should be braced to a purlin running across the rafters above them, and they should form an angle of 45 degrees or greater to the horizontal.

Finally, the struts must support the rafter so that it has no unsupported length longer than its recommended span. If you can't do that and still reach a bearing partition without dropping the strut below 45 degrees, you need to upsize the rafters.

Misaligned bearing walls. In other instances, loads carried by bearing walls or posts must be transferred through floor systems. If the bearing wall or post above doesn't line up closely enough with a bearing wall, post, or beam below, the floor joists in between can be overstressed, causing severe deflection. This can eventually split the joists, as well as cause finish cracking problems.

How closely must they align? Bearing walls supported by floor joists must be within the depth of the joist from their bearing support below (just as with cantilevers), as in Figure 6.

This code requirement applies only to solid-sawn wood joists. Engineered products such as wood I-beams are required to have the loads line up *directly* over each other, and special blocking is required. Special engineering of either dimensional or engineered lumber may allow placing loads at other locations, but you shouldn't try it without consulting an engineer first.

Bringing columns to foundation properly. If you use a column to support a beam or other member, make sure it bears on something that can in turn support it. A common mistake is to rest one on the floor, without extra blocking or support beneath. Doing this can crush the underlying joists. Columns shouldn't rest on unsupported floor joists; they should run contin-

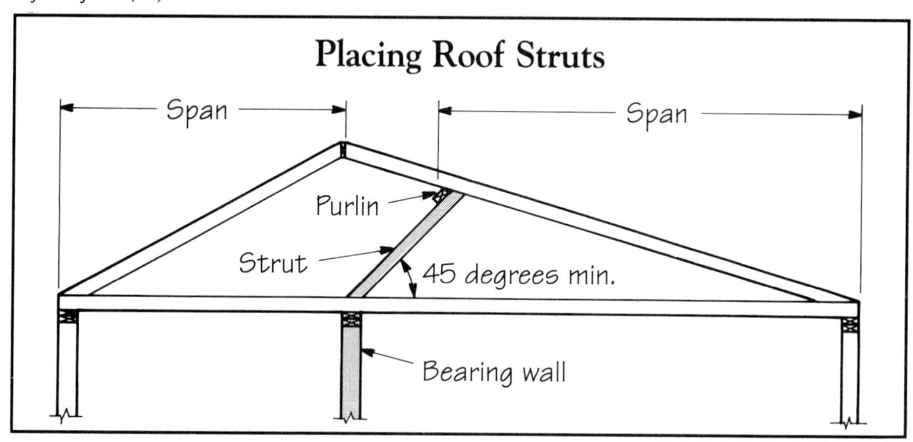

Figure 5. *Struts supporting rafters should always land on bearing partitions. Also, the strut should not drop below a 45-degree angle.*

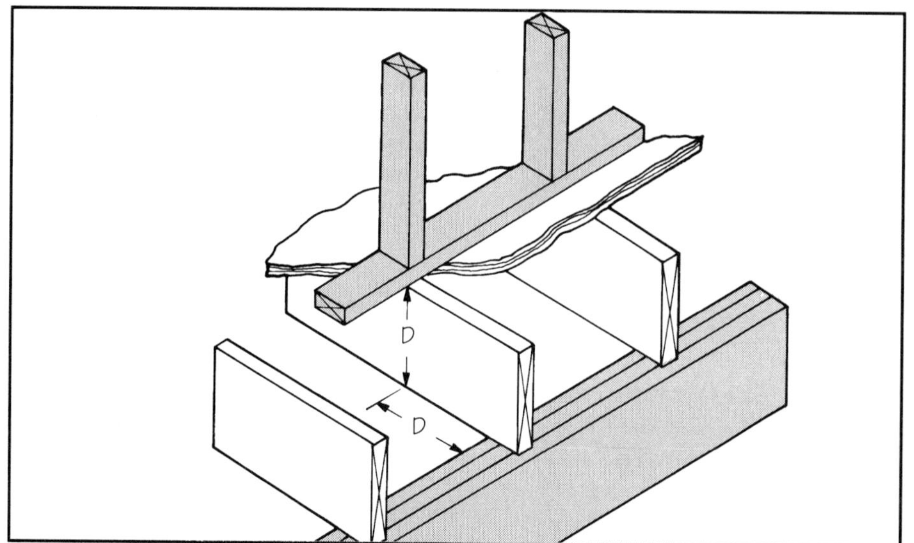

Figure 6. *If a bearing wall doesn't line up with the support below, it should lie no farther away than the depth of the joists (D). If the joists are engineered lumber, the walls and support must align exactly.*

uously to the foundation, or (if you must have a clear space beneath) to an engineered beam or header to transfer the load out to other columns or bearing members.

Columns shouldn't rest on rim joists either, for similar reasons. If you need to rest a column at the rim, add full-depth vertical blocking inside the rim joist the full depth and width of the column base, so that the load is transferred through the blocking to the foundation.

Puny Hangers

A simple but common framing error is hanging a three-member beam (such as three 2x10s nailed together) from a double joist hanger. This usually occurs because triple hangers are hard to find. But if only two of the three members are supported, then only two carry the load. The third member just goes along for the ride. Toe nails or end nails are not going to make it carry the load.

If you're going to use a hanger, use one that holds everything, and use the right size and the correct nails. Undersized hangers and inappropriate nails will weaken the system.

The correct hanger is necessary to carry the vertical load as well as to laterally support the member to prevent rotation. And without the correct nails, the hanger doesn't mean much. Eight-penny galvanized nails or roofing nails won't do. You can buy regular joist hanger nails that are heavy enough to handle the shear stress, yet only 1 1/2 inches long so that they won't go clear through the lumber and possibly cause a split.

Of course, the best way to support a beam is from beneath. When possible, use a beam pocket or a column directly under the end of the beam. Be sure the full bearing surface of the beam is supported clear to the foundation.

Tapering Beams and Joists

It's sometimes necessary (or at least convenient) to taper the ends of ceiling joists or beams to keep them under the plane of the roof, as in Figure 7. But by reducing the depth of the joist or beam, you reduce its load-carrying capacity.

If you must taper-cut the ends of ceiling joists, make sure the length of the taper cut does not exceed three times the depth of the member, and that the end of the joist or beam is at least one-half the member's original depth.

With taper-cut beams, you should also check the shear rating. If you can't meet this criteria, you'll probably have to lower the beam into a pocket so that enough cross-section can be left, after taper-cutting, to carry the applied load.

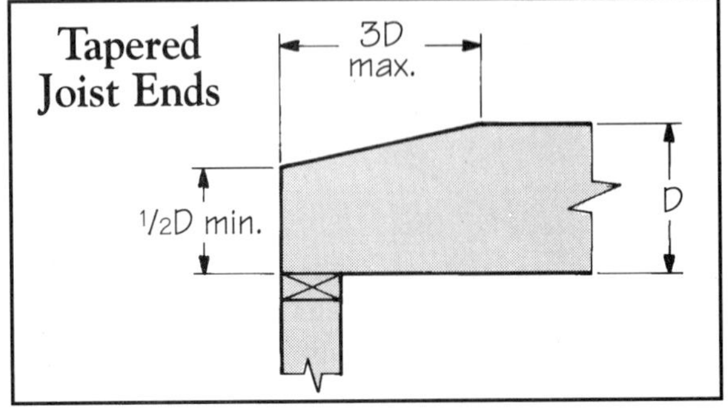

Figure 7. *Overtapering joists to fit beneath roofs creates inadequate joist depth at the plate. A proper cut (right) leaves at least half the depth of the joist.*

Watch Your Mouth

Another area that inspires excessive cutting is the level cut of the seat of a rafter. Many times, especially on low slope rafters, this level cut becomes a long taper cut on the tension (lower) side of the rafter, as in Figure 8. If the bearing point on the rafter is at the heel (interior side) of the cut, there is no problem. But usually these long cuts put the bearing point near the toe. This reduces the effective size of the rafter, producing stresses that can create splits at the bearing point, and eventually a sagging rafter.

To prevent this, cut your rafters so that the heel rests on the plate. This will mean using a slightly longer rafter. It will also give you a few extra inches between the top of the exterior wall and the roof sheathing. This translates into more room for attic insulation to extend over your outside wall, reducing those cold spots that can cause condensation or ice-dam problems at the eaves.

Raising the Rafters

Another way to add room for attic insulation at the eaves is to set the rafters atop a ledger board running perpendicular over the ceiling joists, as in Figure 9. Unfortunately, builders who do this often fail to put in a rim joist or block the ends of the joists to prevent them from rolling over. The resulting design creates, in essence, hinges at the top

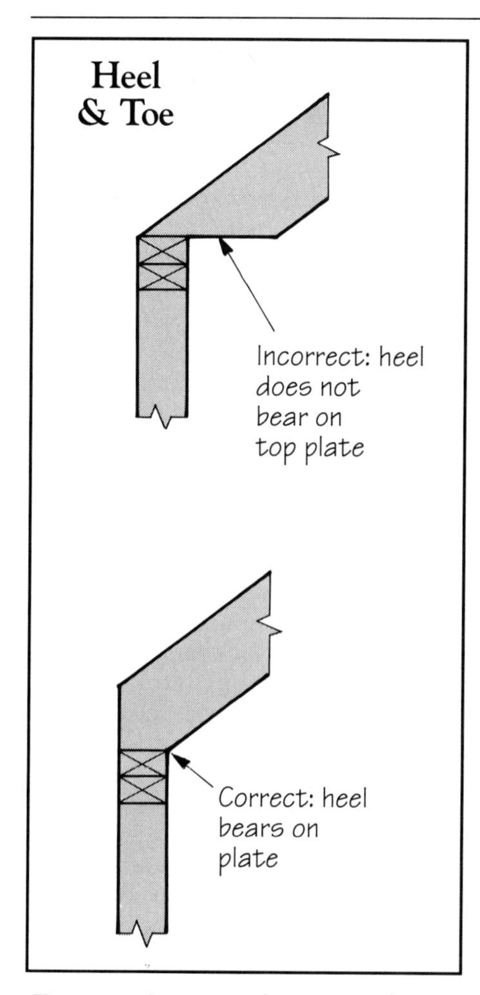

Figure 8. *Setting a rafter's toe on the top plate (top) risks splitting the rafter and causing the roof to sag. The inside edge of the level cut, or heel, should rest on the plate (bottom).*

and bottom edge of each joist. With a strong enough lateral force, such as a high wind or a strong tremor, all the joists could rotate and fall over—bringing ledger, rafters, and roof crashing down onto the now-flat joists.

To prevent this, install full depth blocking between all joist ends or a rim joist nailed against the ends of the joists. Either solution will also provide a baffle to prevent air from penetrating the ends of the batts and keep the batts (or blown-in insulation) from creeping into the eaves.

Blocking is also a good idea where

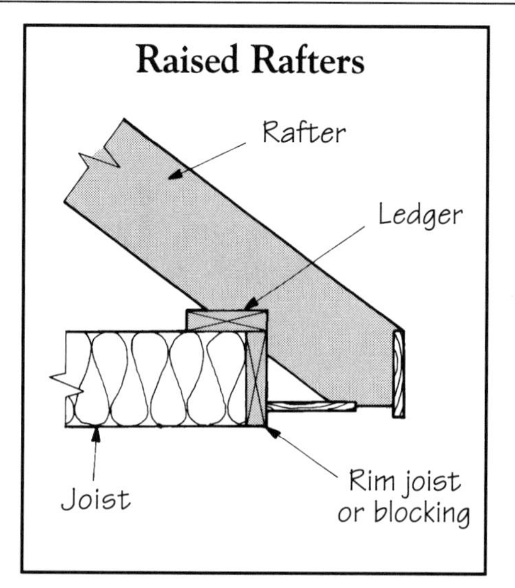

Figure 9. *When nailing rafters to a ledger over joists to make room for insulation, use a rim joist to keep the joists from rotating.*

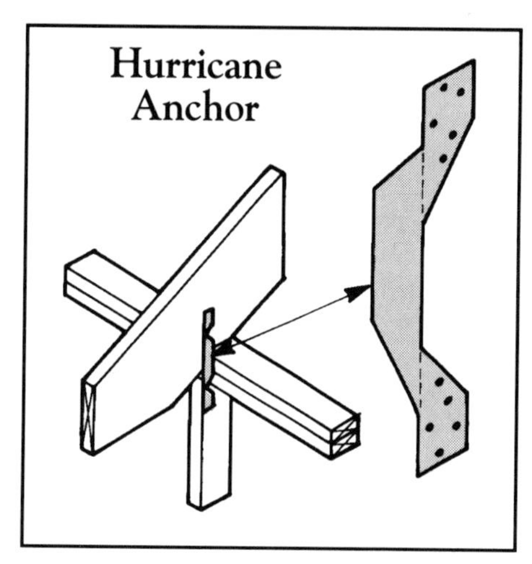

Figure 10. *Nailing rafters to plates, and plates to studs, is not always enough to resist high winds. Hurricane anchors at 4-foot intervals will securely tie rafters to studs.*

joists lap over a center girder at foundation level or over a support wall at second-story level. If the centers are unblocked, the job of keeping the joists upright falls to the nails holding the floor sheathing to the joists. These nails just aren't designed to resist the strong sideways forces created by wind or earthquake. Full-depth 2x blocking over center supports will prevent the joists from rotating in such an event. The blocking also stiffens the floor, since it stops the rotation caused by deflection of the joists under load.

What if a few of these blocks get knocked out by mechanical contractors putting in ductwork or plumbing? That's not usually a problem, as long you don't remove consecutive blocks, so that each joist is blocked on at least one side.

Connecting Rafter to Wall

Conventional construction leaves too little connection between rafters and walls. Nails connect rafter to plate and plate to stud, but do nothing to connect the rafters to the wall itself. Such structures are subject to damage from the high, near-hurricane force winds that sooner or later blow across virtually every roof.

As a result, the building codes are beginning to get more restrictive about how rafters and trusses are tied to the rest of the building. For example, the 1991 Uniform Building Code has added Appendix Chapter 25, which applies to high wind areas. Under its requirements, rafters or trusses must be tied not just to the top plate, but to the studs below at 4-foot intervals. This means using some kind of metal connector to provide a positive tie to the studs.

The answer is the hurricane anchor (see Figure 10). You don't need to face a hurricane to need it–winds of roof-damaging gale force blow in most parts of the country. If you build in an area subject to high winds (or seismic conditions), you should consider using these or other holddowns. ■

David Utterback, of the Western Wood Products Association, provides technical information and gives building code seminars for building professionals throughout the country.

When the Framing Shrinks

Detail for wood shrinkage, and avoid callbacks for cracked plaster, stuck doors, and popped nails

by Henry Spies

Wood is the most common building material in the U.S., but it is also one of the most unstable. Wood shrinkage (and expansion) has been the bane of builders since Noah. Basically, wood shrinks dimensionally as it loses moisture. Most of the shrinkage is across the grain, but there is some along the grain as well. The amount of shrinkage varies with the species of wood and with the specific moisture-content range.

For instance, as douglas fir dries, the cells close and lock the opening, making it difficult for moisture to be reabsorbed. At the other end of the spectrum, Southern-yellow-pine cells are shaped like a bellows, and absorb and give off moisture like a sponge.

The amount of movement is also directly related to the portion of the log from which the lumber is cut. So-called juvenile wood, wood from the first ten or so years of the life of the tree, is more unstable than mature wood. For the same change in moisture content, juvenile wood may expand or contract three times as much as mature wood. As first-growth timber has almost vanished from the scene, the lumber now marketed is from the smaller logs of second- and even third-growth trees, and contains a much greater proportion of juvenile wood.

As an example, consider the common 2x10 floor joist. When it is purchased at 19 percent moisture content, the federal standard for "dry" lumber it is 9¼ inches high. As the moisture content drops to the 9 to 11 percent range that is common in heated buildings, the vertical dimension will shrink as much as ¼ to ¾ inch, and that 2x10 may now be only 8½ inches high! That amount of shrinkage is enough to cause many types of problems. And due to the increasing use of juvenile wood in construction, those problems appear to be on the rise.

Nail Pops

The most common problem associated with wood shrinkage is the nail pop. When a drywall nail or screw is driven into a stud, the drywall is pressed tight against the face of the stud. As the stud dries, the point of the nail stays exactly where it was driven. However, the wood between that point and the surface of the stud shrinks—as much as 1/16 of an inch for a nail penetration of 1 inch. This creates a gap between the drywall and the face of the stud. Because the nail point stays put, the longer the nail or screw the larger the gap (see Figure 1).

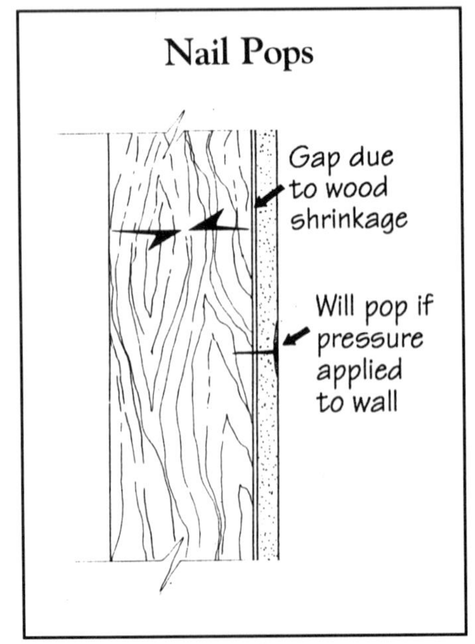

Figure 1. *To avoid nail pops, use screws with ¾-inch penetration into the wood, a screw gun for control, and dry wood. If necessary, use a control joint between the wall and ceiling.*

If anything applies pressure to the drywall, it will slide down the shank of the nail or screw, and the head will put enough pressure on the taping compound to "pop" the surfacing off the head. (The reason there are fewer problems with drywall screws is because it takes more pressure to slide the drywall on the screw shank.)

Nail pops most often appear near the intersection of the wall and ceiling drywall. As the studs shrink slightly in length, they allow the top plate and trusses to move downward, forcing the wall drywall to push against the ceiling. This pressure makes the ceiling drywall slide on the nail shanks and cause the pops on the ceiling. Nail pops can begin with the first heating season, but they can also occur several years later if there was no previous pressure on the drywall to cause it to slide.

Corner Cracks at Openings

Perhaps the second most common problem caused by shrinkage is cracks in the plaster or drywall. These are usually diagonal, extending from the corner of the first opening in from the outside wall to the ceiling in any partition (see Figure 2). As the floor joists shrink in a platform-framed house, the frame of the house is lowered fairly uniformly. However, one end of the floor joist is supported on a concrete foundation, which is stable. The other end usually rests on a wood girder, which shrinks about as much as the joist. The result is a shallow bowl in the floor that causes the diagonal cracks and sticking doors in the partitions.

In some cases, it is necessary to shim the columns in the basement or crawl space to re-level the house frame. The problem will not occur with a steel girder, of course, or if the floor joists are hung on the sides of the girder with joist hangers. The latter approach, however, may lead to a discussion of your ancestry by a heating contractor who is trying to run the ductwork.

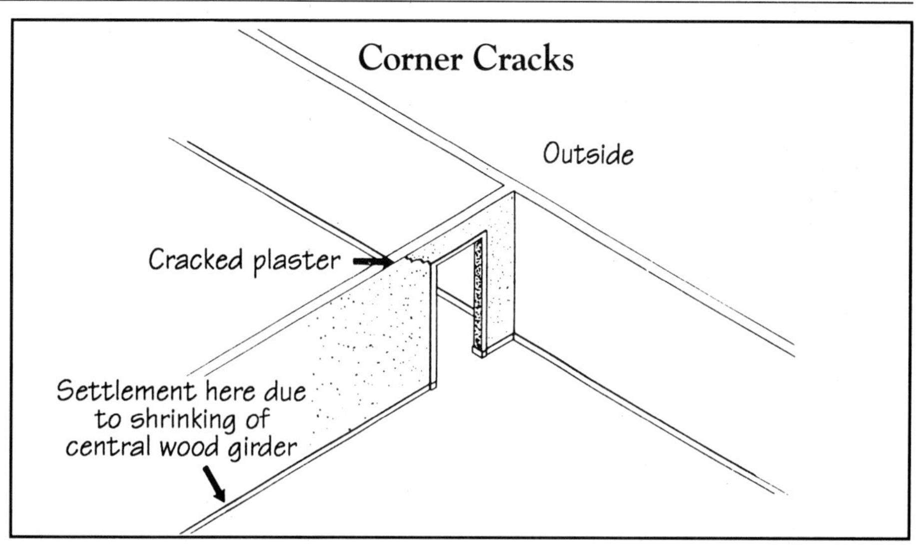

Figure 2. *Differential shrinkage can cause the house to sag over a center wood girder. The solutions are to hang the joists from the center girder or use steel instead of wood. After the fact, you can shim the columns to re-level the house.*

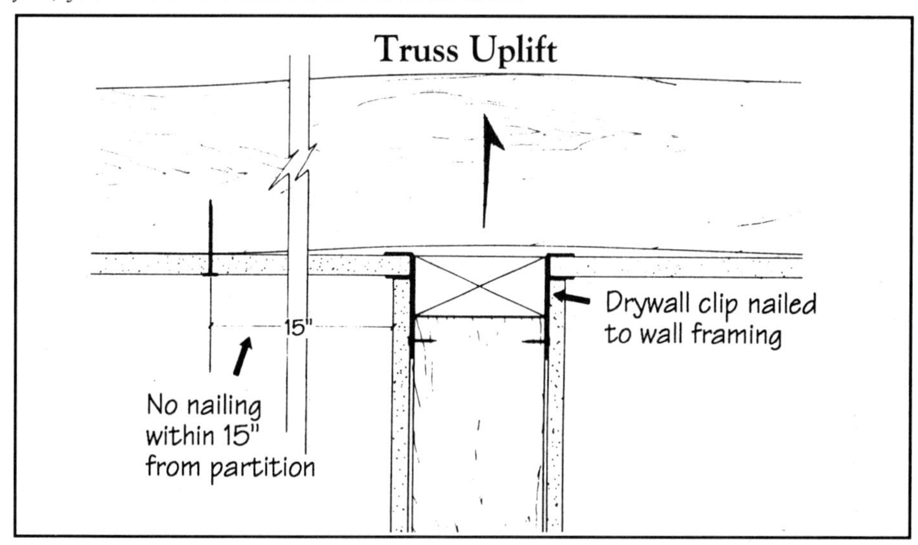

Figure 3. *Well-insulated trusses may rise during the first heating season, creating a gap at ceiling partitions. The solution: Float the ceiling corners and fasten them to the partition with clips.*

Sheathing and Subfloors

As douglas-fir plywood began to disappear from the eastern and midwestern markets, it was replaced by Southern-yellow-pine plywood. Due to the wood's cellular structure, as discussed above, the pine plywood tended to move with variations in moisture content. Each piece of pine plywood was marked with instructions to leave at least 1/8 inch of clearance on all sides, and 1/4 inch in areas exposed to extreme moisture conditions.

If the plywood sheets were spaced apart as instructed and the framing was on standard 16- and 24-inch centers, the joints no longer fell on the framing after the first two or three sheets, and the plywood couldn't be nailed. As a result, most carpenters butted the sheets tightly together, as they had always done with fir plywood. As the humidity increased, the sheets of plywood "grew," and the only way they could expand was to buckle away from the framing.

I have found subfloors that rose off the joists more than 1/2 inch, and roof sheathing that buckled enough for my hand to fit between the sheathing and the shingles. The answer was simple, and is now available — sheets that are cut slightly undersized and marked "sized for spacing."

Truss Rise

Another phenomenon of the energy-conserving age is truss uplift. When insulation is installed so that it completely covers the bottom chord of the truss, that member is much warmer in the winter than the top chords. As a result, it dries more and there is some longitudinal shrinkage.

If one member of a triangular structure—such as a truss—is shortened and the other two remain the same length, the peak will rise. Since the bottom chord of a truss is connected to the peak by the web members, the bottom chord is raised also, thereby lifting the ceiling off the partitions in the center of the house (see Figure 3). This usually takes place during the first winter that the house is heated and, in most cases, occurs only once. If the bottom chord is pine and is juvenile wood, however, it will probably recur.

The forces are rather strong. One builder used metal angles and lag bolts to fasten the truss to the top plate of the wall after the first year's problem. The next winter, the rising truss lifted the wall right off the floor.

In new construction, the best answer is to refrain from nailing the ceiling drywall to the truss for about 15 inches on each side of the parti-

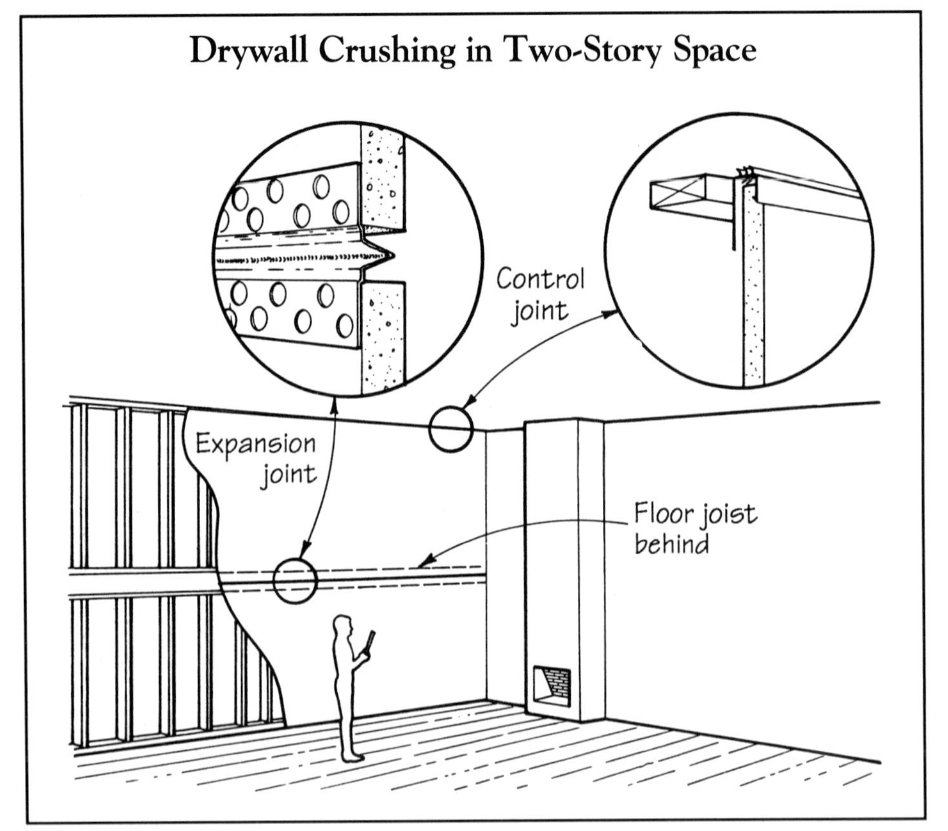

Figure 4. *Drywall in a two-story space is likely to crack if floor joists of an adjoining space abut the wall at mid-height. The cure: Use an expansion joint in the drywall over the joist, and a control joint at the ceiling. USG's P-1 vinyl trim will work.*

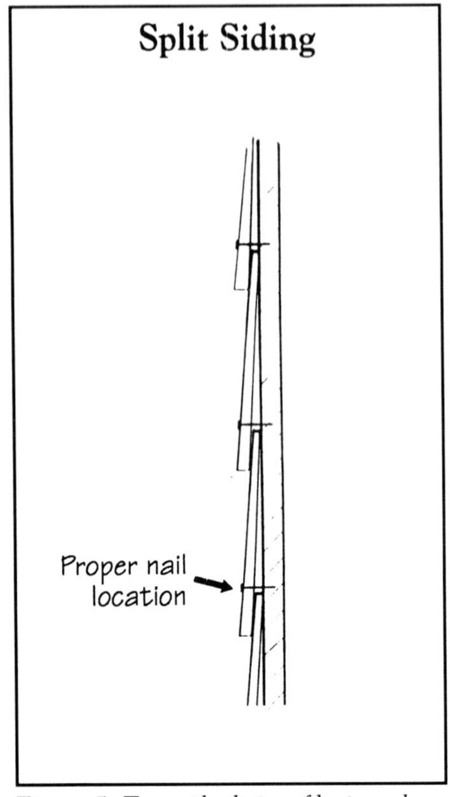

Figure 5. *To avoid splitting of horizontal siding, nails should miss the top of the underlapped board.*

tion, and to support the ceiling drywall with corner clips attached to the studs. Then, as the ceiling rises, the drywall flexes but does not break the taped joint between wall and ceiling. About the only after-the-fact remedy is to nail a cove or other molding to the ceiling, and allow it to slide up and down the wall as the ceiling moves.

Drywall Crushing

In the multistory living areas with cathedral ceilings that are popular today, it is not unusual for the endwall framing of the open area to include a band of horizontal wood—the second-floor joists of the adjoining space. If the drywall extends from floor to ceiling in the multistory area, it will be nailed to studs on both the first and second floors of the adjoining structure. As the floor joists of the second-floor platform shrink, the drywall will buckle and crush in the area of the floor joist (see Figure 4) unless an expansion joint (such as a wood molding over an open drywall joint) is provided.

Another alternative is to apply the drywall to resilient channels, which will allow the studs and joists to move without affecting the drywall. At the ceiling, a control joint rather than a taped joint is almost a necessity.

Siding

The effects of wood shrinkage are just as evident on the outside of the house. The wood siding available today, particularly the cedar coming in from Canada, often has a very high moisture content, perhaps as much as 35 percent. If there is more than one nail across the width of the siding board, splits are inevitable as the wood shrinks. That is why the nail in the face of the siding must be high enough to miss the top of the siding beneath it (see Figure 5). The top of the board should be held by friction, not by the nail.

Also, the amount of overlap should be increased as the moisture content of the siding increases. I have seen siding shrink enough to come unlocked from the piece above.

Veneers

In many parts of the country, it is common practice to add a belt of masonry veneer about four feet high across the front of the house for the sake of appearance. The veneer is capped with a stone coping or a brick cap course, and the siding is brought down to the veneer with a metal flashing that extends from the sheathing out over the cap.

To make a neat joint, the builder typically fits the siding carefully to the cap course and, when the floor system shrinks the inevitable 1/2 inch, the siding is pushed down, perhaps even behind, the cap and forces it away from the house (see Figure 6). At a minimum, the flashing will be bent, forming a channel to hold water against the bottom of the siding.

To avoid this problem, the siding should be stopped a full inch above the cap, and the flashing installed accordingly. Then, when the wood shrinks, the joint will close to a permanent position with the appropriate clearance. Incidentally, if vertical siding is used above the cap, the ends should be beveled (not square cut) to provide a drip edge on the outside surface.

Perhaps the most spectacular and disastrous example of this type occurred when someone built a three-story, wood-frame apartment building with full brick veneer. After the three floor platforms had shrunk during the first winter, none of the casement windows on the third floor could be opened, because the windows no longer matched the opening in the brick veneer. The through-the-wall air conditioners were all tipped inward, causing the condensate to drain inside rather than outside. It was an expensive demonstration of the folly of neglecting the effects of wood shrinkage in design and construction. ■

Henry Spies is a building consultant formerly with the Small Homes Council-Building Research Council of the University of Illinois.

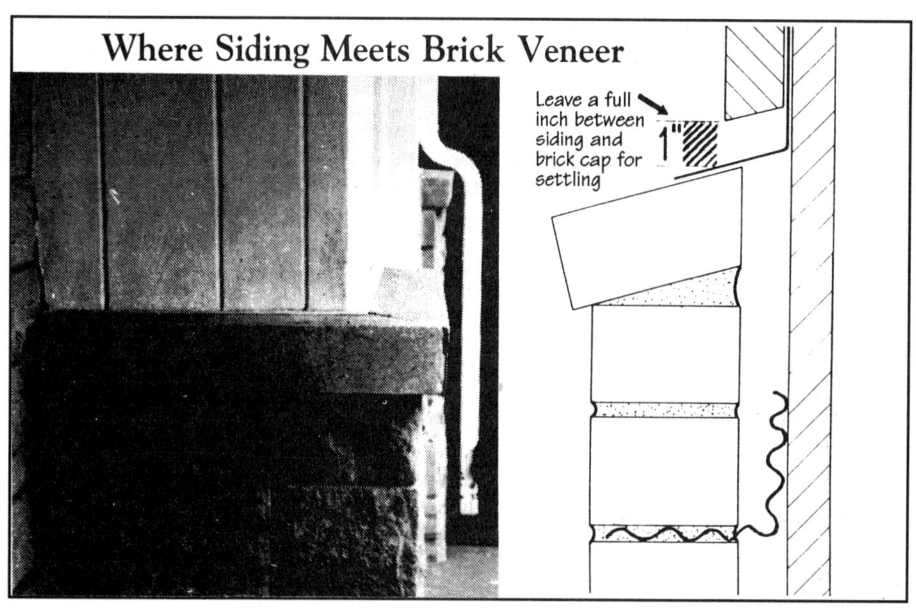

Figure 6. *Some houses have a masonry band at the base about 4 feet high. When the floor joists shrink, the siding will fall and can force the brick cap away from the building. The cure: Leave room for movement in the flashing.*

The Cure For Rising Trusses

by Henri de Marne

Thousands of houses built with roof trusses are experiencing a very noticeable crack in the drywall finish at the joints of interior walls and ceilings.

The separation generally begins to occur in the fall and remains throughout winter and early spring. It often disappears entirely during the late spring or summer. This phenomenon is called by a variety of names: truss uplift, rising-truss, ceiling-partition separation, etc., but I prefer the name "rising-truss syndrome" coined a number of years ago by the NAHB Research Foundation. Somehow it seems to give it an extra dimension that is appealing.

Why does it occur? Is there a remedy? These questions I am frequently asked by builders and homeowners as well.

A truss is a building component made of several parts that work with each other to transfer loads, thus permitting longer spans with smaller members.

When trusses are manufactured, stacked, stored, and installed, all their parts are in temperature and moisture equilibrium. But once the ceiling is insulated, the rules of the game change, particularly since levels of insulation are considerably higher now than they were 15 years ago.

During the warm and humid weather of late spring and summer all parts of the truss absorb moisture and expand. They are in a state of equilibrium similar to that present at the time of installation.

But as fall rolls on and the heat is turned on, the bottom chords, cozily nestled in fiberglass, cellulose, or rock wool insulation, begin to release the stored moisture at a faster rate than the top chords and webs, which are exposed to colder air.

Thus, the bottom chords shrink in both length and width, while the other members remain in a cold and humid environment.

During the heating season, moisture from the living spaces is also slowly finding its way into the attic through joints and holes in the vapor retarder. It migrates into the open attic space resulting in a sharp increase in relative humidity.

Consequently, the truss members not covered by insulation are bathed by cold and very moist air, which prevents their drying. The upper truss members may even absorb additional moisture. Thus the top chords retain their dimension or can even lengthen slightly while the bottom chord shrinks in length and width. The result is the familiar arching which causes the ceiling and wall to separate.

This problem is particularly acute in the first year of a house and may be less so in succeeding winters.

It is not a structural problem. The movement is actually very small, seldom exceeding a fraction of an inch. However, it distresses homeowners.

Truss uplift varies in degree depending on the type of wood used and how dry it was when the trusses were manufactured. Quality wood is not as prone to longitudinal expansion and contraction as juvenile or compression wood. Nor is kiln-dry wood, but it does occur to some degree with all types of wood.

What Not To Do

Under no circumstances should you attempt to prevent this arching by attempting to mechanically fasten the bottom chords of trusses to the plates or studs of interior partitions. The result would be far worse than the small cracks that normally develop. The rising truss can pull up the top plate and damage the drywall at the top of the walls as the fasteners are pulled up. Or the entire interior wall can be pulled up causing a crack at the floor.

To repair the separation, do not shim the space between the bottom chord of the truss and the top plates. This may cause the interior walls to become bearing (without proper support underneath) when the trusses go down in the spring. And this in turn can cause the ends of the trusses to pull off the exterior bearing walls, cracking the joints there.

And certainly do not cut the webs to release the arching pressure.

Prevention

The easiest and surest way to eliminate truss uplift is to omit all insulation in the attic—or at least reduce it considerably. This of course is impractical, so we'll look at other ways to minimize the problem.

- If trusses come with an upward camber built in, do not force them down to fasten them to the interior

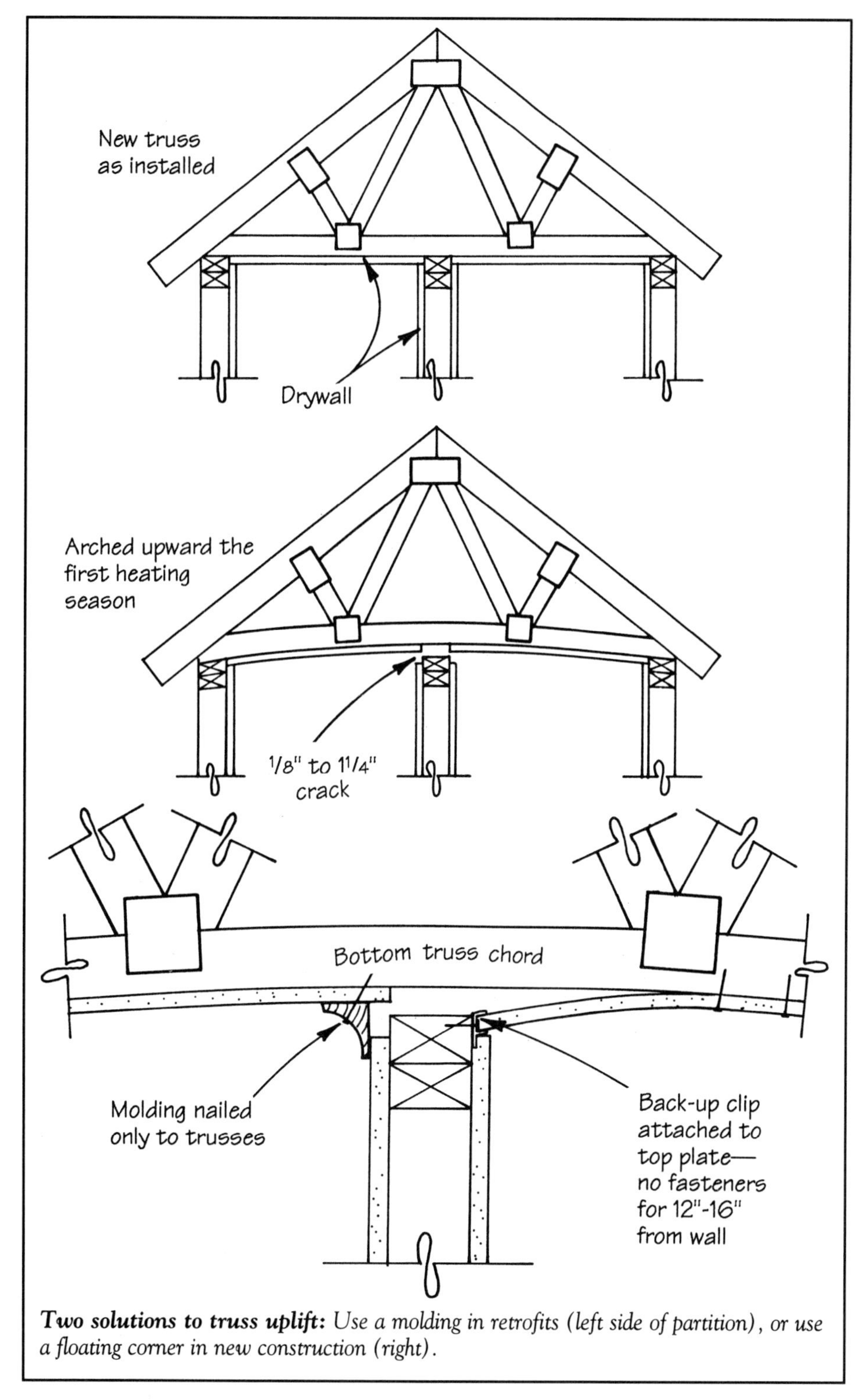

Two solutions to truss uplift: *Use a molding in retrofits (left side of partition), or use a floating corner in new construction (right).*

wall plates as this would increase the stresses later on. Shim them instead.

- When trusses are delivered, cover them up and, once erected, paper the roof immediately.
- Be as careful as you can installing the ceiling vapor retarder and seal all holes around pipes and wires to reduce convection of interior moist air into the attic. Some builders insist that there should be no ceiling vapor retarder so moisture can escape into the ventilated attic, a practice no authority I am aware of recommends. It is best to install the ceiling vapor retarder prior to erecting the interior partitions, instead of stapling it to the top partition plates, so it will not risk being torn if the trusses rise.
- Follow the Recommended Specifications for the Application and Finishing of Gypsum Board by the Gypsum Association (write for them at 810 1st St. NE, #510, Washington, DC 20002).

The Association recommends floating interior corners to minimize or eliminate nail popping caused by shrinkage and settlement of lumber and buildings.

To accomplish this, do not fasten ceiling drywall within 12 to 16 inches of the interior walls. Instead rely on the taped corner joint to hold the drywall together, allowing the ceiling boards to flex as the bottom chords of the trusses arch. Better yet, use drywall clips to hold the drywall flush with the top plate–thus relieving the strain on the taped joint.

- Provide ample attic ventilation by means of continuous soffit and ridge venting with unobstructed air circulation at eaves. The air space at the eaves should be at least 1 1/2 inches deep.

Cure

Where these precautions were not taken and damage has already occurred, probably the simplest permanent solution is to hide the crack with a slip-joint molding.

Nail (or glue with a construction adhesive) the bed, crown, or cove mold to the bottom of the trusses only. The molding will follow the truss up and return down with it in the spring. The crack is still there, and, unfortunately, convection of warm moist air into the attic will still occur, but the crack will be hidden.

Look First

Finally, don't jump to conclusions. Examine all cases of separation to determine whether they are due to the rising-truss syndrome or perhaps a more serious cause such as frost heave or expansive soil movement. ■

Henri de Marne is a building consultant in Waitsfield, Vt.

Section 7. Roof Trusses

Rob Swanson

Fast Framing with Trusses

These engineered giants can do wonders—but only if ordered right and handled well

by Daniel Marrazzo

When trusses are set with a crane, both speed and safety depend on crew and crane operator being in sync.

Roof trusses may offer speed, economy, and a lot of open interior space, but many residential contractors view them as mysteries rather than old friends. Everyone has heard stories of truss disasters—I have one of my own—but following some simple guidelines for ordering and installation will eliminate most of the risk and give you a roof in no time at all.

Ordering

This is one of the most important phases and one of the most neglected. The big problem is procrastination. Many fabricators look for a minimum of four weeks lead time. This means ordering your trusses about the same time you pour the foundation, not when the walls are going up.

As with any other building product, you should shop around for trusses. And as with many other products and services, the lowest price often isn't the best choice. Quality varies widely, and you should talk with other contractors and erectors about which fabricator they like and what to watch out for.

Number one on my list is a fabricator that will deliver when they say they will. If you order four weeks in advance, you should expect to get your trusses within two or three days of the date agreed upon. There's nothing worse than having a crew standing around waiting for the trusses to show.

Although you can literally order trusses over the phone from the blueprints, you'll risk fewer costly errors by talking things over with the truss company representative on site and having him help you measure. Reps can often figure ways for you to save a buck. And they are trained to notice things that may escape your eye, but could come back to haunt

The Book on Bracing

I suspect that most builders secretly enjoy a little risk in their lives or they would have grown up to be engineers, who clearly have a loathing for it. But when it comes to erecting trusses, what appears to be a casual risk can turn into a large-scale game of pick-up or much worse.

Builders who are used to working with rafters and joists sometimes forget that truss roofs are really engineered systems that rely heavily on proper bracing for integrity right from the start.

Risk prevention in this case is simple: Follow the rules when it comes to temporary bracing. They can be found in a manual published by the Truss Plate Institute (583 D'Onofrio Drive, Suite 200, Madison, WI 53719; 608/833-5900). It's titled *Recommended Design Specification for Temporary Bracing of Metal Plate Connected Wood Trusses*, but at the institute it's called DSB-89.

The manual won't ever make it to the best-seller list, but it does allow you to look up very exact specifications for a huge variety of truss types and sizes. And it's written by engineers, so it probably takes into account both the stuff you've thought of, and the summer squall or winter snowfall you weren't counting on.

But as you can see from the drawings we've included here, the specs call for a lot. A lot of labor. A lot of 2x4s. A lot of trouble to go to.

But then again, cutting it *too* close can also buy you a lot of trouble — including serious injuries. The risk, and the decision, is once again yours.

Some Definitions

Temporary bracing is just what it sounds like, but in many cases you can satisfy the *permanent bracing* requirements shown on the plans by installing temporary bracing to those specs and leaving it in place.

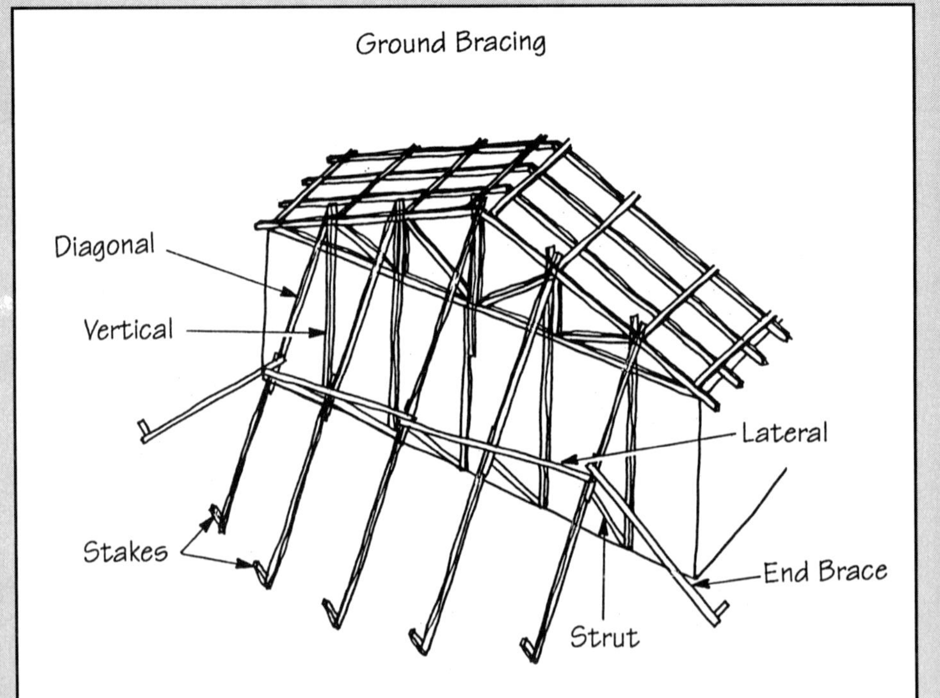

Ground bracing not only stabilizes the gable-end truss, but helps keep the entire complex of trusses plumb and in place. Diagonal and vertical ground braces should be located in line with lateral top-chord bracing.

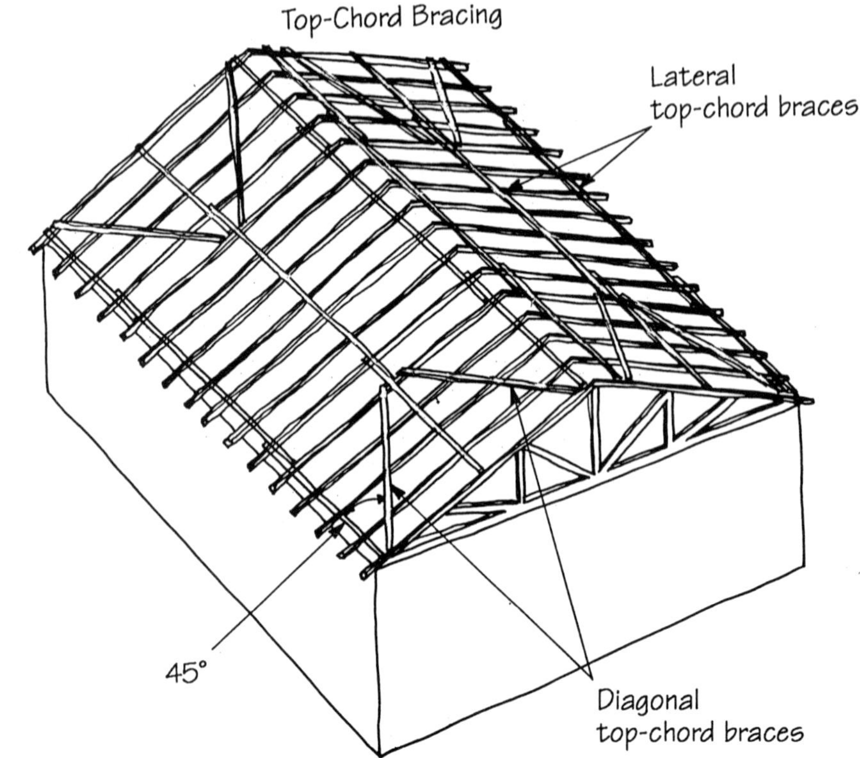

Top-chord bracing, both lateral and diagonal, establishes proper spacing and keeps the tops of the trusses from moving around until the sheathing is applied. This bracing is necessary even if trusses are installed with short "spreaders."

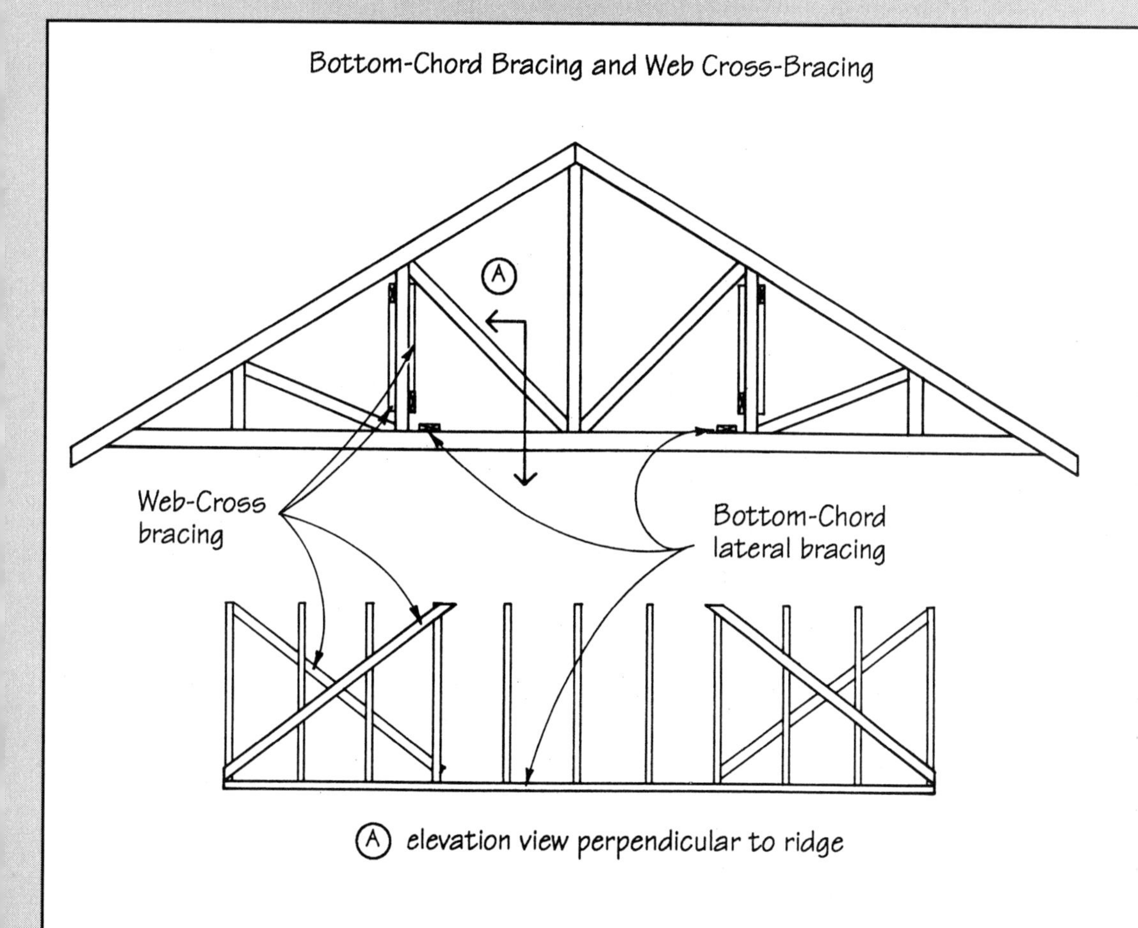

These braces should be installed according to the permanent bracing plan and left in place. Bottom-chord bracing provides continuous lateral support; pairs of web cross-braces triangulate the trusses on each end of the building and every 20 feet in between.

Bracing falls into four categories depending on where you install it: *ground bracing, top-chord bracing, bottom-chord bracing*, and *web bracing*. When you're setting trusses, you are often most concerned with top-chord bracing (specifically *lateral* bracing—the continuous runs of 2x4s perpendicular to the trusses) because it not only provides an initial sense of stability, it establishes the necessary spacing. Bottom-chord lateral bracing does the same for the ceiling members. But it's the diagonal bracing (at 45 degrees to lateral bracing) and web cross bracing (nailed to the webs in pairs) that form an X that *triangulates* the trusses to prevent racking. And finally, it's ground bracing that holds up those first few critical trusses upon which the whole premise rests.

A Few Guidelines

Although the TPI bracing requirements are specific to the type, pitch, and size of the truss you're using, here are some of their guidelines that apply in most situations:

- Use grade-marked 2x4s at least 10 feet long for all braces, and at least two 16d nails at each intersection.
- Short 1x or 2x *spacers* or *spreaders* can't ever be used in place of continuous lateral bracing.
- The 2x4 bracing members must overlap at least one bay (two trusses).
- Get your spacing right the first time, so you don't have to remove lateral bracing prematurely to fix things.
- Most bracing should be placed as near to the panel points (the intersection of webs) as possible.
- Don't rely on scabs nailed to the end walls to restrain the gable-end truss.
- Web cross bracing and bottom-chord diagonal bracing are typically required at the ends of buildings and every 20 feet in between.
- Most roofs require web cross bracing and bottom-chord lateral bracing for permanent bracing.
- Permanent top-chord lateral and diagonal bracing is often satisfied by the sheathing.
- Even when you are fully braced, don't stack all your sheathing in one spot–but scatter it.
- Leave your ground bracing in place until you are fully sheathed and permanently braced.
- With the sheathing in place you can breathe easier. It acts as a diaphragm that transfers lateral loads to sidewalls and end walls.

–Paul Spring

Figure 1. *Pick up trusses from two points, not from the peak as in this photo. Reason: Unbraced trusses are unstable and easily damaged under their own weight.*

you when the trusses are delivered.

But even with their advice, you will ultimately be asked to sign a work order for a particular number of trusses of specific lengths and pitches. Make doubly sure and measure again before you sign—you only get one chance. And remember later on as you inspect foundation forms and framing or masonry, that the top plate is going to have to be parallel and exactly where you predicted or you're going to have a serious problem. Trusses aren't like rafters–there's no cheating to adjust a soffit.

Preparation

The safe and efficient installation of trusses begins before they ever arrive. Because crane time is expensive and truss erection is dangerous, you need to prepare as you might for a big concrete pour—double check your layouts, clear the decks, and make sure everyone on your crew knows where they're supposed to be.

Work with your excavator so that you have a flat, compacted perimeter of at least 20 feet around the building for access by machines and crew, and make sure it has been cleared of all scraps. Have blocks cut to length and 2x4 bracing set out nearby (12-footers are easiest to handle, but some longer braces will be required).

Stud walls should be straight, plumb, and heavily braced; block walls should be plated and secure. The layout—typically 2 feet on-center—should be marked on the top plate so that it is easily visible to a carpenter who is walking the plate and dealing with the crane operator. The first truss is placed 22½ inches from the gable end. All the trusses that follow should be installed 24 inches on-center from this first layout line.

Most truss erectors around here use 25-inch long 1x3s (or sometimes 5/4 stock) called *spreaders* to space the top chords of the trusses. This spreader is nailed across adjoining trusses; the next spreader is alternated either above or below the end of the previous one. Spreaders are usually cut the day before, nails started in each end, and then stacked in bundles so they are out of the way yet handy.

Delivery

Often you can coordinate the delivery of the trusses so the company's boom truck or mechanical arm can load the roof directly. But if you do have to store them on site briefly, pick the spot carefully. Sticker them either flat or vertically so that they aren't racked and can't slide or fall and get damaged or hurt someone. Cover them with a tarp; you don't want to handle a frozen truss, and you don't want to have to deal with one that's been sun-baked either.

Many truss manufacturers can also provide a truck-mounted crane when you're ready to erect. Whether you provide for a crane this way or find your own, planning ahead is the key. Many crane companies are booked even further ahead than are truss fabricators.

Shop crane companies both for price and experience. There is nothing more aggravating or dangerous than a crane operator who is not familiar with the crew's needs during erection. Trusses can also be ruined by an operator who doesn't know what he's doing (see Figure 1). Trusses must always be lifted vertically from two points equally distant from the peak. This is often done with big "truss hooks" and a spreader bar. Ask around for a company that sets a lot of steel and trusses. But also remember that crane companies charge travel time home port to home port—so shop locally.

Cranes in my area typically rent for $300 to $600 per day; there is usually a half-day minimum. If you think you will require more than one day's time to install your trusses, it's a good idea to reserve a second day. It beats trying to negotiate an extra hour out of a reluctant crane operator at quitting time.

Although crane installations are most common, you can load some jobs by hand or use a high-lift, job-site forklift fitted with a truss boom. Only low-slope, short-span trusses such as those used for a 24-foot garage with a 4/12 pitch should be considered for manual loading. Even then, it's taking a risk with both your crew and the trusses.

If you have a skilled operator and a high-lift forklift on site (they're very popular with the masonry trades around here and are less expensive than a crane), it does let your crew find their own pace instead of having to keep up with a crane. A high-lift forklift is also very handy after the trusses are up—for loading sheathing and even roofing materials. You may have trouble setting to the center of your building with one, however, and it takes knowledge and skill to manage loads as fragile as trusses at this height.

The Right-Sized Crew

Too many or too few can be dangerous in setting trusses, but there isn't an exact formula. Start with a carpenter on each outside wall to toenail the trusses and install spreaders and bracing. On spans over 30 feet you may want a third crew member in the middle. Last, beside the crane or lift operator, you'll need someone on the ground to rig the trusses for the crane and supply the crew with spreaders, etc.

Setting the First Trusses

The first truss set is a gable truss. Typically, a gable truss has vertical members every 16 inches for the siding and costs a good deal more. If your design will allow it, you'll find it's cheaper to order two extra standard trusses instead, and then nail vertical 2x4s to the face of these trusses to create gable trusses. You can even sheathe them down on the ground to save yourself having to do it later from scaffolding.

Bracing this first truss, which in turn will be braced in several ways to the next three or four, is the key to the success and safety of the installation. And you'll repeat this same procedure on the other end of your building with the last four or five trusses.

Typically, the first truss is braced with 2x6 (or wider) verticals that extend from the ground to well up on the gable end. These keep the first truss upright. The temporary verticals are nailed all the way up the wall and gable end and are backed by diagonal braces that run back to the ground where they are staked. It's a lot tougher on a second story, where the bracing has to be anchored to the floor on the interior of the building.

Once the gable truss is set and braced, then the next three trusses can be installed on their layout lines and toenailed through their metal truss plates. But don't expect the trusses to line up automatically on the top plate. To keep them aligned and prevent a wavy fascia, string a line down the tails. This will keep the overhang even, and can make a real difference when you're trimming out later on.

As each truss is installed, a spreader should be nailed from the previous truss to the new one. Although these stiffen things up considerably while keeping the layout, continuous 2x4 bracing across the top chords is recommended.

Once the first four trusses are set, you should install a run of 2x4s on the top edges of the top and bottom chords. This lateral bracing will help to permanently support the trusses and it is important that it is laid out and nailed uniformly.

Next, you should set diagonal 2x4 braces in the plane of the webbing. These are extremely important in keeping the trusses from moving laterally. We attach these near the lowest panel point (an intersection of web members) on the fourth truss and then the highest panel point on the gable truss. Depending on the trusses and their span, this is often a good time to install two or three sheets of sheathing to lock in these four trusses.

The Remainder

The rest of the trusses can be installed two at a time if the crane operator is skilled and your crew is working together well. It saves a lot of wear and tear on the operator and a lot of trips back to the truss pile.

It works like this. Keeping the cable taut, the operator lowers the two trusses into position so the first one can be toenailed at the plate and anchored with spreaders. Once it is secure, the cable is slacked slightly to allow the hooks to be slipped off the first truss. The remaining truss can then be raised just slightly and slid into place.

Sheathing

Once the final gable truss is set and braced, it's common to relax and assume that the trusses are now secure. That's a mistake. Let me give you an example from my very recent past:

After raising fifty-one, 60-foot trusses over a masonry warehouse and installing four courses of plywood on one side of the roof, we were caught by an unexpected rainstorm and packed it in for the day. After a night of wind and rain, we were greeted the next morning by only half the trusses we had raised—the rest had collapsed and were inside the building along with the top few courses of masonry and some interior walls. Apparently, strong winds caught hold of the sheathed side of the roof and overloaded the system. This was obviously the hard way to learn about truss failure.

I now make very sure of my bracing and run two or three courses of plywood on both sides of the roof. These sheets can usually be handed up from inside the structure and can be nailed off later with a nailgun. Then I complete the sheathing as soon as possible. This isn't much different than sheathing rafters, although you may be required to use 5/8-inch plywood and clips along unsupported edges.

But even after you're sheathed, you need to remember that you're dealing with engineered members. Unless you are making a small correction on the tails to straighten a fascia, don't even think about cutting, drilling, or modifying a truss. And watch for subs who are used to notching a joist with a reciprocating saw to accommodate a pipe, wire, or duct. If you allow this to happen to a truss, you're almost guaranteeing a failure.

One last thought that illustrates the kind of thought and precaution that trusses require: Remember to warn your clients (you can even state it at the end of your contract where they'll see it) that these are engineered systems that won't tolerate interference. This will seem a little extreme until the day you return to a house you built to complete a punchlist and find a pull-down stairway has been installed and a portable flea market loaded on the lower chords. ■

Dan Marrazzo is a residential and light commercial contractor from Yardley, Pennsylvania.

Stacking a Roof: The Production Approach

There's no time for a crane or a wasted move when setting trusses in the tracts

by Paul Spring

Production building doesn't always have the best reputation when it comes to quality, but no one can fault it for its economy of time and motion. If you want to know how quickly and efficiently you can get from A to B, spend some time watching tract piece workers. Their profit and pride are based primarily on speed, and that comes from specializing and planning every step of the job in detail. Setting trusses is one of those disciplined efforts.

A Matter of Balance

In some of the hot building markets in the West, experienced crane operators and their machines can run nearly $1,000 a day—an unnecessary expense in the mind of most production builders. Their framers have learned to *roll* trusses of all sizes and types like joists. The only crane involved is one on the delivery truck, and it only spends a few minutes stacking the banded bundles of trusses on top of the walls. In fact, the tract term around here for setting trusses is *stacking a roof*.

Yet a typical crew of two (in piece work, even a lone carpenter stacking isn't that unusual) will still be ready for the sheathing guys by the end of a long day.

The key is careful planning–*everything* is laid out, precut, has nails already started, and can be found hanging within bob-and-weave distance from the plates before a single truss is loaded.

Prepping

While one of the crew is laying out the top plates of the interior walls, another starts on the cut list. Most production framers rely on frieze blocks (finished blocking between rafters on the top plate) and vent screens of the same length for attic ventilation to hold the layout on exterior walls. Unless frieze and ridge

Prepping the gable end. *Working on the plates, this carpenter has preset toenails on the bottom chord of the gable truss and is cutting notches for the outriggers. Note the frieze blocks, vent screen, and outriggers hanging from the 1x4 ledger, which keeps these "accessories" in reach.*

The "stack." *It looks haphazard, but the bundles left by the delivery truck are carefully ordered for the stacking crew. They will tip up the trusses by hand, starting with the gable truss on top of the first bundle.*

blocks are supplied by the truss fabricator, they'll be on the list. They are cut 1/8 inch short to account for the thickness of the gang nails they butt. Drywall backing should also be cut to fit unless drywall clips are used.

In fact, even barge rafters (rakes) can be plumb cut on either end, and joined if they aren't too cumbersome. Their *outriggers* (gable-end lookouts) can also be cut to length and laid out. Nails should be started in all of these "accessories."

To keep blocks, backing, and outriggers handy but out of the way, they are quite literally hung by their toenails on a 1x4 ledger or 16d nails on the inside of the exterior walls just below the plates.

Bracing too is cut to length, laid out, and leaned against walls, so that very little thinking, measuring or sawing takes place once the trusses start going up.

To support the gable trusses when they are set, 2x verticals are nailed securely to the end walls. A central vertical is used if the trusses are relatively tame in size. But if they extend more than 6 or 7 feet above the plate, then two or more verticals will be used as high as a hammer will reach.

Last, a catwalk should be built down the center of rooms or garages if they involve long spans or vaulted trusses. This serves as a walkway for a *point man*, who can help raise and steady trusses too large to be handled by the carpenters on the exterior walls.

Loading and Scattering

With luck, the truss yard has banded the trusses in the correct order and the driver is already familiar with the temperament of the production carpenter whose roof has been badly stacked.

Still, the speed and safety of the job depend on how well a crew anticipates and deals with the stack in moving and distributing the trusses on the plates. First in importance are the gable-end trusses. Because they require some preparation and are key components in bracing, they can't end up at the "bottom of the deck."

When working with moderate-

Tipping up. *Working alone, this piece-work carpenter tips up the first 38-foot common truss, relying on careful preparation and skillful positioning. When the truss is standing, he will temporarily brace it by nailing it to the outriggers jutting out from the gable truss.*

Nailing in frieze blocks. *Vent screens and frieze blocks not only finish off the space between the truss tails in this open-soffit design; they also maintain the 2-foot on-center spacing. The plywood "special" hanging from the header will be used to keep ceiling insulation from blocking the vent.*

Installing outriggers and sway braces. *With the four-truss sway brace in place, the last of the outriggers is nailed up. Once bracing is completed and a straight line snapped on the rafter tails, the roof will be ready for sheathing.*

pitch trusses in the 30-foot range, most crews will scatter them along the plates, resting the trusses on each other, like a stack of coasters that have spilled. Bigger trusses are typically left stacked and then carefully walked to their layout—these monsters are just too big and tippy to be moved twice.

Moving about on the plates while wrestling with these monsters in tandem is tricky at best, but more than strength or daring, it requires anticipation, concentration, and above all, teamwork.

Gable-End Trusses

Because the studs in gable-end trusses are often just stapled in place, production framers will sometimes nail a temporary 2x horizontally across the truss. This makes the gable-end studs safe to lean against.

This truss also has to be notched for its outriggers. At least two of these outriggers are typically nailed in place, and toenails are started along the lower chord of the truss for quick nailing into the top plate.

As soon as the truss is raised, one carpenter will steady it while the other slides it back and forth until it is positioned correctly on the plates. Then the uprights can be nailed to the truss, and the toenails driven into the top plate.

While one of the crew pulls up the drywall backing for the end wall and installs it, another can be nailing a ridge block to the peak of the next truss. After each installs a frieze-block or vent-screen on their top plates, the next truss can be raised. It will be face-nailed to the ends of the outriggers and toenailed to interior walls and exterior walls.

At this point, a 1x4—already marked with the layout—is pulled up and nailed as high up on the top chords as possible. On large spans with steep pitches, a second row of this lateral bracing is typically used about a third of the way up from the frieze blocks.

Finishing Up

After four trusses are erected, a 2x4 or 2x6 sway brace is installed both to triangulate the roof and to plumb the gable end. The 45-degree heel of this diagonal is toenailed to the drywall backing just inside the gable-end truss, and then spiked to the face of the third ridge block out once the gable truss has been brought into the same plane as the wall below. A *choker* is often used for extra security. This horizontal 2x sits just below the ridge block and is often nailed to the vertical webs. The top of the sway brace is nailed to the choker with two 16d nails.

With these trusses braced, the barge rafters can be nailed to the outriggers that are in place, and the remaining ones can be installed.

If the trusses have been scattered or stacked in the right order, then raising the remainder will not hold any big surprises. Once all the trusses are set, one worker can install permanent bracing according to the truss manufacturer's specs, while the other *backnails* the trusses (toenails his way back down the plates in the opposite direction) and strings and cuts the truss tails for a straight fascia. ■

Paul Spring is a former editor at The Journal of Light Construction.

Shopping for Custom Trusses

If you haven't considered roof trusses recently, you'll be surprised at their versatility and economy

By Frank Paul

Gary Carlson

Trusses can compete with conventional framing even on complex roofs, given advances in computer-aided design and manufacturing, and the growing scarcity of larger timbers and skilled labor.

If you bought trusses 25 years ago, you were a pioneer. And you were probably a spec builder putting up simple houses with straight gables.

If you're buying trusses today, you are just as likely to be a small custom builder. And you're probably using truss types even *we* couldn't imagine back then: valley and hip systems, girder trusses, energy trusses that can be stuffed with insulation, and trusses that create vaulted ceilings, shed roofs, flat roofs, roofs with load-carrying attic space built in, and more.

Costs Dropping

Between standard truss design software and in-house designers, full-service truss companies can now create a system for almost any roof drawn. And they can do it with increasing economy. In fact, some of the trusses we sold five years ago for $150 now sell for $120.

One of the reasons our truss prices have dropped is automated assembly. The next step for us is computerized

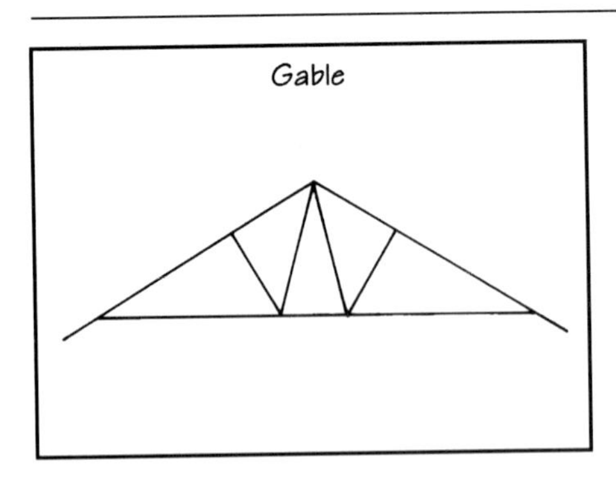

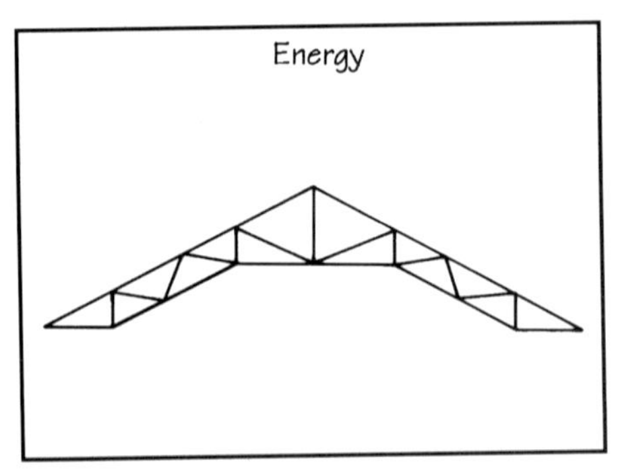

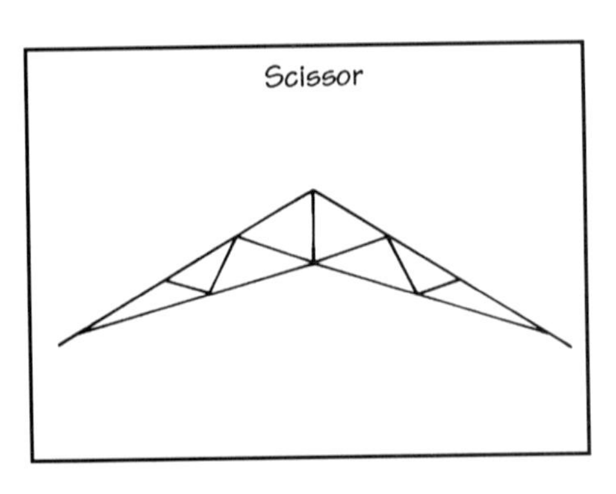

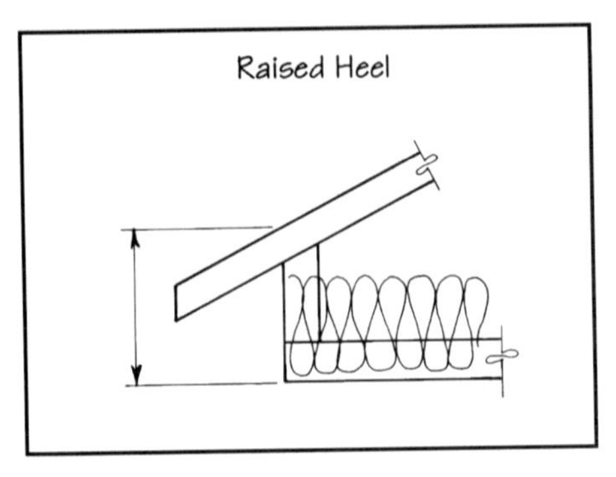

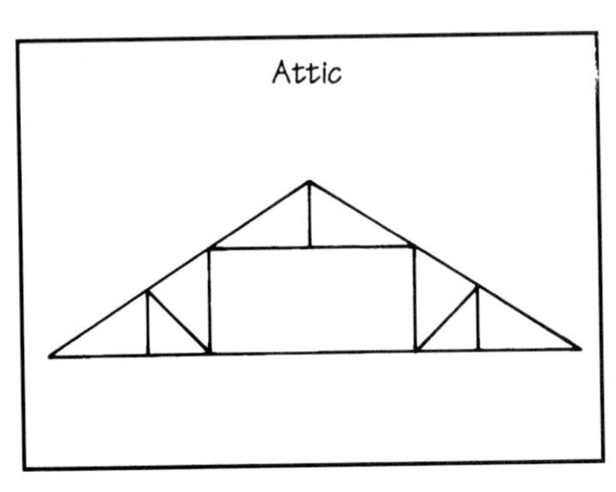

cutting; we estimate that it will triple the speed of that part of production.

Another factor reducing cost is increased design capability. What was once a challenge to this industry is now old hat to designers with 10 and 20 years experience. Aided by CADD systems and the increasing use of machine-stress-rated lumber, these professionals can even take on unusual, one-of-a-kind projects.

Truss Advantages

The best known advantage of wood trusses is their clear-spanning ability. Truss manufacturers routinely produce 50- to 80-foot trusses for commercial and agricultural applications, but even in residential work, a clear span roof can eliminate expensive engineering, simplify foundations, and offer much greater freedom in floorplans. (For a glossary of truss terminology, see "Truss Talk," page 157.

Another advantage with trusses is reduced labor. Because they are spaced 24 inches on-center and are already cut and formed as units, trusses are much faster to install than standard rafters. They're also a lot harder to steal.

The greatest savings using roof trusses is still with a simple gable roof. Take a 26- x 40-foot ranch as an example. Our truss package for this house runs about $800 delivered. The material for conventional framing, figuring 2x8 rafters and 2x6 joists, would run nearly that much before cutting and installation. Also, the conventional framing would require nailing rafters and joists, as well as sheathing and ceiling finish, at 16 inches on-center rather than 24 inches.

At the other end of the spectrum is a 40-truss roof with 15 different truss types. Like all manufacturing processes, small quantities drive up unit costs significantly because of setup expenses. However, even in this case, it would be worth pricing out trusses (see "Pricing and Ordering Trusses," facing page). If two-thirds of the 40 trusses were stock gables, there would probably be significant savings in using them. You could still conventionally frame the portions of the roof where that is more economical.

Truss Types

If you haven't considered trusses in the past few years, you're probably not aware of the many standard types now being produced. Here are the most popular, with some notes on price and sizing.

Gable. These are the most common truss type; we produce them from 16 feet to 80 feet long in slopes from 3/12 to 12/12. Stock sizes run from 16 feet to 40 feet at a 5/12 slope.

Scissor. Scissor trusses, used to create cathedral ceilings, run only 10 percent to 20 percent more than gable trusses in lengths up to 30 feet. Stock sizes at our company run from 24 to 30 feet in a 5/12 slope. The bottom chord of these trusses runs at $2\frac{1}{2}$ in 12. In most of our scissor truss designs, the bottom chord slope cannot exceed half the slope of the top chord.

One concern with scissor trusses, particularly in spans over 40 feet, is *horizontal deflection.* As scissor trusses deflect vertically with loading, they move horizontally where the heel of the truss rests on the bearing surface. Allowance for this movement—we limit it to $\frac{3}{4}$ inch with our scissor trusses—should be made at one bearing wall.

There are several variations on the scissor truss. One is the *modified scissor*, which uses a horizontal bottom chord on one part of the span with a scissor on the remainder. This produces a flat ceiling in rooms on one side of the house with a cathedral ceiling on the other. *Vaulted ceiling* trusses are similar to scissor trusses except that they include a horizontal section of bottom chord in the center.

Attic frame. This is a highly flexible design for creating living or storage space within the truss framework. Attic trusses do require a considerable slope: At 26 feet you'll need to be up around 8/12 to get usable living space at minimum ceiling heights.

The cost of attic trusses also tends to be high because of their construc-

Pricing and Ordering Trusses

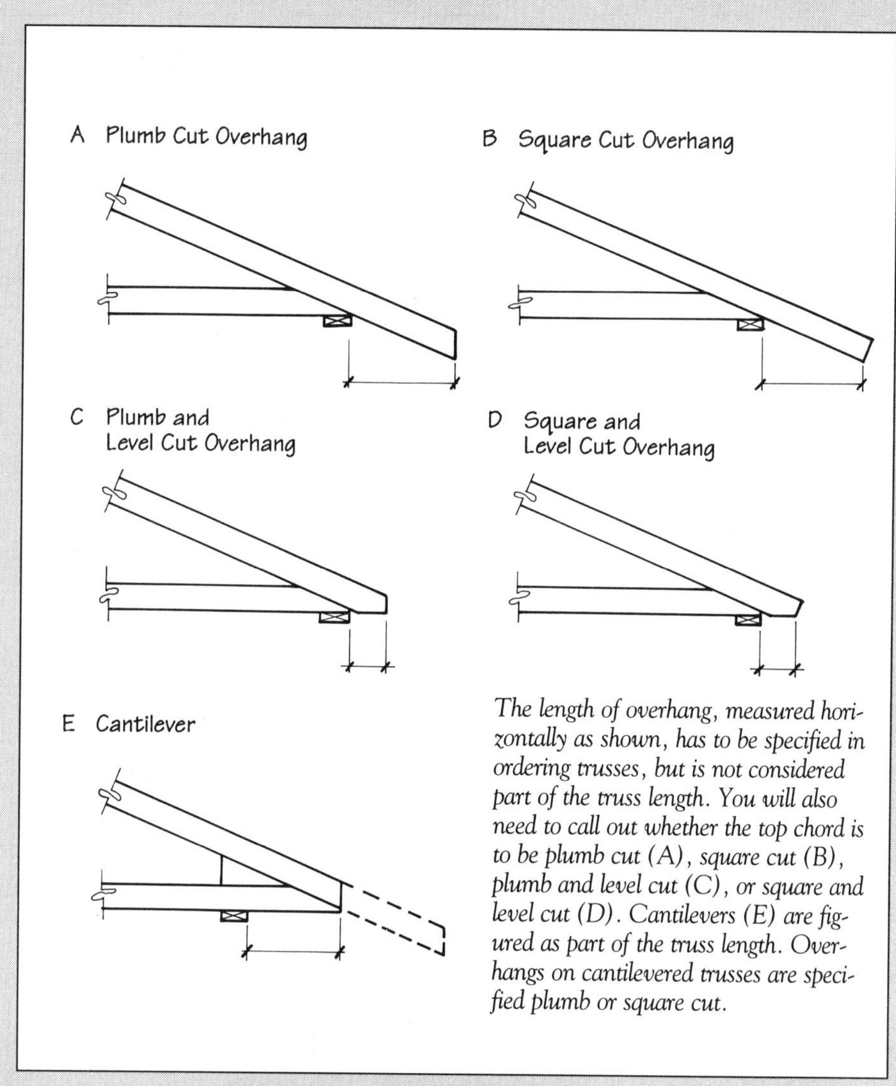

The length of overhang, measured horizontally as shown, has to be specified in ordering trusses, but is not considered part of the truss length. You will also need to call out whether the top chord is to be plumb cut (A), square cut (B), plumb and level cut (C), or square and level cut (D). Cantilevers (E) are figured as part of the truss length. Overhangs on cantilevered trusses are specified plumb or square cut.

Getting a price quote and ordering trusses doesn't require a lot of knowledge or experience; it's the manufacturer that has to do most of the thinking.

Look For Experience

The first step is to find a truss manufacturer in your area with a lot of experience. Most manufacturers now use a computerized design program, but to make full use of this requires an experienced designer on staff.

You should make sure they carry sufficient product liability insurance. Not all companies do.

You should also look for good service, whether you are dealing with a wholesale manufacturer that sells through a network of retail lumberyards, or a retail manufacturer that will take orders directly. Either arrangement is fine, as long as the manufacturer will work with you before you place the order and after if you need it. Ask around to see how other contractors have been treated, particularly if things didn't go smoothly for some reason.

Reliability is also vitally important. Look for a company that guarantees delivery dates, since late delivery can really mess up the schedule. Also find out whether the delivery is to "plate line" (up on the walls) or "ground," in which case the builder needs to provide a crane.

You should also ask about lead time, which can vary from a few days to several weeks. And you might consider getting a price from more than one manufacturer, particularly if you're dealing with an unusual roof design. Not all manufacturers carry the same lumber and connectors, so their approach—and price—for a given job may be different than the next guy's. Also check to be sure that the quote includes delivery.

Simple to Order

The requirements for ordering trusses are quite simple. You'll need to know what the loading requirements are for wood trusses in your local jurisdiction, and you'll need the building dimensions, the slope of the roof, the size of the bearing walls (2x4, 2x6, etc.), and the length and type of overhang and cantilever you want.

Often the easiest way to present all this is to submit your plans for a takeoff. But if the framing plan isn't too complicated, you can typically get a price over the phone. At the other extreme, if you're dealing with a complex roof, you might want to sit down with the manufacturer. Large truss makers will send a rep to visit your office or the site.

Most truss manufacturers will get back to you promptly with a price. If the manufacturer can't be competitive with conventional framing, some will include any options that will help the builder keep the price down.

If the job is complex or you request it, the truss manufacturer will draw up the job to confirm the loading requirements, dimensions, etc. Then they'll set a delivery date. The only requirement after that is to make a date for a crane if required, and to make sure the top plates match the dimensions on the plans.

— F.P.

tion. The strength of trusses is based on a system of triangles, or *panels*. Because of the open space within an attic truss, a simple 24-foot span that would normally use 2x4 top and bottom chords requires 2x8s or even 2x10s, depending on the slope, room size, and loads.

Most of our attic trusses are used in residential work and are designed for a 40-pound live load. The typical application is garages, where a clear span is vital. In other areas of the house, the high cost usually has the builder looking for ways to provide intermediate bearing. If the area is just for storage, we often suggest less expensive Fink-type trusses with a center panel designed for light live loads.

Hip system. Although a fully hipped roof framed with trusses has a high initial cost, a properly designed system can make up for much of this in labor savings.

The most common hip system uses step-down trusses and a prime hip girder that extend the length of the roof (see Figure 1). The hip package is completed with trussed hip end framing and hardware, or the builder can choose to finish the framing conventionally. Field labor costs are the ultimate consideration here.

Valley sets. The costs of valley systems (see Figure 2) are high because there is usually only one of any given size in a roof plan. Smaller valleys in the 12-foot to 30-foot range are typically framed with conventional rafters even when trusses are used elsewhere on the roof.

"Energy" truss. Sometimes called a "super" truss, this is part of a truss system developed in the late 1970s by Roger Beaulieu of Roki Associates, and is used primarily by builders of superinsulated homes. It can accommodate roof R-values up to 60. It is relatively expensive compared to gable trusses, but is particularly useful with two popular regional styles: capes and salt boxes.

Raised heel. For most other architectural styles, a raised-heel truss will accommodate increased levels of insulation and still provide good airflow between soffit and ridge vents. Essentially a gable truss with an added riser at the heel, this feature adds about 10 percent to the cost. The height of the riser should be specified based on the amount and type of insulation being used.

Others. There are many other common types of trusses whose uses are relatively obvious from their shape. A so-called *mono* creates a shed roof, while a *dual pitch* is a gable with two different slopes. A *flat* is just what it sounds like, and a *sloping flat* is a single slope with both ends truncated. A *stub* is a gable truss with

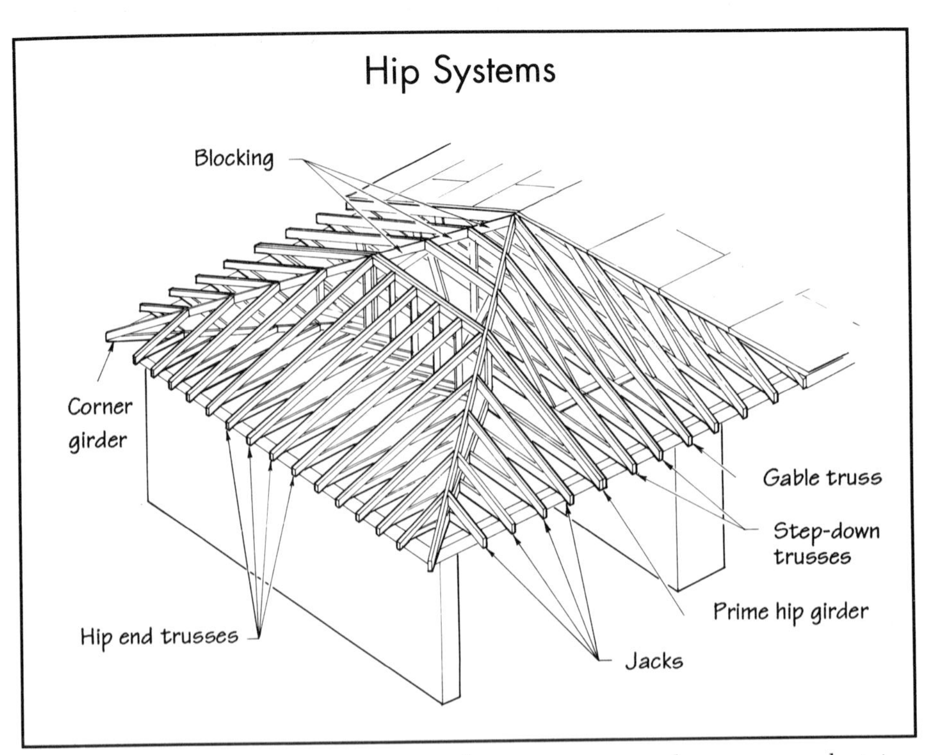

Figure 1. *Hip systems.* *The most common hip system uses step-down trusses and a prime hip girder that extend the length of the roof.*

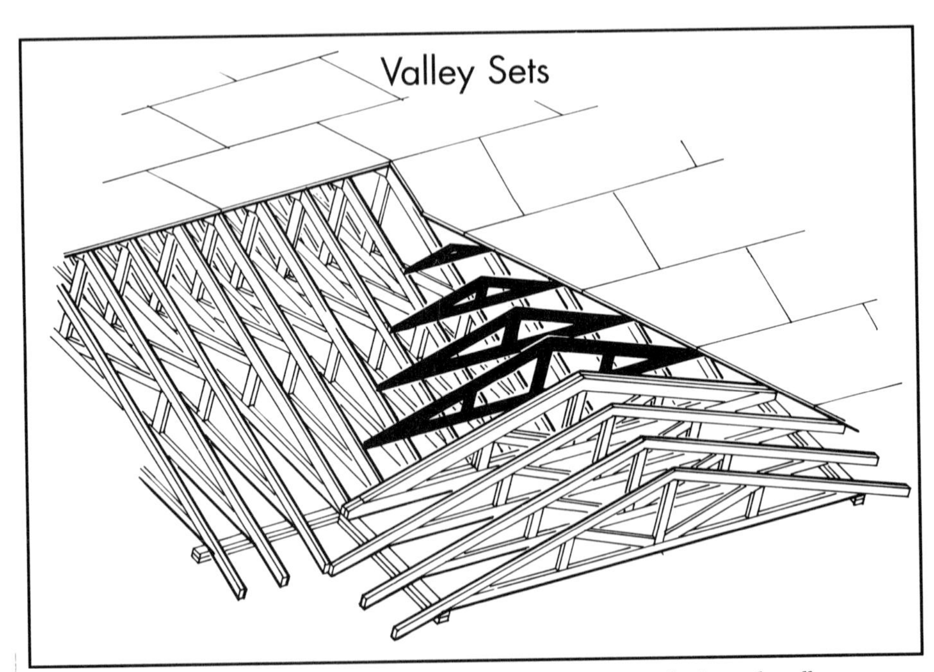

Figure 2. *Valley sets.* *Although valleys are usually conventionally framed, valley truss systems may prove economical if the same size valley is used more than once in a roof plan.*

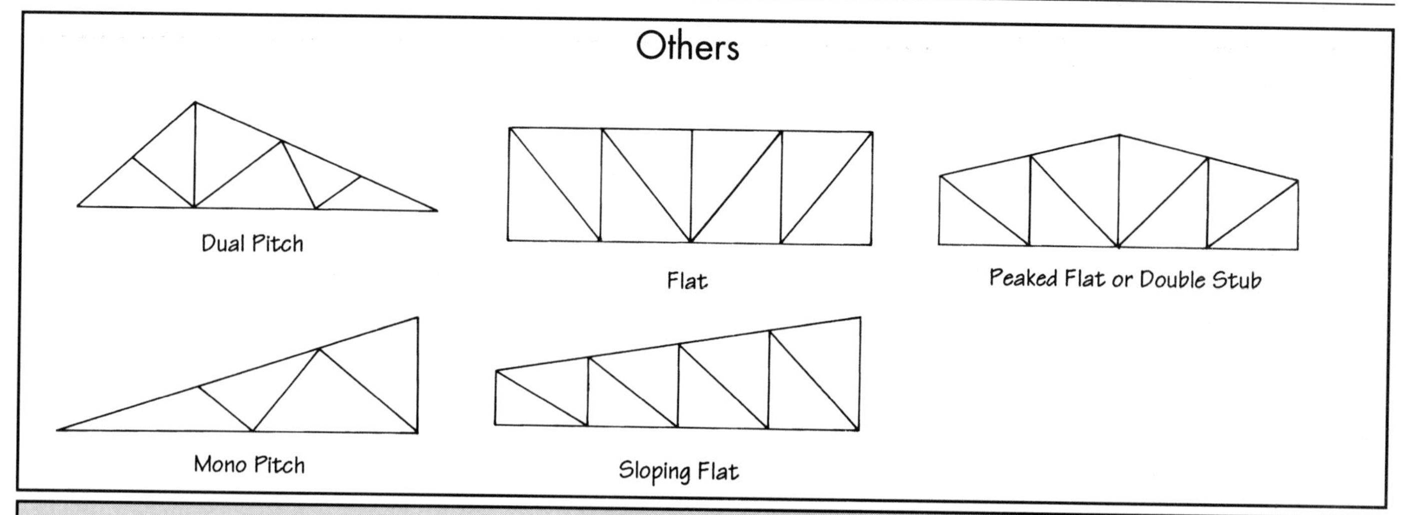

Truss Talk

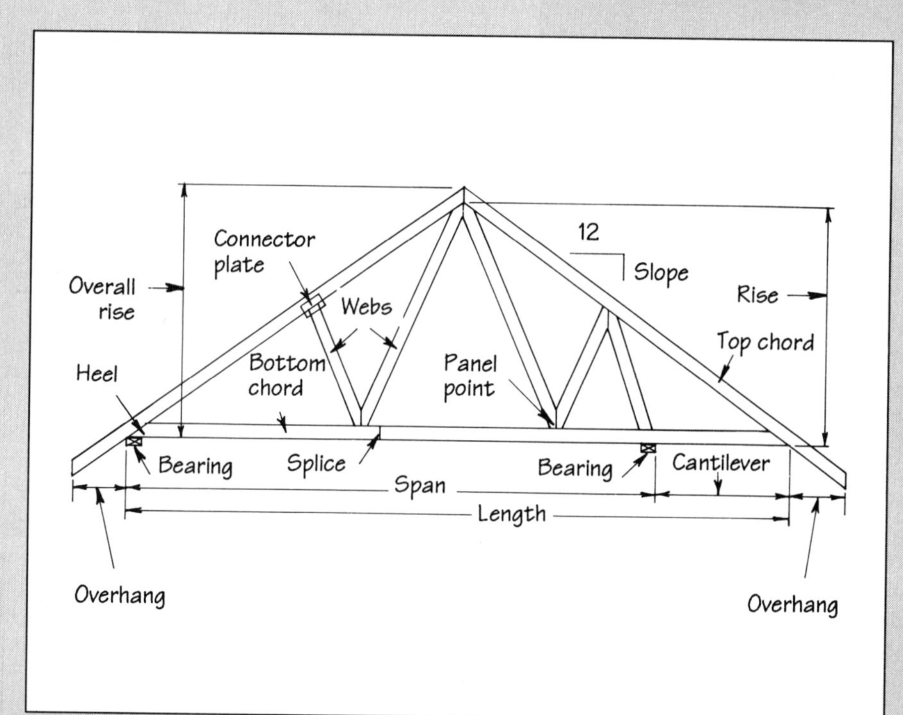

Many of the engineering terms used to describe trusses are also used in conventional framing. Others are unique to truss construction. Here are the common ones that are worth reviewing if you haven't dealt with trusses in the past.

Bottom chord: The horizontal or inclined (on a scissor truss) member at the bottom of a truss. In a conventional system, the ceiling joist.
Butt cut (heel cut): Slight vertical cut at the ends of the truss bottom chord made to ensure uniform span and to provide clearance for sheathing.
Camber: An upward crown built into a truss bottom chord to compensate for deflection caused by the dead load.
Cantilever: Extension of the bottom chord beyond its support, exclusive of overhang.
Clear span: The ability of a truss to span the distance between the exterior walls of a building without requiring any interior support.
Connector plate: Pre-punched, metal-toothed connectors used to join two or more members in a truss. They are mechanically embedded in the wood.
Heel: Point where the top and bottom chords intersect.
Length: Overall measurement of the bottom chord.
Overall rise: Vertical distance from the bottom edge of the bottom chord to the uppermost point of the truss.
Overhang: The extension of the top chord of a truss beyond the heel, measured horizontally.
Panel: A chord segment between two panel points.
Panel point: Point where a web (or webs) intersect a chord.
Plies: Refers to the number of identical trusses joined together to form a girder, as in "2-ply."
Span: Horizontal distance between the outside edges of support (bearing).
Splice: Point at which two chord members are joined together to form a single member.
Symmetrical truss: Truss with the same configuration of members and design loading occurring on each side of the truss centerline.
Top chord: The inclined or horizontal (on a flat truss) member that establishes the upper edge of the truss; rafter in conventional framing.
Webs: Members that join at the top and bottom chords to form the triangular patterns that give a truss its inherent strength.

— F.P

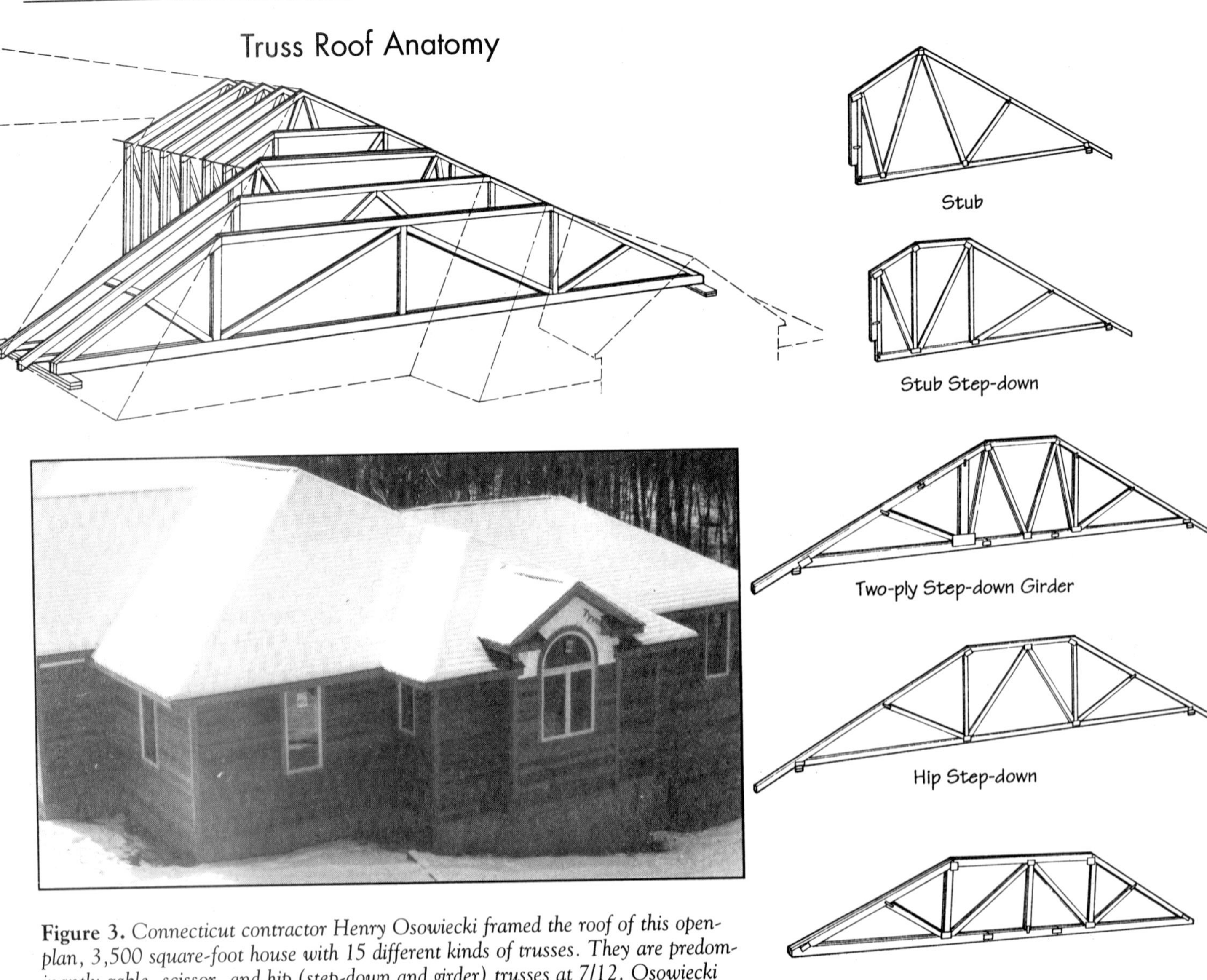

Figure 3. *Connecticut contractor Henry Osowiecki framed the roof of this open-plan, 3,500 square-foot house with 15 different kinds of trusses. They are predominantly gable, scissor, and hip (step-down and girder) trusses at 7/12. Osowiecki opted for trusses because they eliminated the expense of engineering and kept the floorplan free of bearing walls. He was able to save substantially on the overall price of the trusses by conventionally framing valleys and hip ends.*

one truncated end. A *peaked flat* (*double stub*) creates a pitched roof with two vertical ends.

Putting Them Together

A recently built single-family home we supplied in Thomaston, Conn., is a good example of how a variety of truss types combined with some conventional framing can make sense for a small, quality-conscious builder.

When Henry Osowiecki started considering the roof on the 3,500-square-foot house he was building (see Figure 3), he wasn't worried about his crew's ability to frame it. Although the 7/12 roof has a vaulted section and a number of hips, valleys, and other complications, there was nothing that unusual. However, the open floor plan made no accommodation for bearing walls or beams. The time and cost that would go into engineering and building the necessary bearing prompted Osowiecki to get a price on trusses.

Working with a field rep, he identified the roof planes that would be easy enough to frame conventionally and that would be expensive to cover with trusses. These included valleys and hip areas below the prime hip girders. This produced a savings of between 30 percent and 40 percent over using trusses exclusively.

He ended up with a roof composed of 58 trusses of 15 different kinds, and six different areas that required conventional framing. He didn't have to alter his plans to accommodate bearing walls or beams, and finished the framing phase of the house ahead of schedule. The cost of the trusses delivered to the site was $7,800. ■

Frank Paul is a vice president at Wood Structures Inc. of Biddeford, Maine, a leading supplier of trusses to the New England area.

Truss Collapse

Domino-style truss failures can be catastrophic. Good bracing and quality control by the fabricator are essential.

by Raymond A. DiPasquale

The wood truss is a building workhorse. It is economical and efficient, easy to fabricate and erect, and allows clear spans with a minimum of material and weight. Yet roof trusses can fail. Failures are often reported in the Northeast during the winter, when heavy, wet snow loads push roof systems beyond their ultimate capacity.

In addition to overload, other factors can cause a wood-truss system to fail. The case study below illustrates some of these factors.

The Domino Effect

A system of 60-foot-long, clear-span roof trusses collapsed about an hour after the last truss had been erected, and after the workers had left the site. The trusses involved were the lower sections of two-part, piggyback-type triangular trusses that had a final pitch of 7/12. They had been erected with a crane and spreader bar, and were spaced 24 inches on-center. The trusses were long and high, with unusual stability problems.

Truss failures of this kind usually have several contributing causes, but only one trigger that sets the system in motion. Since roof sheathing is not yet in place and temporary bracing between elements is only lightly attached, the system cannot resist the momentum of the progressive collapse. Once in motion, the domino effect typically causes complete destruction. An examination of the pieces as they lie is the only reliable way to determine the probable cause.

This massive failure (above) occurred an hour after the last worker cleared the site. The evidence pointed to a weak joint as the trigger. In general, the trusses suffered from wood defects and badly installed gusset plates (right). Better bracing might have prevented a total collapse.

In this case, it was possible to pinpoint where the collapse had started on the basis of the collapse pattern. The direction and angle of both the horizontal and diagonal temporary bracing between the trusses also influenced the failure pattern. The investigation showed that a specific truss was the trigger. The truss had serious material defects at one end.

A Weak Joint

On the "trigger" truss, there were two knots within inches of the first

Truss Inspection Checklist

All trusses should be inspected before they are erected. Things to look for:

- Use of inferior lumber. Since trusses are generally hidden from view, the fabricator may be tempted to use material with excessive knots or splits, or of a stress grade below that assumed in the design.
- Installation of the gusset plates. These have to be large enough to embrace all members that meet at a joint. They must be centered on the joint so the lugs or nails are fastened to all members in proportion to the forces transferred at the joint.
- Use of warped lumber or inadequately dried lumber. This can result in joint eccentricities or shrinkage distortions, which can produce secondary stresses during service loading.
- Improper joint fit. If individual members are not cut to the correct length, then the truss will be distorted in order to make the connection, or the gap will be too great under the gusset. This also results in eccentricities and secondary stresses.
- Knots in the vicinity of gussets. Such trusses should be rejected.
- Damages during handling, shipment, and erection. Trusses are extremely flexible in the direction perpendicular to the plane of the truss, and can be deformed during shipment on a flatbed. During handling they are often lifted in the weak direction, which puts excessive stress on the joints and members. Consequently, joints can fail before the truss is ever installed. Look for loose lugs or nails, or deformed or bent gussets.
- Repairs made to damaged members. Reject such trusses.

Check out the fabricator and his operation. Since trusses are relatively easy to make, many lumber dealers may be willing to fabricate them, but without having the quality-control procedures that would assure a good product. When a manufacturer is under pressure to fabricate a large order in record time, quality control can suffer.

Specifications for a truss are usually of the performance type, and do not dictate how to make the truss—only that it perform satisfactorily under service conditions. However, the performance specification should also insist that the truss be fabricated under the direction of a licensed design professional. We've always insisted that the shop drawings be sealed by a professional, but having a professional direct the fabrication is something new.

The performance specification should be explicit about the roofing and ceiling materials, and the loads to be applied to both the top and bottom chords. Information is sometimes sketchy, and all the loads anticipated are not delineated. Failures have been caused by such under-design.

The potential for failure can begin during the very early life of a wood truss. Units that do not meet standards should be returned to the fabricator for replacement. Once the truss is incorporated into the job, it is extremely difficult to replace.

Bracing trusses during installation is critical, yet often done in a haphazard manner. Most wood-truss failures during erection are caused by instability–and bracing is what prevents it.

There are guidelines, published by the Truss Plate Institute (TPI), which spell out in detail the recommended procedures for lifting, installing, and bracing wood trusses (see Resources).—R.A.D.

joint at one end of the lower chord. The knots were about 1½ inches in diameter, and about 2½ inches apart. Directly above them, but located in the web member that comes into the same joint, were two more knots of the same size and configuration. The failure appears to have started near these two sets of knots.

In general, knots decrease strength because their grain is at a large angle to the grain of the member, and the grain around them is distorted. When lumber dries and shrinks, checking can occur around knots. The weakening effect of knots is greatest when members are in tension and torsion, rather than in compression. Where shear stresses are present, the knot will reduce strength, primarily because of the diagonal tension that results from the shear.

The stress configuration in wood trusses is anything but precise. At the joint in question, the bottom chord was theoretically in tension, and the diagonal web and the top chord were in compression. But because gusset plates restrain movement at the joint, there were also local bending and shear stresses in the plate and in the members that met there.

In addition, trusses are subjected to all kinds of stresses while being transported and lifted into place. These include stress reversals, torsion, localized stress concentrations, and—invariably—out-of-plane bending. Since trusses are totally unstable in the direction perpendicular to their plane, gussets can loosen during erection. In rare instances the metal itself will yield, but in most cases the wood yields because it is softer. This occurs at the points where the connector lugs are driven into the wood. The connector is thereby loosened and, as was observed in the wreckage in this case, is easily dislodged under slight impact.

Bad Bracing or Bad Trusses?

Material defects in prefabricated wood trusses can be traced to the manufacturer. But material defects don't necessarily lead to total failure

Resources

The booklets *Bracing Wood Trusses: Commentary and Recommendations* and *Handling and Erecting Wood Trusses: Commentary and Recommendations* are available, along with *Quality Standards*, for $3.50 each from the Truss Plate Institute, 583 D'Onofrio Dr., Suite 200, Madison, WI 53719

Two other booklets, *In-Plant Quality-Control Procedures for Metal Plate-Connected Wood Trusses* and *Code of Standard Practice*, are available from the Wood Truss Council of America, 111 E. Wacker Dr., Suite 600, Chicago, IL 60601

Will They Meet the Load?

Wood trusses are designed and made to order. The architect or builder specifies the overall profile and the live loads (generally, snow load). Orders are usually processed by lumber-supply companies and fabricated either by them or by a specialty manufacturer.

Several consulting firms specialize in the analysis and design of wood trusses for the many lumber companies that fabricate them in their own facilities.

The basic structural analysis and design approach assumes that all the joints are pins (free to rotate like a hinge), and that loads are applied only at the joints. The members are then designed as either in pure compression or pure tension. Most of the analysis and design work is done by computer these days, since it is fast and accurate and can handle secondary effects. Computers can produce a graphic diagram of the truss configuration, the size of each member, and notes about the connections.

Even though the typical design approach assumes that each joint in a truss is free to rotate, in actual fabrication the connection is made with a gusset plate, which gives the joint some rigidity. This restraint of movement results in some secondary stresses when the trusses are loaded. Only some of the computer programs account for this in designing the members and joints.

Another assumption—that all loads, whether dead or live loads, are applied at the joints—is also not completely true, since the sheathing is nailed to the top chord, and load is applied continuously along its length. This introduces local bending in the top chord. The same is also true if a ceiling is attached directly to the bottom chord.

Only the most sophisticated computer programs fully consider these secondary stresses but it is important that in the end, user, designer, and specifier understand that these loading conditions exist. The conditions could, by themselves, over-stress the trusses and cause a failure. —R.A.D.

of the system. If there had been more redundant load paths in the system, the defective member could have failed without taking the rest of the roof with it. Lack of structural redundancy is a particular problem during the early phases of construction. A small glitch at that stage can cause extensive damage.

A more substantial temporary bracing system might have prevented the total collapse of the roof in this case, but the inadequacy of the bracing system was not the primary cause of the failure. There was evidence that the contractor followed many—although not all—of the manufacturer's guidelines for bracing during the installation.

Although inadequate bracing is often the major cause of roof-truss collapse during construction, this investigation turned up several other areas that did not conform to good practice and standard specifications.

Some of the defects noted were:

- Undersized or badly trimmed lumber at highly loaded joints.
- Knots located in gusset-plate area.
- Gusset plates not centered on several joints.
- Gusset-plate lugs not adequately embedded in the lumber.
- Defective lumber. An attempt had been made to repair split lumber with a plate; truss should have been rejected at plant.
- Poor fit of members at truss end and joints.
- Undersized connectors.

In light of the material defect that triggered this failure and the many other deficiencies in the fabrication of the trusses, the truss manufacturer bore the burden for this incident.

Starting Over

Because the truss ends were only nominally anchored to the wood at the time of failure, the wall had remained intact. The fix was therefore easy. New trusses were fabricated and installed, but in the rebuilding, the contractor followed the recommendations of the Truss Plate Institute (TPI) for bracing the system.

The fabricator supplied the new trusses and the contractor paid for the erection, since there was some question as to whether additional bracing would have confined the failure to a local area. No legal action was taken. ■

Architect and structural engineer Raymond A. DiPasquale is a failures investigator in Ithaca, N.Y.

Section 8.

Engineered Lumber & Hardware

MaryLee MacDonald

Engineered Beams & Headers

Engineered lumber can support greater loads and longer spans, but good performance requires proper handling and detailing

by Clayton DeKorne

MacMillan-Bloedel

Many new homes boast wide open floor plans and all sorts of architectural bric-a-brac—cantilevered decks, wild roofs, lofts, balconies, popout dormers. These features create complex loading conditions that are often difficult to support with solid-sawn timbers or beams built up from dimensional lumber. As a result, more builders are turning to the higher strength and stiffness of engineered structural wood products, including glulams, laminated veneer lumber, and Parallam.

Practically all the manufacturers of engineered lumber products provide technical support to both the distributor and the guy on site. This support includes span charts and installation details, computer software for sizing beams, and a staff of on-line engineers and technical field reps.

To use these materials successfully, however, requires more than just the right engineering specs. It also requires knowledge of ordering, storage, handling, and connections. Each material is managed very differently on site. Here's a look at several beam materials, the recommendations of the manufacturers, and the insights of a number of builders.

Glue-Laminated Timber

Glulams seem to fall into two categories for residential builders—exposed and enclosed members. Many builders think of glulams primarily as replacements for solid-sawn timbers in exposed applications, such as ridge beams, purlins, and headers. But according to Jim Walsh of Bohemia, a glulam manufacturer, 85 percent of the glulams they sell to residential contractors are used for concealed floor girders and garage door headers.

Glulam grades. Glulams come in three appearance grades—industrial, architectural, and premium. Builders often confuse these grades with structural grades of lumber, but they have nothing to do with the strength of a glulam, only its finished appearance.

Industrial-grade glulams are planed to a uniform dimension after coming off the press, but otherwise are left unfinished. This grade is sometimes referred to as a "non-appearance grade." The wood has checks, knot holes, and press marks, and some may have one re-sawn face.

Bohemia

Glulams are made from carefully selected lumber glued face to face with a rigid, exterior glue. The laminations are finger-jointed to just about any length, and stacked to just about any height—making big beams that will carry almost any load.

Architectural-grade glulams have been sanded on all four sides, the edges have been eased over, and large blemishes have been filled with putty. These beams have been sanded, but they may need touching up with a belt sander to take out dings and banding marks. If the member is used as a column or is at eye-level, you can go over it with an orbital sander to eliminate sanding marks.

Premium-grade is something of a misnomer, because it is easily confused with premium-grade lumber. In glulams, "premium" signifies that every defect down to the smallest check has been filled with putty. If you expect a clear, unblemished, unputtied finish, you'll need to custom-order the stock from the manufacturer and pay a substantial premium.

Appearance-grade glulams are delivered in a watertight wrapper to protect the finish. Most manufacturers recommend slitting the underside of the wrapper to allow condensation and any water that gets in to escape, but otherwise leaving the wrapper on until the building is closed in.

But builder Dennis Hunt of Argyle, N.Y., cautions against leaving the wrapper on at all. He explains that any water that leaks in will be held by capillary action against the wood, which will raise the grain and stain the surface. He also warns that the wrappers are slippery to climb around on.

Glulams are manufactured with lumber dried to a 12 to 16 percent moisture content, which is where wood in buildings usually equalizes. As a result, glulams are dimensionally very stable. Some seasoning checks may open up on the ends, or parallel to the grain on the sides of beams, after finishing. Seasoning checks won't compromise a beam's structural performance, but they might not look too good. Manufacturers offer these suggestions to reduce checking and preserve the finished surface:

- Keep beams covered but allow the wood to breathe.
- Keep the beam off the ground, even if the wrapper is still on.
- Keep beams out of direct sunlight to prevent tanning.

Making Connections

Trus Joists Corp.

Beam pockets. Codes require you to protect untreated foundation beams from concrete. They specify either a 1/2-inch separation from all concrete surfaces or some kind of vapor retarder between the wood and the concrete. Beams should never be grouted into the pocket. If there is good bearing on the foundation, a floor girder will be held securely in place by the compressive load. Rotational forces are resisted by continuous bracing along the top edge. In most cases, this bracing is made of joists that either lap over the beam and are toenailed to the top edge, or, in a flush beam, butt each face of the beam. In areas with severe wind loading conditions, a steel anchor strap may need to be buried in the concrete and secured to the beam ends.

Header connections. Headers should be supported by enough trimmers on each side to provide the correct bearing length. In addition, the top plates should run continuously over the header if possible, and lap the wall by at least 32 inches to tie the header and wall together.

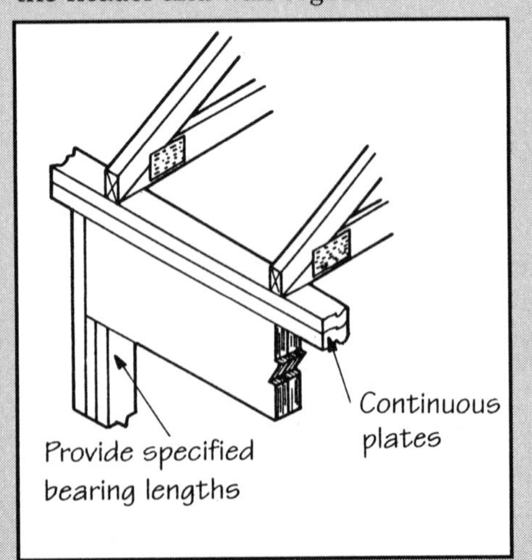

• Keep beams from extreme or rapid drying. This may mean protecting them from strong drying winds and not storing them near heaters or on hot paved surfaces.
• Avoid sudden humidity changes. This may mean you have to dehumidify a space when plaster or joint compound is drying.
• Seal the beams as soon as possible after unwrapping, and seal any new cuts immediately.
• If possible, condition the beams by allowing them to slowly acclimate to the interior of the building. This may take weeks or even months, depending on the size of the beam.

Residential size glulams. Recently, manufacturers have been marketing glulams designed to replace conventional header stock. At least three manufacturers, Bohemia, Unit Structures, and Weyerhaeuser, make glulams sized to fit standard framing widths (3½ and 5½ inches) and carry residential loads. (Design values are set at about Fb = 1,700 psi and E = 1,800,000 psi.) These beams are available in standard lengths up to about 24 feet.

By contrast, standard glulams have a higher stress rating (Fb = 2,400 psi and E = 2,000,000 psi) and come in stock sizes that are narrower (3⅛ and 5⅛ inches) than 2x4 and 2x6 wall framing.

Camber. Other than residential headers, most big glulams are built with a *camber*, or slight upward bow, for strength. It's usually a very slight curve with about a 10,000-foot radius. On long members it's only as noticeable as a small crown on a rafter. For most applications, the camber requires no special alterations to the rest of the framing.

It is important, however, to pay attention to which side is up. A glulam manufacturer will often put a premium grade of lumber for added strength on the tension side of a cambered glulam. The top is usually well labeled on the outside of the wrapper or stamped on the top edge of the beam.

Connection details. Most connec-

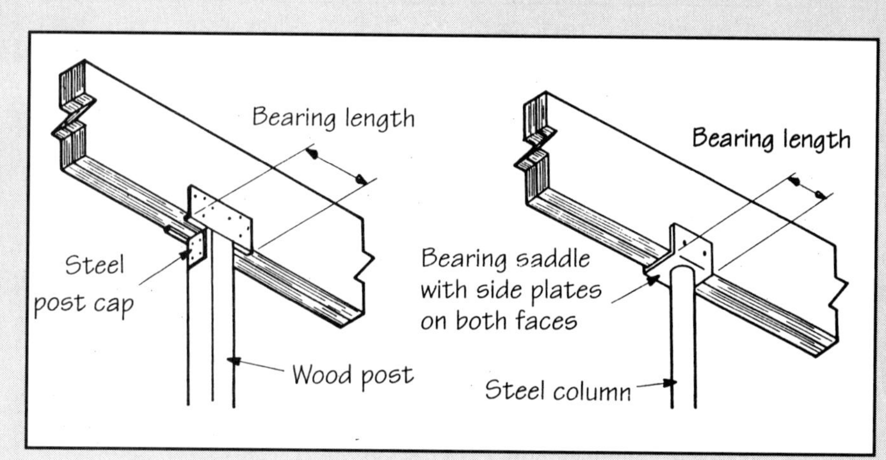

Post caps. Load-carrying beams must have a minimum bearing length at each support. For this reason, most manufacturers of engineered lumber recommend steel post caps for wood posts or a steel post with a bearing saddle for support. At the very least, wood posts should be toenailed into the beam to keep it from sliding over if it gets bumped later on.

Splices. In general, avoid splices whenever possible—you usually can with engineered wood. If a splice is necessary: (1) don't position the splice just off the bearing point, which is the point of maximum shear, and (2) make sure you follow the recommended nailing schedule.

Splices between plies should be staggered and whenever possible the splice should fall directly over the supporting post. For splices that must fall in the midspan, there are two schools of thought for designing a beam. One school recommends that splices should fall at the quarter points along the span, where the splice will have the least effect on bending stress. In this case, make sure the laps are well nailed. Another school recommends that the splices should fall at the center of each span. This leaves plenty of room for nailing the laps, but the splice falls at the point of maximum bending stress. In this case, the beam should be slightly oversized to account for this loss of strength. —*C.D.*

Splice directly over bearing point

Alternate: splice centered between bearing points

Bearing points

tions for engineered lumber beams aren't much different than conventional framing connections. The difference lies primarily in the size of the connector and the strength of the steel. But because the loads are often much greater in an engineered lumber beam, the need to get the details right is more important.

Glulam connections are usually made with through-bolted steel gussets and brackets, or off-the-shelf nail-on clips and hangers.

Nail-on connectors are most common for enclosed framing, while exposed beam connectors are often custom-made. In either case, it's worth sketching out your plans with a technical rep for help with hardware specifications.

Hunt points out that distributors can sometimes offer hardware at prices way below a custom metal shop. A few hangers for a simple roof might easily cost $35 to $40 apiece from a metal shop. But a retailer ordering in bulk can get the same hangers made for $3 or $4 each, and he'll probably pass some of the savings on to you if you buy his beam stock, says Hunt. In addition, distributors might also offer a pre-cut package, including pre-drilled holes and hardware.

When concealed connections are needed, *splice plates* or *split-ring con-*

Sizing Up Beams

The chart to the right shows three typical beam applications in residential construction—a window header, a garage door header, and a floor girder. For each load condition, we've compared four options—dimensional lumber , glulam, LVL, and Parallam.

For clear spans up to 12 feet long under roof loads up to about 420 pounds/linear foot, built-up lumber beams are a reasonable choice. For the longer *clear* spans, engineered lumber is the only option, since dimensional-lumber beams require a center post.

The costs in the chart do not include labor for laminating multiple lumber pieces, or for shimming thin pieces.

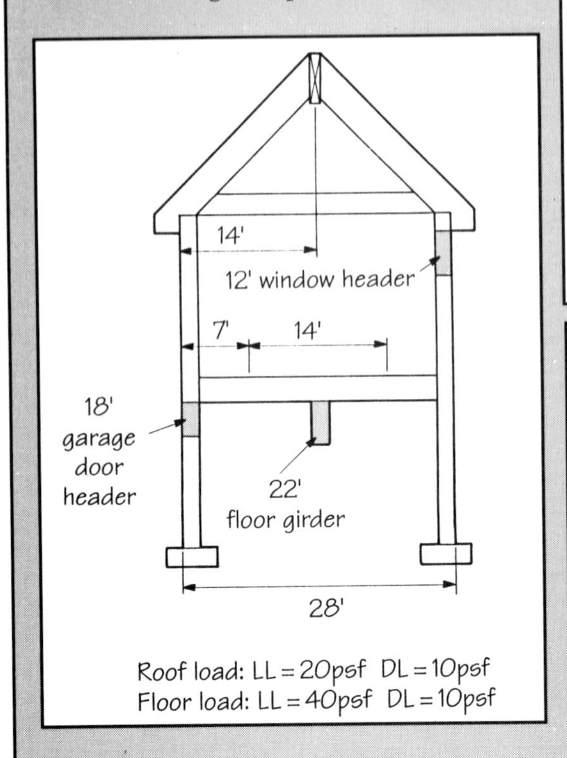

12' WINDOW HEADER (420 PLF)

	Beam Size	*Cost/LF**	*Remarks*
Doug fir (1,200F, 1.3E)	triple 2x12	$3.75	Clear span
Glulam (2,400F, 1.8E)	$3\frac{1}{8}$x$10\frac{1}{2}$	$4.00	Industrial-grade, clear span
LVL (2,800F, 2.0E)	double $1\frac{3}{4}$x$9\frac{1}{2}$	$4.40	Clear span
Parallam 269 (3,100F, 2.1E)	$2\frac{11}{16}$x$9\frac{1}{4}$	$3.45	Clear span

18' GARAGE DOOR HEADER (770 PLF)

	Beam Size	*Cost/LF**	*Remarks*
Doug fir (1,200F, 1.3E)	quadruple 2x12	$5.20	Requires two 10-foot lengths with center post
Glulam (2,400F, 1.8E)	$3\frac{1}{8}$x$16\frac{1}{2}$	$8.00	Industrial-grade, clear span
LVL (3,100F, 2.0E)	double $1\frac{3}{4}$x18	$12.20	Clear span
Parallam (2,900F, 2.0E)	$3\frac{1}{2}$x18	$8.70	Clear span

22' FLOOR GIRDER (700 PLF)

	Beam Size	*Cost/LF**	*Remarks*
Doug fir (1,200F, 1.3E)	quadruple 2x12	$5.00	Requires two 11-foot lengths with center post
Glulam (2,400F, 1.8E)	$5\frac{1}{8}$x18	$11.00	Industrial-grade, clear span
LVL (3,100F, 2.0E)	triple $1\frac{3}{4}$x16	$15.70	Clear span
Parallam (2,900F, 2.0E)	$5\frac{1}{4}$x16	$11.50	Clear span

*Material cost only. Average national pricing, no minimum order

nectors are often used (see Figure 1). Hunt recommends splice plates if you aren't familiar with split-rings because they can be difficult to align.

Toenailed lag screws are not recommended. Lags are not strong enough in the end-grain, or diagonal to the grain, to hold the loads a carrying beam usually sees. Lags, however, are sometimes used with face-plates screwed straight into the face grain.

Heavy members. Large glulams are heavy. Hunt claims a 5-inch-thick, 12-inch-deep beam that's 25 feet long is the most his crew can maneuver by hand. Above this size, a backhoe or crane is needed to set the beams into place. To avoid this complication, more and more builders are turning to thinner material.

Laminated Veneer Lumber

On most residential job-sites, laminated veneer lumber (LVL) has displaced both steel and glue-laminated

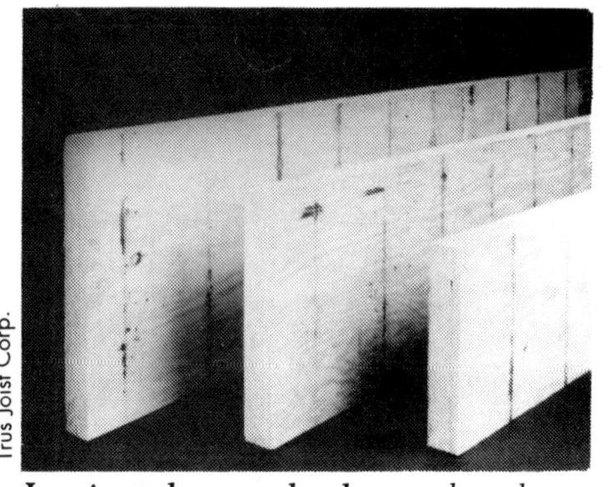

***Laminated veneer lumber**, such as these Micro-Lams, is made from 1/8-inch-thick veneer plies glued together. The grain of each ply runs absolutely parallel along the length of the beam, giving the wood higher strength values than dimensional lumber.*

timbers in "big span" applications. The advantages are obvious. You don't need a crane or a welder. The most common material comes in thin, 1 3/4-inch plies that are lightweight and easy to handle. The clips, hangers, and other connectors are readily available and behave much like conventional lumber hardware. And, as demand increases, more lumberyards are stocking LVL.

Specific sizing. While LVL handles a lot like conventional lumber, manufacturers are quick to caution that it's not the same stuff. First, it must be sized for specific load conditions. It can't be easily substituted on site or used as a standardized material. "If changes or substitutions are made, a builder has to keep going back to the span charts, or, better still, come back to the distributor for engineering support," says John Dawick of Louisiana Pacific's Gang-Lam division.

That recommendation might disappoint builders who want a material they can develop an "intuition" for. But at least the manufacturer isn't asking builders to "consult an engineer" with the expectation that they will have to bear an additional cost. That service is readily available from the distributor and has already been covered in the cost of the material.

Moreover, as Mike Baker of Alpine Structures cautions, "LVL isn't the 'super wood' it's sometimes made out to be." According to Baker, overspanning can be a problem, and he claims to have seen numerous installations where the load carrying capacity was greatly exaggerated in the field.

At this point, most of the LVL products on the market have similar design values (Fb = 2800 psi and E = 2,000,000 psi). Gang-Lam brand LVL has slightly higher values (Fb = 3100 psi and E = 2,100,000 psi). In the future, Baker claims, manufacturers will be offering LVL with lower values as they "fine tune the resource base." As this happens, it will become increasingly important to know what you've got.

Laminating on site. According to several manufacturers, the most overlooked LVL connection is between

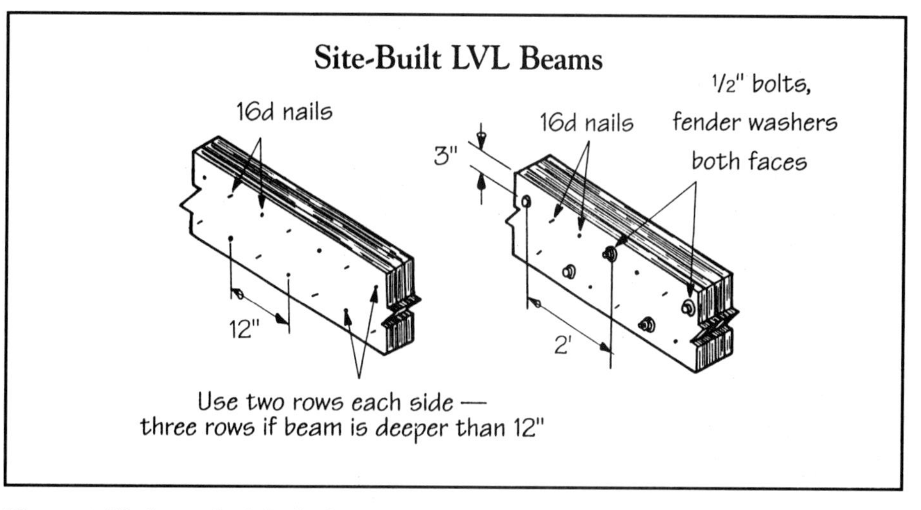

Figure 1. *Nailing schedule for laminating LVL.* *For laminating up to three 12-inch-deep plies (left), nail two rows of 16d nails spaced 12 inches on-center. The rows should be nailed from both sides with the spacing staggered. For 14-inch or deeper plies, use three rows of 16d nails spaced 12 inches on-center. For laminating four or more plies (right), use through bolts staggered every 2 feet, in addition to the regular nailing.*

plies. Manufacturers are very specific about the nailing schedules for fastening plies, as shown in Figure 1.

Note that the nailing between plies doesn't rely on any glue. Manufacturers are reluctant to recommend onsite glue bonds because it's too difficult to control the quality with structural glues, such as resorcinol or phenolic resin. And elastomeric adhesives, such as PL 400, will deform under heavy loads.

The manufacturers I surveyed all claim that the most common error they see with LVL is insufficient nailing when the beam is *sideloaded*. Sideloading occurs when joists are hung on only one side of the beam. If the plies are not well-connected, most of the load will be carried by only one ply. When the recommended nailing schedule is followed, this isn't a problem.

Another common error is drilling and notching beams. LVL should never be drilled or notched for electrical and plumbing pass-throughs. For that matter, neither should any girder, header, or other single-member beam.

Figure 2. *LVL is manufactured at a low moisture content and is prone to cupping if one face gets significantly wetter than the other. Preventing this requires careful storage in the yard and on site.*

MacMillan-Bloedel

Parallam *is made from hundreds of matchstick-like strands of Douglas fir and southern yellow pine that are covered in glue and then pressed into a huge "parallel strand" billet. The billet is then sawn into different dimension beams. Pressure-treated Parallam is also available for exterior applications.*

Compared with repetitive members such as joists and rafters, the shear forces are more concentrated in a single-member beam. As a result, the center portion of the beam section is extremely important.

Dimensional stability. As manufacturers are quick to claim, LVL has greater dimensional stability than common framing lumber. But while this is true in a general sense, the most common complaint I heard from builders had to do with cupping. Of the builders I spoke with who had used LVL, seven out of ten reported some incidence of cupping.

LVL cups most often when it is stored on the ground and picks up moisture from the soil (see Figure 2). Therefore, manufacturers stress that LVL must be stored on blocks and covered from weather. Builder Bill Smith of Great Barrington, Mass., recommends ordering a roll of house wrap to use as a material cover before it goes on the house. Smith recommends leaving the ends of the pile open to allow any wet material to dry out.

Smith and others have reported that LVL is often received wet. In this case, the material starts to cup as soon as the sun comes out and one face begins to dry and shrink. Even though manufacturers send out wrapped pallets of material, these get opened by distributors. Stocking yards, in particular, are notorious for delivering wet material.

Smith recommends that if your supplier is careless about covering material, you should insist on selecting your own beam stock and should pull sticks from the middle of the bundle. But, he points out, this defeats the purpose: One of the reasons he likes using engineered lumber is that its consistent quality is supposed to eliminate this time-consuming trip to the yard.

Cupped LVL is especially troublesome when nailing together multiple plies. John Dawick of L-P's Gang-Lam division insists that the worst thing you can do is try to draw cupped pieces together with clamps or through-bolts. He recommends waiting several days until the wood has a chance to dry out and come back to its original shape before you wall it in. Dawick says LVL is fairly forgiving stuff, and a few rain showers before the building is enclosed won't be a problem.

LVL manufacturers now make a more stable, $3^1/_2$-inch-thick material; other manufacturers have turned to entirely different technologies to produce dimensionally stable beam stock.

Parallam

Parallam is available in thicknesses from $1^3/_4$ to 7 inches, so depending on which size you spec, it's handled either like LVL or glulams.

Parallam seems to solve several problems associated with the other materials. Builders report it comes off the truck more dimensionally consistent than glulams and LVL, and remains much more stable than LVL. Out of eight builders I surveyed who had used Parallam, no one reported any evidence of cupping or twisting. And because it is a solid material, side-loading is not an issue.

Distributors I called unanimously agreed that they don't worry about callbacks with Parallam the way they do with LVL. Those who sell other engineered products in addition to Parallam report that Parallam sales are now outnumbering LVL sales by as much as four to one.

I had to really press builders to find anything wrong with Parallam, and turned up only two minor gripes. First, cutting Parallam dulls steel blades faster. Carbide blades are recommended. Second, builders say that large Parallam beams are heavier than equal size glulam or solid-sawn timbers. The manufacturer responds to this complaint by offering Parallam in a variety of sizes, so builders have the option to use several thinner plies. But, as with dimensional lumber and LVL, multiple-ply beams are usually fastened together on the deck and lifted into place in one piece. If this is the case, it makes sense to use a solid dimension material. ■

Clayton DeKorne is an associate editor with The Journal of Light Construction.

Roof Framing With Wood I-Beams

For greater speed and longer spans, wood I-beams make good sense

Two men easily position a 24-foot I-beam rafter. The wood I-beams are lightweight and straight enough that they don't need crowning.

Wood I-beam joists are becoming a much more common sight in residential floor systems for reasons that are well advertised by their manufacturers. Even builders who haven't used these engineered members are becoming more comfortable with the idea.

But using the same components for roof rafters is met with some skepticism. For that reason, we went to Mike Hoch of Sunbuilt Homes in Champaign, Ill. He and his partner Mark Hieronymus build both spec and custom homes. They began using wood I-beams in their roof framing a couple of years ago, and they have become loyal fans of those manufactured members. Here's what Hoch had to say about them.

JLC: *Why did you start using wood I-beams in your roofs?*

Hoch: We decided to try them because our house designs often call for rafters over 24 feet. At that length you pay a real premium for Douglas fir, and it's a special order—no yard stocks 2x10s or 2x12s that long. And that means hoping they come in on time and in good enough condition to use.

JLC: *Is the quality difference that noticeable?*

Hoch: This is the fourth house where we've used wood I-beams for rafters, and we think we've got the best looking roofs in town. We've only used Trus Joist's TJIs (see list of major suppliers at the end of the article), but they're the best framing material we've ever seen—dead straight and the same dimension each time. We don't even crown them. And we've never sent one back for defects.

With fir rafters we inevitably ended up sending three or four back to the yard because of a 2-inch crown, loose knots, or checks. Then we twiddled our thumbs waiting for the new material to arrive.

JLC: *But in a direct cost comparison, doesn't solid stock come up cheapest?*

Hoch: The labor we save handling the lighter I-beams helps outweigh any price difference. On the house we're building now, the TJIs cost about $500 more than solid 2x rafters. But one of these 9 1/2-inch-deep I-beams weighs about half what a 2x10 douglas fir rafter does. This means one man can easily handle a 30-footer. We figure that'll save us two days labor for our crew of three on this job alone—plus a lot of Doane's pills.

JLC: *Why not just sub out the framing?*

Hoch: Quality is an important component in the houses we build, so we don't want to give away that control. And if we're doing the framing, we can hammer out last minute design details as we go. What the switch to I-beams has meant to our small crew is that we can wrap up the framing phase of the houses we build with the speed of production framers.

JLC: *When don't you use wood I-beams on a roof now?*

Hoch: We'll look at fir when rafter lengths get below 20 feet if the lumberyard has a good supply of straight stock. But we no longer think of wood I-beams just in terms of big jobs.

We recently did a hip-roofed room addition with a 22-foot horizontal clear span. We didn't want posts, and the house was designed for lots of skylights. We would have had to double up 2x12 rafters to get the strength needed. Instead, we used 11 1/2-inch wood I-beams, 2 feet on-center.

JLC: *Why not just use trusses?*

Hoch: If we had a one-story, gable-roofed house with no complications, we'd probably consider trusses. But generally, they're not a good option for us because of the kinds of houses we build—we like to tuck a second story under the roof.

I suppose a truss engineer could come up with a workable design, but that kind of engineering can cost a small fortune, and we still would end up with a 12/12 pitch on the outside, and a 6/12 sloping ceiling on the inside. You can really see the transition between the two around skylights.

But we also like using wood I-beams because they're an efficient use of our forest resources—they can use all the lumber in the tree to make them.

JLC: *Is there any trick to ordering wood I-beams?*

Hoch: We have to adjust our schedule a little to make sure that the yard has time to make up the order. It takes a little less than a week for them to get beams when that's necessary, but they keep a large supply on hand.

Our yard buys directly from the manufacturer, so the beams all come in as 60-footers. When we give the yard a cut list (we round off a couple of inches long), they literally go out there with a chainsaw to fill the order. This can take a day or two if we need a lot of different lengths.

But there are also some advantages to this system when it comes to price. With wood I-beams you don't pay a premium for longer lengths—it costs the same per foot for a 10-footer as it does a 50-footer—and you only pay for the length you need rather than ordering in two-foot increments as you do with solid sawn lumber.

JLC: *What questions did you have on the first job where you used I-beam rafters?*

Hoch: The usual: How were we going to tie them together, how much blocking were we going to have to do, and how much of a pain in the neck would cutting them be?

Blocking turned out to be no big deal. You just fill in on either side with 3/4-inch material—usually plywood or OSB—and then treat the beams like 2x dimension lumber. We sometimes use 2x blocking just to use up scrap pieces that would otherwise be wasted.

JLC: *What about cutting?*

Hoch: That requires a little more effort, but I devised a jig that works well both for commons and compound angle cuts (see Figure 1, next page). I make it out of a 12-inch-wide, 2-foot-long scrap of 1/4-inch lauan plywood. I screw a stop along one edge at the bottom to hold against the I-beam like a Speed Square, and another on top at the angle of cut to use as a guide for the shoe of my saw.

You can also use a radial arm saw or Sawbuck if you have a lot of rafters to cut. Either way, safety glasses are a good idea—wood I-beams throw a lot of splinters.

JLC: *Where did you go to get your questions answered when you were starting out with I-beams?*

Hoch: We've gotten excellent support from our area technical rep, and most manufacturers' literature is very helpful. They also hold seminars for builders, architects, and engineers from time to time.

When questions came up we called our rep. When he couldn't deal with something over the phone, he came out to the site. We've found his suggestions very helpful in solving engineering questions, particularly on our first job.

JLC: *Does your rep get involved with the roof design before you submit the plans?*

Hoch: On the first house where we used wood I-beam rafters, the framing was very tricky, and we had an engineer design the entire package. But on subsequent houses, we've gotten all the necessary information from the technical manual and our rep. On this house, he came up with a modification to the roof framing

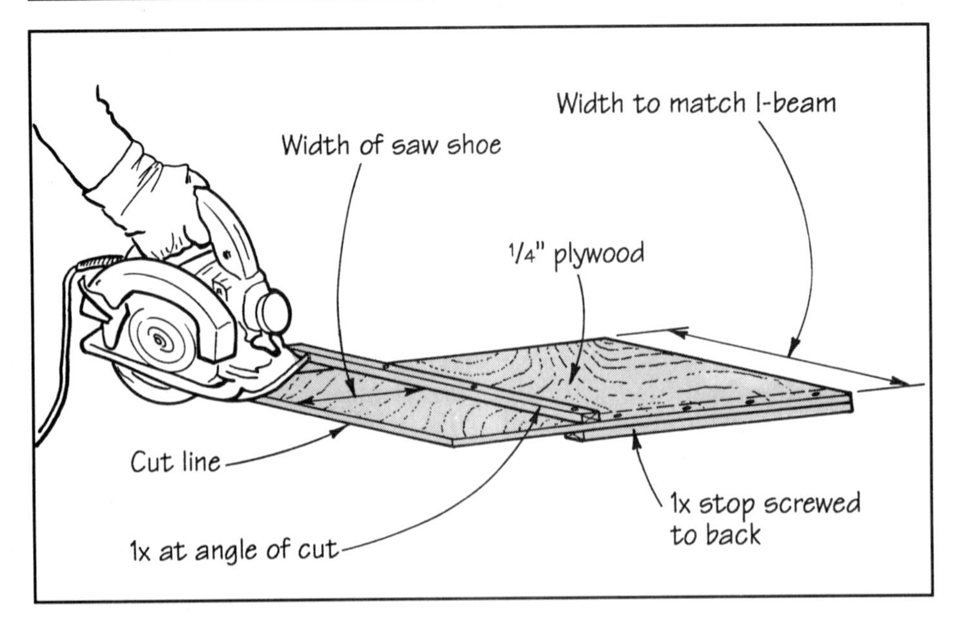

Figure 1. *Hoch came up with this jig (top) to simplify marking and cutting I-beams. Because these members splinter badly, he recommends using glasses when cutting (above).*

plan that saved us in the neighborhood of $1,000.

JLC: *Any reluctance on the part of building departments to accept this engineering or wood I-beams in general?*

Hoch: The building and safety departments are reluctant to approve anything new. We're the only ones in the city using wood I-beams, so we've had to answer all the questions.

The inspector's main objection had to do with fire safety—he was concerned that I-beams burn faster than 2x material (although not too many firemen are going to be climbing around on a 12/12 pitch). He also wanted to require cross bracing at the plate in between the rafters. The manual shows this bracing, but when you read the fine print, you realize it's only necessary with 14- and 16-inch deep I-beams. Although we were able to convince the inspector that the rafters won't tip once the plywood is on, we do apply cross bracing if it's windy, just in case.

JLC: *Speaking of wind, wood I-beams have a reputation for acting like sails in a stiff breeze.*

Hoch: They do flex a lot, but then anything 30 feet long and a foot wide isn't going to be much fun when it's blowing.

JLC: *Doesn't that "flexibility" become a problem when you cut through the flange for a birdsmouth?*

Hoch: How you handle birdsmouths is a critical detail with wood I-beams. You can cut into the web slightly, as long as the bottom flange of the beam bears completely on the top plate, and you're not making any other cuts into the I-beam along its length. If you need another birdsmouth, then you have to use cross bracing.

JLC: *How do you handle knee walls then?*

Hoch: It's easiest to rip a continuous wedge and nail it into the top plate of the knee wall. This provides full bearing for the rafters.

JLC: *How about connections at the ridge?*

Hoch: The most common way—which produces a finished interior—is to plumb cut the I-beams at the ridge, and then tie them together over the top with a metal strap (see Figure 2). But on that first house, our clients didn't mind the ridge beam dropping down into the room, so we butted the rafters on top of the ridge and connected them with gussets. This detail worked out well for us and we've stayed with it. As far as the ridge beam goes, we either box it in or use stain-grade material.

JLC: *What do you use for gussets?*

Hoch: We cut a one-piece, V-shaped gusset from 3/4-inch OSB or plywood; it fits between the flanges and should run at least a foot down each side of the ridge. Each rafter pair requires two gussets—one on each side.

We mass produce them by cutting three or four patterns that are 1/4 inch or so smaller than the area they cover. (You don't want to have to use a saw or a sledge up there to get them to fit.)

Once we've checked the pattern pieces, we lay them out on three

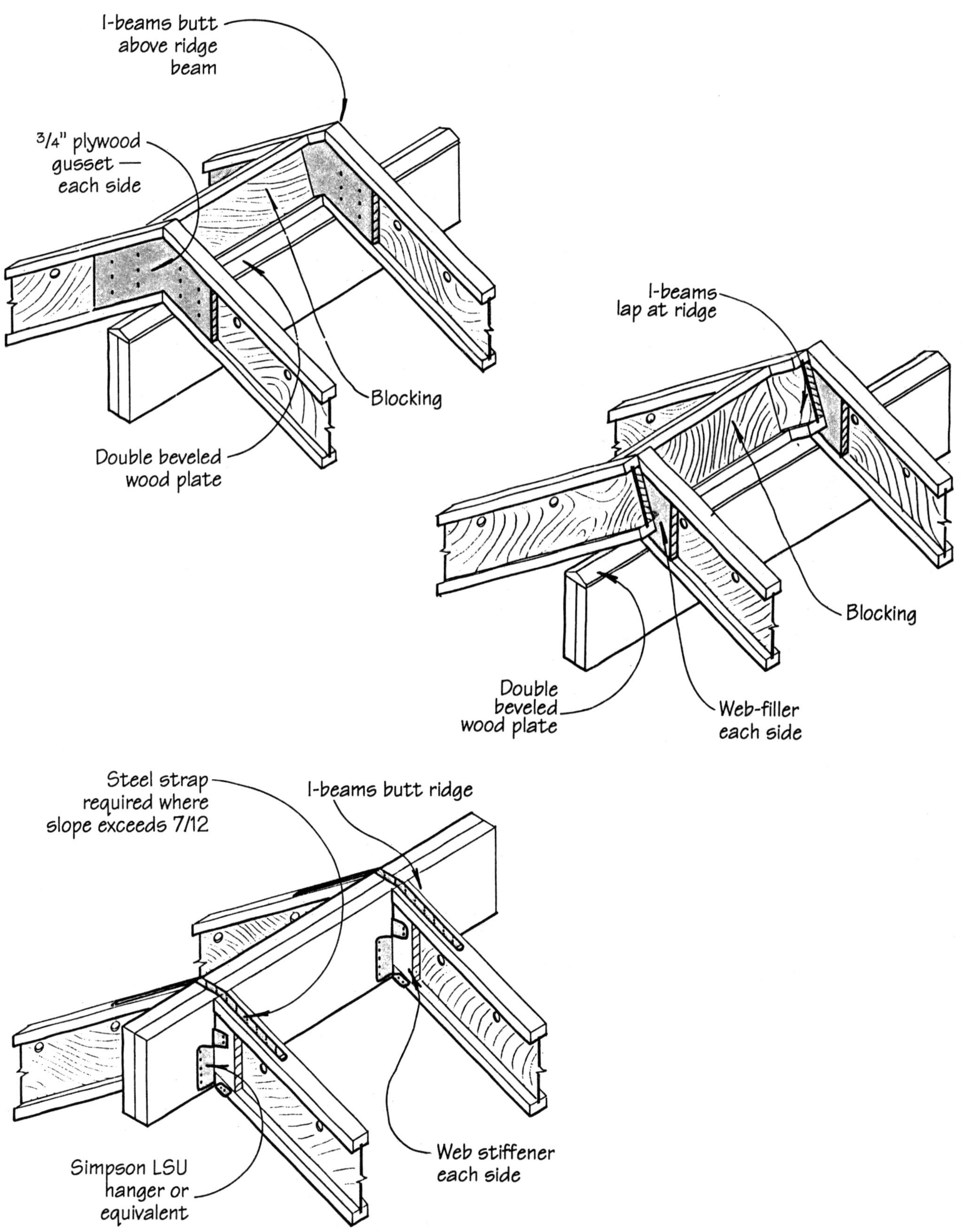

I-beams butt
above ridge
beam
3/4" plywood
gusset —
each side
Blocking
Double beveled
wood plate
I-beams
lap at ridge
Blocking
Double
beveled
wood plate
Web-filler
each side
Steel strap
required where
slope exceeds 7/12
I-beams butt ridge
Simpson LSU
hanger or
equivalent
Web stiffener
each side

Figure 2. *There are several common ways to detail wood I-beams at the ridge. Because of the crucial role that the bottom flange plays in an I-beam, only one birdsmouth—at the exterior wall plate—is typically cut on each rafter.*

Figure 3. *This I-beam jack rafter is blocked with a scrap 2x to increase its bearing on the LVL valley beam. Hoch saves the blocking work for rainy days after the shell is complete.*

pieces of plywood or OSB screwed together and use a circular saw to cut 50 or so pieces at a time. We nail each one with a dozen "eights," which are clinched on the back side. We've found it's easier to install all the rafters first then come back and nail the gussets, unless it's windy.

JLC: *Backing up a bit, what's your framing sequence and how do you deploy your crew when setting I-beam rafters?*

Hoch: We start by stringing a line and setting our ridge beams. Once these are in place and braced, we set scaffolding on the second floor deck so my partner Mark has a solid place to work at the ridge.

Once we nail a double-beveled wood plate to the top of the ridge beam, we're ready to go. I cut rafters on the ground and hand them up to Marty, who's positioned on the second floor deck near the plate. He's armed with a Senco pneumatic nailer and drives two 10d or 12d nails through the flange into the plate at the birdsmouth. We set about a third of the rafters then Marty goes around to the other side and we set that side.

JLC: *How are the rafters nailed at the ridge?*

Hoch: We use two 16d sinkers at the ridge—one on each side of the web. These are driven by hand; having a nail gun at the ridge is more trouble than it's worth. Either way, though, you need to nail the rafter in the right spot the first time because pulling nails out of the flanges isn't an easy task.

JLC: *Do you ever use hangers?*

Hoch: No, we haven't found it necessary. But the hardware is widely available now.

JLC: *How do you handle hips and valleys?*

Hoch: Basically, the same as in conventional framing, with the exception that web stiffeners are required on both sides of the jack rafters where they meet the hip or valley rafter. This blocking can be installed after all rafters have been set; in fact we save this task for a rainy day after we've run the sheathing (Figure 3).

For longer hip and valley rafters we use Micro=Lams. These laminated veneer lumber members are always straight and strong. On smaller houses or for short valleys, regular framing lumber works fine too.

JLC: *How about dormers?*

Hoch: Again, the procedure isn't much different than with conventional lumber. We double-up the I-beams on either side of the dormer, then frame conventionally with 2x6s.

When connecting wood I-beams with conventional framing, you must fill in between the flanges with solid blocking to stiffen the I-beam and create a flush nailing surface. Once you have framed a few houses with wood I-beams, this becomes second nature.

JLC: *How do you handle overhangs?*

Hoch: On our gable ends we typically have small overhangs—just 8 inches. But the technical literature does give a detail for outriggers.

At the eaves, the technical manual shows a lot of cross bracing and blocking needed for fascia support. We've modified their detail somewhat and found it a lot more practical for our 24-inch overhangs (see Figure 4).

We start by cutting a bevel to match the pitch of the roof on the top edge of our 2x6 subfascia. Then we nail through the sheathing into

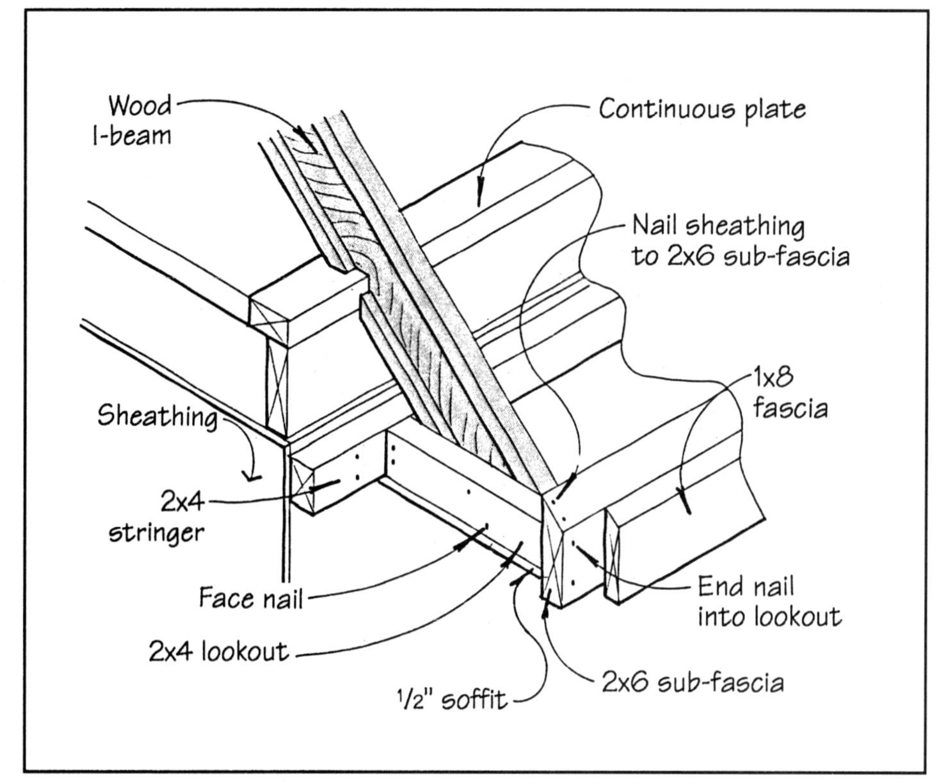

Figure 4. *This detail—modified from the technical manual—creates a very stable 24-inch overhang without a lot of crossbracing and blocking.*

the beveled 2x so that it hangs from it. If you're not using a nail gun, this requires another pair of hands to *buck* the underside of the 2x6 with a hammer to absorb some of the bounce.

Next, we nail a 2x4 ledger to the house with 16s and cut lookouts to fit between it and the subfascia. We toenail the lookouts to the ledger, facenail through them into the rafters, and then nail through the subfascia into the endgrain of the lookouts. This creates a very stable overhang, ready for soffit and fascia.

Hip and valley overhangs are handled just like conventional framing: You cut small rafters out of 2x6s as needed to support the sheathing and to provide nailing for the soffit material.

JLC: *How about temporary bracing?*

Hoch: Wood I-beams are not very stable laterally before the sheathing goes on, so you don't want anybody up on the rafters until you've completed the blocking and temporary bracing. We use 2x4s every 8 feet unless we can get it sheathed—typically 1/2-inch OSB with clips—the same day.

JLC: *Any problems with subs adjusting to the new material?*

Hoch: Not really. Our electrician likes the knock-outs because it saves him time when running his wires. And the drywall hangers like them on cathedral ceilings. They're flat, and nails seem to hold better so they don't get many complaints about nail pop. ■

Manufacturers/Suppliers of Wood I-Beams

Alpine Engineered Products, Inc.
P.O. Box 2225
Pompano Beach, FL 33061
800/735-8055

Boise Cascade
Laminated Veneer Lumber Div.
P.O. Box 2400
White City, OR 97503-0400
800/232-0788

Georgia-Pacific Corp.
133 Peachtree Street NE
Atlanta, GA 30303
404/521-4000

Louisiana Pacific Corp.
111 S.W. 5th Ave.
Portland, OR 97204
503/221-0800
800/547-6331

Trus Joist MacMillan
9777 W. Chinden Blvd.
P.O. Box 60
Boise, ID 83714
800/338-0515

Weyerhauser Company
Corporate Headquarters
Tacoma, WA 98477
206/924-2345

Willamette Industries Inc.
Structural Wood Products Div.
2550 Progress Way
Woodburn, OR 97071
503/981-6003

New Members, New Connectors

A guide to the new generation of hangers and anchors designed for trusses, LVL, and wood I-beams

by William Loeffler

Up on the top plate, a worker sets a scissors truss into a specially designed anchor. The anchor (inset) prevents uplift while allowing the truss to slide back and forth as it deflects.

The days are waning when a crew setting roof trusses sent a young, agile member skyward to walk the plate and toe-nail them on the layout. Plated trusses, wood-web I-beams, and other manufactured structural members have become sophisticated components capable of generating tremendous forces at bearing points and demanding creative solutions where they intersect. All of this puts a real emphasis on the connecting hardware—and your familiarity with the new generation of hangers and anchors designed for these members.

Why Not Just Toenail' Em?

Builders have worked with trusses for years without metal hardware, so why the emphasis on connectors now? Especially with roof systems, where computers are making truss design more and more complex, metal hardware provides an inherently stronger connection that doesn't rely as heavily on the installer as does toe-nailing. I have a lot of respect for the care used by most builders, but structural hardware provides greater strength with less demand for judgment and precision. I'm not being critical about the amount of care and skill used by most builders, but when you look at the numbers, it just isn't worth taking the risk.

Here is what I mean. A perfectly executed toenail is driven at an angle of 30 degrees to the face of the wood, at one-third the nail's length from the end of the member. Under ideal conditions, the nail has five-sixths its normal lateral load value: 78 pounds for a 10d common nail. This same nail installed in a steel hanger has 118 pounds of lateral strength—a 50 percent increase.

The Right Stuff

In some parts of the country, truss fabricators are supplying hangers with each job, or at least marking plans with part numbers. Wood I-beam and LVL (laminated veneer lumber) suppliers commonly stock the popular hangers and can order specials. But in many cases, you're still on your own.

The temptation is to grab something off the shelf at the lumberyard. That works fine for joists—a 2x joist hanger can generally carry all the load the joist can carry.

But plated trusses, wood-web I-beams and LVL are wolves in sheeps' clothing—the strength engineered into them far exceeds what you're used to for the dimensions involved. (And even the dimensions are different from minimal lumber when it comes to smaller wood I-beams and LVL.)

This means you have to pay attention to the loads these members will have to satisfy and their connection points. Here's an example. Let's say you need to hang a truss with an end reaction (load) of 1,200 pounds from a girder truss (typically two or more trusses ganged together) whose bottom chord is a 2x6. A 2x6 joist hanger off the lumberyard shelf can only handle about 700 pounds—well short of capacity. How about a 2x10 joist hanger? This can carry 1,200 pounds and could be nailed to the vertical web on the girder truss (see Figure 1).

This 2x10 joist hanger is also a bad choice however, because at least two of the hanger nails would have no support value. Here's why. When nails are loaded parallel to the grain as would be in the case of nailing to a web, the nails have to be spaced 15 to 20 times their diameter from the end of the member and from each other. That means a 10d nail would have to be at least $2^1/_4$ inches from the web end.

Two hangers that *would* work well are a 2x6 hanger that has a double row of nails in each flange (giving it a capacity well beyond the 1,200 pound requirement), or a strap hanger, that would provide nailing farther up the web.

Another tempting, but mistaken, application is the use of glulam hardware to hang one truss from another. Glulam hangers will meet the load requirements, but they generally use 20d or 40d nails on close centers. This is no problem for a meaty glulam, but you're likely to split the 2x bottom chord or webs of a truss with this nailing schedule.

Ordering Connectors

Most major manufacturers of metal connectors (see "Sources of Supply") print separate catalogues that feature hardware for engineered wood products. It makes sense to pick one up and get familiar with what's available. If you're having trouble figuring out your needs, call one of the manufacturers listed in the sidebar for advice.

When it comes to ordering for special situations, be specific. Here's what a hanger salesman will need:

- end reaction requirements and uplift requirements, if any;
- a description of the member;
- a description of the header.

The header (used to denote whatever the hanger attaches to) description is particularly important in the case of trusses hanging on trusses. The size and number of webs at the panel point (intersection of webs) of the girder truss determines the style

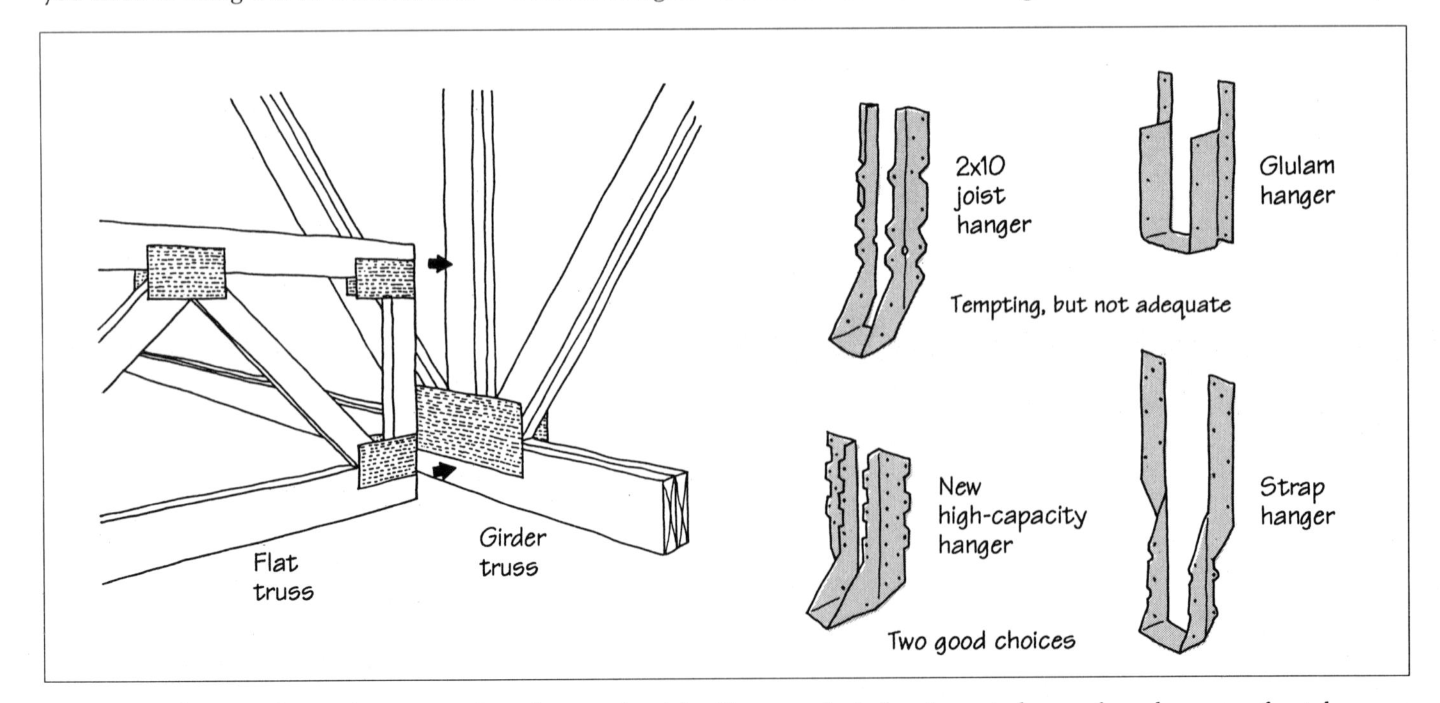

Figure 1. *Choosing a hanger for engineered members can be tricky. For example, in hanging a single truss from the center of a girder truss (left), neither the 2x10 hanger nor the glulam hanger shown at right will work, even though they are rated to carry the 1,200-pound load. The reason: The bottom chord and vertical web of the girder truss don't provide the right nailing.*

Managing Metal Connectors on Site

by Jack Stuart

When a job calls for nothing more complicated than joist hangers, a builder can have a couple of boxes thrown on the truck with his lumber order and not give the whole thing too much more thought. But heavier connectors—which cost more and are a lot harder to get—require careful planning to ensure that they end up at the right place at the right time.

So if you have a large interest in seeing the job run smoothly and economically, it is imperative that you have a hand in the take off, ordering, and supervision of installation. Here is the take-off system I use and some thoughts on the management and installation of different kinds of connectors.

Takeoffs. The key is producing a master print and take-off list that can be distributed to the related trades and suppliers. Then if changes have to be made, an amended plan and list can be sent out so that everyone is working with the same specs.

If there are special conditions or manufacturer's specifications that are often ignored by installers, take the time when you're doing the takeoff to draw a detail of how things should be handled. These drawings are vital when two or more trades are involved.

Light wood-to-wood connections. These 12- to 20-gauge, stamped-steel connectors—joist hangers, post bases and caps, framing anchors, wall braces, and straps—are pretty straightforward, but you still need to make sure that they are installed with the right number and kind of fasteners. It is not uncommon for a sharp inspector to check the nailing, because if this condition isn't met, the hardware can't handle the load for which it was engineered.

Heavier wood-to-wood connectors. These stamped or welded connectors for posts, beams, and trusses are commonly used in light industrial and commercial buildings that have large open areas. Because of the many different configurations available and their size and cost, most suppliers don't carry them in inventory. This means a careful takeoff to ensure that you'll have what you need when you start setting beams, because much of this hardware has a top flange that means you can't go back later to install it.

These connectors are usually detailed on the structural plans and care must be taken to get the right connector with the right nails or bolts in the right spot. As a superintendent on commercial jobs I always tried to get the engineer to do a walk-though at some point after the framing was up (but still exposed) to take a look at all the heavy connections. In my experience, they have always been happy to do it even though it wasn't always required.

Wood-to-concrete connections. These connections—embedded beam seats, post bases, holdowns, etc.—are critical because the tolerances for error are a lot tighter and mistakes are a lot costlier to fix.

Although some people advocate "wetsticking"—embedding the hardware as the concrete is being screeded—I've learned the hard way that it is worth taking the time to fasten all embedded hardware to templates. This braces them securely before the pour.

It helps to have a concrete subcontractor with an appreciation of what these framing connectors do. The sub needs to know how to install them so that the next trade can deal with them properly. I've had problems in the past because I didn't check with the concrete guy to make sure he knew what the connectors were and how to install them properly. Sometimes it works better if you hire someone other than the foundation sub for this job.

However, even when the foundation sub is well informed, someone needs to check his layout (location, alignment, and elevation) before the concrete is placed. Depending on the job size, this could be the owners' representative, the superintendent or a surveyor hired for the purpose. I also typically invite the engineer to take a look, and I have found it advantageous, when possible, to have the framer check things out. All of this takes time and careful coordination, but it's worth the effort.

Steel-to-concrete or steel. These connections involve a weld plate or bolts that are used to mount beams and columns (fabricators typically supply templates for column base plates). Embedding these kinds of connectors requires the same precautions described above, but there are several things to be aware of when it comes to welding.

On-site welding usually must be done by a certified welder, and there is usually a torquing requirement on a bolted, steel-to-steel connection. An independent inspector is sometimes required on site while the welding is going on. It gets expensive to have this inspector on site. It helps to schedule the welding or bolting for a specific time frame.

Some thoughts on bolts. Most of the heavier connectors require either bolts, or lengths of threaded rod, and washers and nuts. Take care to double-check the various lengths and sizes for each connector. Although the company that supplies your connectors will typically carry bolts, you can often save a buck on large orders if you shop around.

Bolts have a tendency to get scattered and lost unless they are in someone's charge. And because it's easy to mistake bolts of similar size and length, particular care should be taken during installation to make sure that the right bolts are getting used in the right places. Although some bolts can be added later, it's often more difficult and less efficient, or in the case of a column-to-beam connection, just plain dangerous. ■

Jack Stuart is a construction superintendent in Bedminster, N.J.

of hanger that will work.

The type of wood in the header member should be mentioned too. Most hanger load ratings are based on nails or bolts into southern yellow pine or Douglas fir. The hanger rating must be reduced to 80 percent if spruce-pine-fir is used. On the other hand, header material like LVL or glulam beams can accept heavier shank nails with less chance of splitting.

Also consider the depth of your header member and the carried member. Describing both members is the *only* way to be certain your hardware will work.

Hangers For All Seasons

Manufacturers have developed a wide range of hangers and anchors to handle the special needs of engineered wood products. Some of the more common applications and hardware solutions are described below.

Floor trusses. *Strap hangers* up to 22 inches long are the most popular connectors for floor trusses. The 16-gauge steel straps are formed in the field over the top of the carrying member and give these hangers a rating of about 2,000 pounds. Strap hangers for floor trusses cost about $3 apiece when bought in volume.

Be wary of strap hangers touting 3,000 to 4,000 pound loads because they may require a 22-inch-high solid header to obtain the nailing required.

Topmount hangers (Figure 2) cost a bit less, but because the straps that lip over the header are factory-formed, you have to order different hangers for trusses of different depths.

Floor openings framed with nominal lumber are often two-ply (3 inches wide), which means you can use either a two-ply hanger or a 3½ inch hanger with a ½-inch shim.

Wood I-beam joists. The first caution is that wood I-beams in small residential sizes won't fit a 2x (1½ inch) hanger. Depending on the beam manufacturer you're using, you'll need widths of 1.6 or 1.75 (1¾) inches.

The most popular I-beam hangers by far are *topmount hangers*. They install quickly because the top flanges set the depth and provide hanger bearing, cutting down on the number of nails to be driven. The joist flanges extend high enough to trap the top chord of the I-beam from rotating, and eliminate the need to add web fillers. Topmount hangers for 10-inch and 12-inch I-beams run about $1.50 each.

Facemount hangers (Figure 3) cost a little less than topmount hangers, but are a bit more labor intensive to set at the correct height. They work well if the header extends above the I-beam, and two-ply I-beams work nicely with facemounts because they fit standard 4x (3½ inch) hangers. In all cases, if the side flanges do not reach the top chord, you'll need to add web fillers to prevent rotation.

Facemount hangers also have an advantage over topmounts in situations where there are uplift requirements of over 200 pounds because they rely on nails driven into the face of the header. Web fillers would be needed, though, to accept the extra joist flange nails.

Dealing with masonry construction. If you have used some of the new masonry hangers, you'll be amazed at how easily they're installed and at the cost savings available on townhouse construction with masonry firewalls.

Saddle-style hangers (Figure 4) straddle a wall, and are merely placed on centers by the masonry crew. End walls require a *hookover style*, which is essentially half of a saddle-style hanger. The hangers can be made to carry floor and roof trusses, or wood I-beams.

Top-chord-bearing truss hangers (Figure 5) are also available in both saddle and hookover styles. Some planning is required since two to three week delivery is common for most masonry hangers.

Laminated veneer lumber. LVL is so strong it presents a challenge in selecting the right hanger. The key is

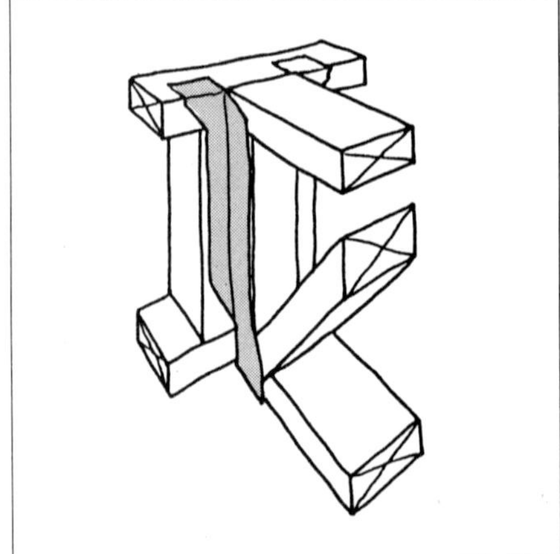

Figure 2. *Topmount hanger.*

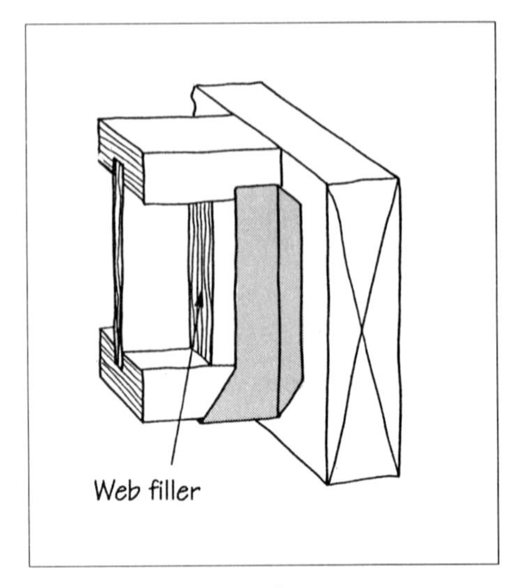

Figure 3. *Facemount hanger.*

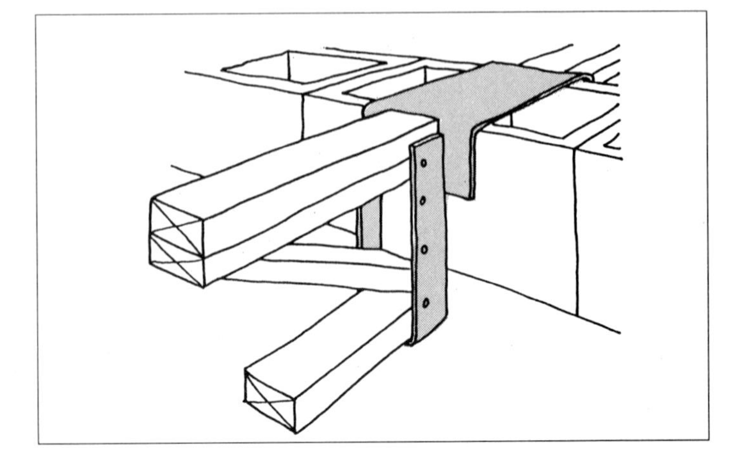

Figure 4. *Welded masonry saddle hanger.*

the end reaction number for your LVL beam.

An 18-gauge stamped hanger made for wood I-beams will bow, sag and could fail under LVL load conditions. With a lot of LVL being doubled up (two-ply) for headers and beams, matching the capacity of the hanger to the beam becomes even more critical.

Facemount hangers can carry 2,000 to 3,000 pounds depending upon their size and nailing schedule. Twelve-gauge *topmount hangers with a deep seat bearing surface* (Figure 6) can handle 3,000 to 5,000 pound loads. For heavy two-ply loads, or three-ply loads and ledger plate applications, a *welded hanger* (the masonry hanger in Figure 4 uses a welded seat) is necessary.

Wood I-beams as roof rafters. This application of I-beams has special hardware requirements. At the bottom end of a sloping rafter, *variable pitch connectors* (Figure 7) provide bearing and anchorage to the wall; the pitch is determined by the elevation at which the connector is nailed to the plate.

At the ridge, consider a *light-sloped U connector* (Figure 8). At about $3, they are less expensive than pitched seat hangers and can be adjusted to the slope on the job site; some can be site-adjusted for skew as well, if that is needed.

Most of these connectors require a web filler with wood I-beams, and you may also need cross-bridging to prevent rotation. Metal bridging is available in longer lengths like 30 inches, 36 inches, 42 inches, and 56 inches for this application.

Roof trusses. Hanger manufacturers are literally unveiling new alternatives each month, and it's now possible to purchase plated truss hangers with 10,000-pound capacities from stock.

At the low end, a 16-gauge, *facemount hanger* can handle less than 1,000-pound reactions but do it for less than a dollar.

Newer style *facemount hangers with wide flanges* of 14-gauge steel that take several rows of nails per flange will get you to the 2,000-pound reaction level at a cost of $3. Often a *strap hanger* at $2 can handle about

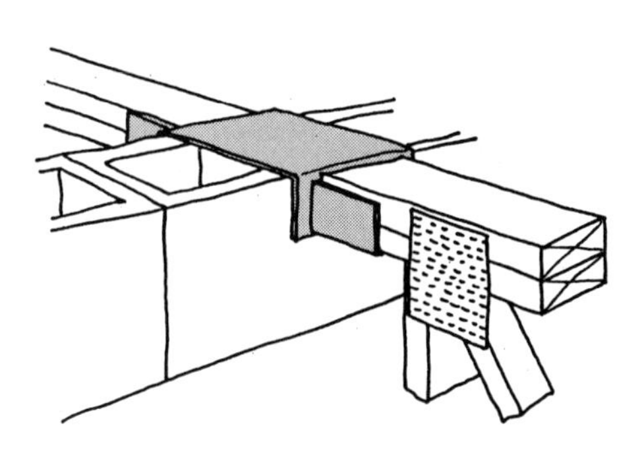

Figure 5. *Top-chord-bearing hanger.*

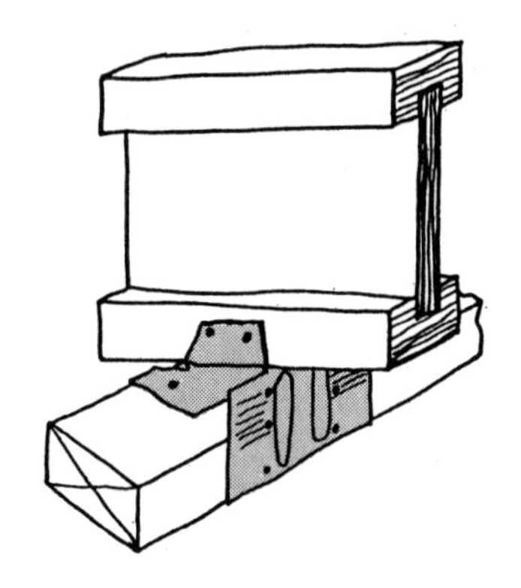

Figure 7. *Variable-pitch connector.*

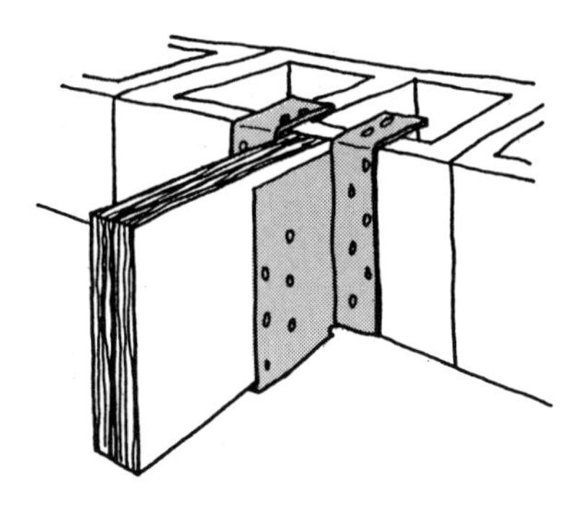

Figure 6. *Deep-seated topmount hanger.*

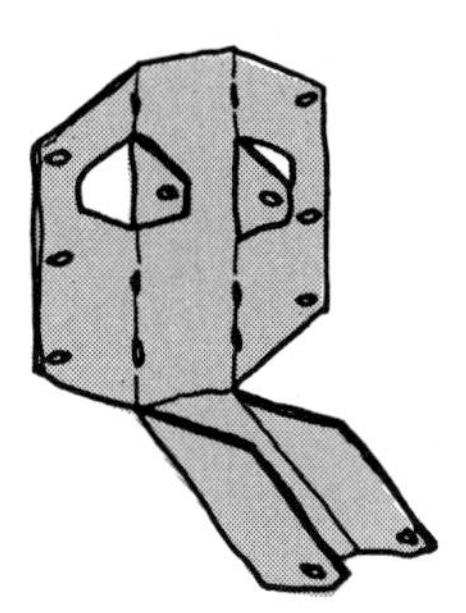

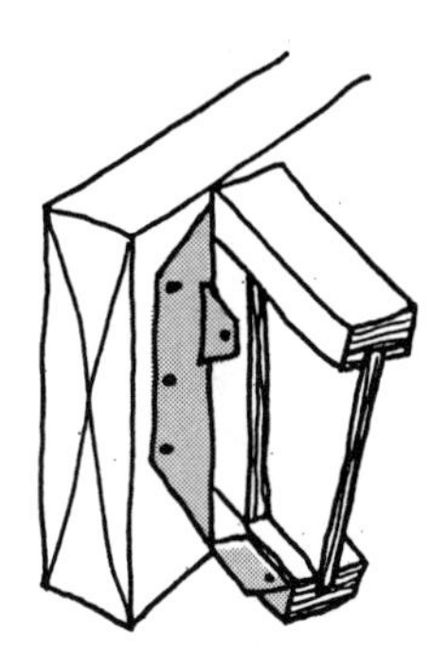

Figure 8. *Light-sloped U connector.*

the same capacity.

Above this range, you'll be choosing between a *bolt-on girder hanger* (Figure 9) or a welded style with top flanges that straddle the webs. The bolt-on hanger can cover 3,000 to 10,000 pounds depending upon the number of bolts. Expect to pay from $30 to $90 each for these. Be aware of the size of the web required if the bolting face has a double row of bolts.

Welded-style top-flange truss hangers (Figure 10) range in capacity from 3,000 to 5,000 pounds and in price from $15 to $30. The lower the price, the less metal in the hanger and the more restricted you are in application. For instance, some welded hangers with top flanges can straddle only a 2x4 vertical web. A larger web or any diagonal webs precludes its use.

For angled connections, there are several alternatives beginning with the adjustable connectors mentioned under wood I-beam rafters. Most

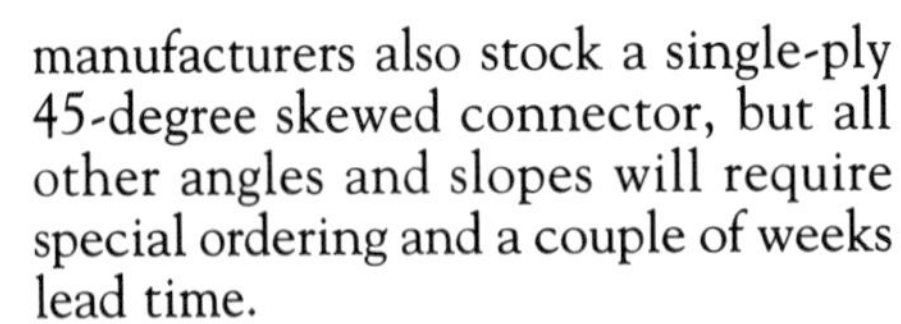

manufacturers also stock a single-ply 45-degree skewed connector, but all other angles and slopes will require special ordering and a couple of weeks lead time.

Hips and jacks. A step-down hip roof system with a corner or hip truss hanging from a girder truss is a good

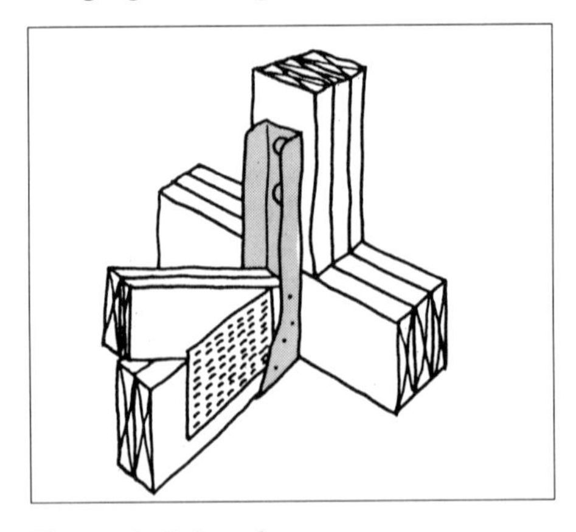

Figure 9. *Bolt-on hanger.*

hanger candidate. This is one of the connections a building inspector is likely to look at first.

Again, the choice of connector depends on the numbers, and the longer the hip member the greater the end reaction. But one hanger can take care of both the hip and the end jack, since they hang at the same panel point. Some hangers have individual seats for hip and jack, while others have a common pocket that eliminates the need for left and right hangers. Some styles depend on bolting while others lip over the bottom chord of the girder truss.

For short setbacks of 7 inches to 11 feet, a 2,000-pound *facemount hip and jack hanger* (Figure 11) will run about $10.

For end reactions up to 3,300 pounds with single-ply hips and 5,500 pounds with double-ply hips, a *heavy-duty hip and jack hanger* (Figure 12) that lips over the bottom chord of the

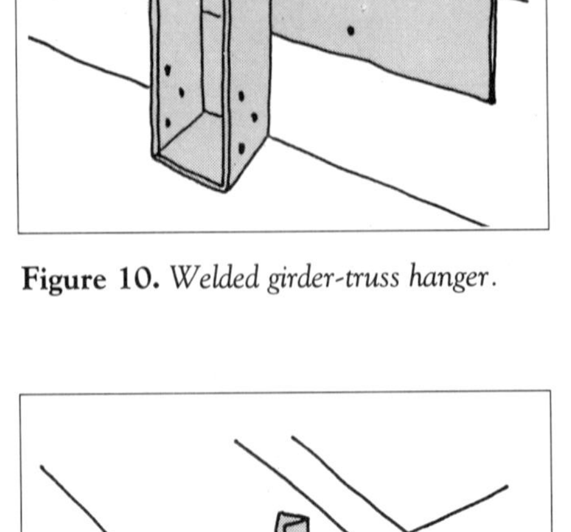

Figure 10. *Welded girder-truss hanger.*

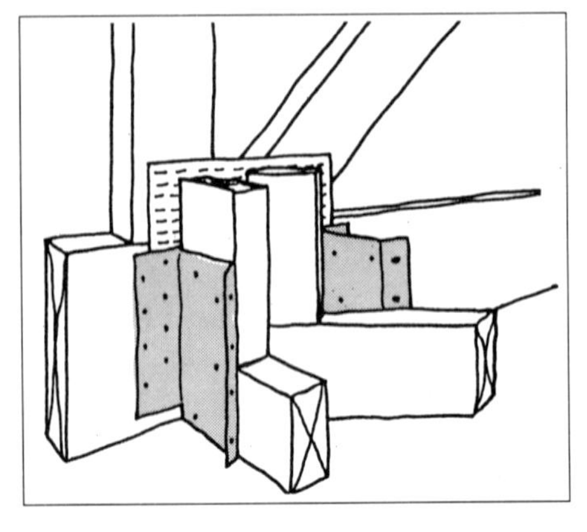

Figure 11. *Facemount hip and jack hanger.*

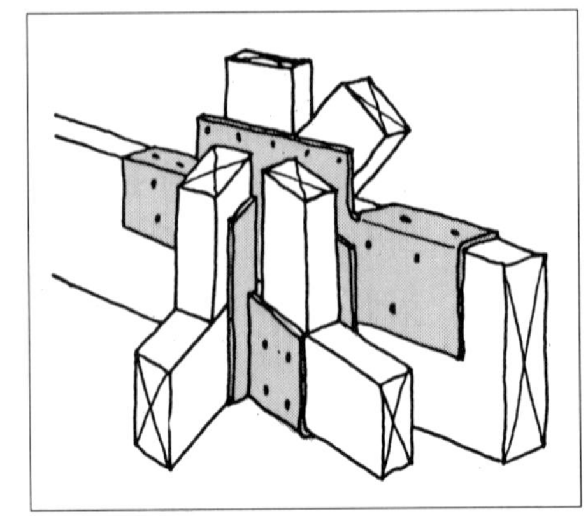

Figure 12. *Welded hip and jack hammer.*

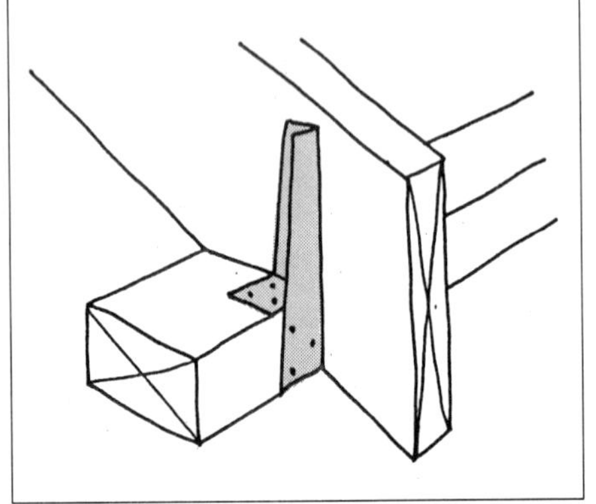

Figure 13. *Truss to single top plate anchor.*

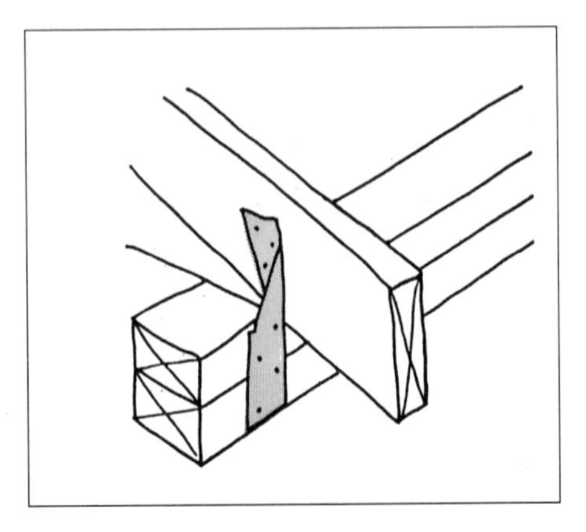

Figure 14. *Truss to double top plate anchor.*

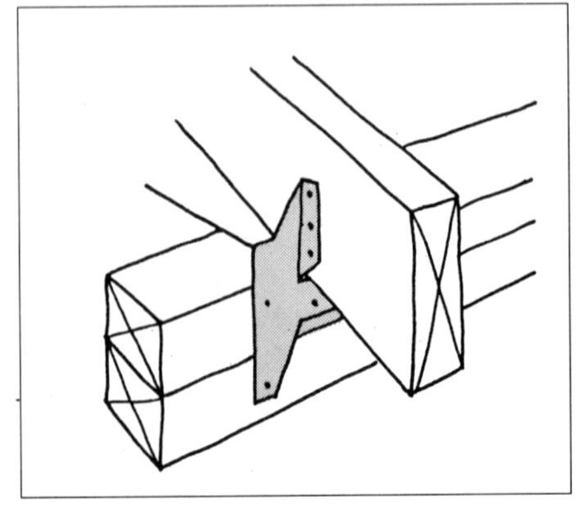

Figure 15. *Straddle-type truss anchor.*

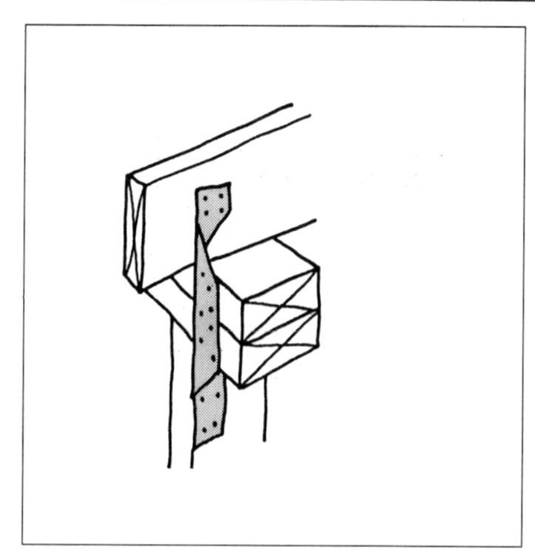

Figure 16. *Extended-truss anchor.*

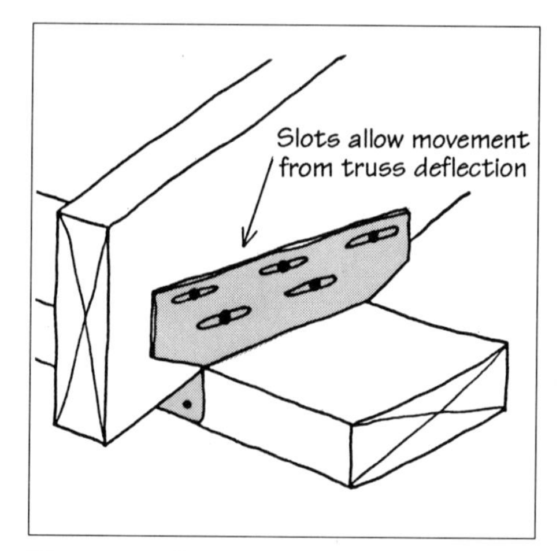

Figure 17. *Scissors-truss connector.*

girder truss at $40 to $50 works best.

King-sized skewed hangers that bolt on to webs can handle three-ply hips at 7,500 pounds at a price tag of $100 each. They are similar to bolt-on girder hangers as in Figure 9, only their seats are skewed.

Truss anchors. Many builders particularly on the coasts, are in the habit of anchoring roof trusses to their bearing plates with hardware, and there is a growing demand among building inspectors for these connectors.

Anchors that attach to *single plates* (Figure 13) and *double plates* (Figure 14) only cost about 20¢ each. Their uplift rating depends upon the number of nails into the plate and into the truss. Generally, three nails are worth a little over 300 pounds and four nails about 400 pounds, but check the manufacturer's rating since the numbers vary widely.

For uplift requirements of 500 pounds, look for a five-nail anchor with a pair of flanges that *straddle* the truss (Figure 15). At 30¢ each, they beat using a pair of the garden variety anchors.

Extended-truss anchors (Figure 16) tie the truss to a stud. They also have a 300- to 400-pound rating, except they transfer the uplift force directly to the stud and bypass the plate.

Anchoring a scissors truss firmly to a wall may cause the wall to buckle because, by their nature, scissors trusses deflect under load conditions. They push outward and then return when the load is decreased. A *scissors-truss connector* (see Figure 17 and lead photo) can help reduce the possibility of wall buckle. This connector has slotted holes that permit the truss to move outward as it deflects while still counteracting uplift forces.

The base of the connector reduces the friction between truss and plate which, in turn, reduces the lateral force a wall must exert before static friction is overcome and the truss slides. Available for 4-inch, 6-inch, and 8-inch nominal plates, these scissors-truss connectors cost about $1.50 on the average. ■

Bill Loeffler has been the manager for 20 years at Cleveland Steel Specialty Co., which has manufactured hangers for 65 years. He is responsible for engineering, testing, tooling, and drawing up new products and special applications.

Sources of Supply

Cleveland Steel Specialty Co.
14430 South Industrial Avenue
Cleveland, OH 44137
800/251-8351;
800/686-8351 (in Ohio)

Lumberlok
1029 Whipple Road
Hayward, CA 94544
800/221-7905;
800/221-7906 (in California)

Simpson Strong-Tie Co., Inc.
1450 Doolittle Drive
San Leandro, CA 94577
415/562-7775

Teco Products Co.
12401 Middlebrook Road
Germantown, MD 20874
800/438-8326

United Steel Products Co.
703 Rogers Drive
Montgomery, MN 56069
800/328-5934;
800/642-4762(in Minnesota)

Section 9. **Panelized Systems**

Gary Mayk

Building With Structural Foam Panels

Strength, speed, and energy efficiency make foam panels a close competitor to wood frame construction

by Steve Andrews

Builders who use structural foam panels cite a number of advantages: They say panels are stronger, more energy efficient, go up faster, and offer higher, more consistent quality than stick-built houses. Enough homeowners and contractors have been attracted to foam panels to bring regular annual increases in sales of about 25 percent to the industry, even during the recent slow times.

Yet the vast majority of houses are still stick-built, in part because builders remain skeptical about foam panels. Is this skepticism healthy? Or are those who steer clear of structural stress-skin panels missing a smart building option?

After researching and writing about structural stress-skin panels for several years, I've found no single answer to this question. For some builders, panels are a great boon; for others, they're too limiting or not worth the extra expense. To decide for yourself, you need to consider your receptiveness to new materials, the types of houses you build, and the ease with which your crew can build stick-built systems that compare in energy performance to panels.

Sandwich Panels vs. Unfaced Panels

There are two types of foam panels designed to carry residential and light-commercial structural loads: sandwich panels, and unfaced panels with integral wood or steel framing members. (This article does not cover cladding or curtainwall foam panels, which are routinely applied over post-and-beam framing.)

Sandwich panels. Structural sandwich panels account for most of the market. They usually consist of two wood facings—typically 7/16-inch OSB—adhered to foam that is 3 1/2- to 11 1/4-inches thick. Sometimes the inside is faced with wallboard, either alone or over the OSB.

This sandwich panel functions as an I-beam, with the facings serving as flanges and the foam as web. Despite the inherent strength of this system, roughly one in four manufacturers add integral studs, usually at panel

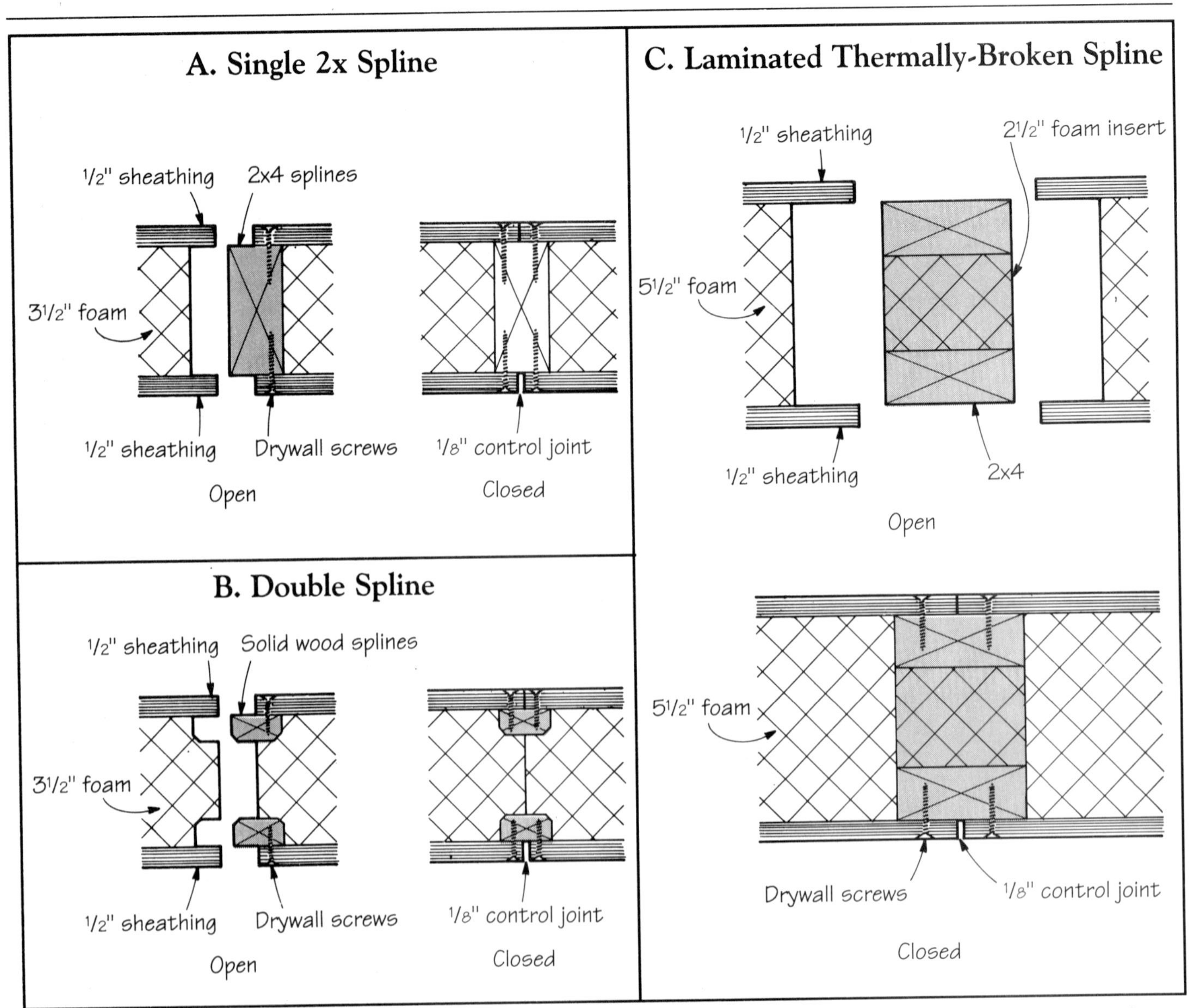

Panel joints. *Manufacturers have developed roughly a dozen ways to connect panels, but three are most common. The single 2x is the simplest and cheapest (A), but slightly compromises insulation value. Thermally-broken double splines (B and C) add a little cost up front, but save energy down the line. All are easy to work with.*

joints; this approach is often taken to avoid battles with local code officials wary of the panels' lack of framing members. There are probably between 75 and 100 sandwich panel manufacturers in the U.S., and most produce about 30 homes a year.

Unfaced panels. Unfaced structural panels consist of slabs of precut foam notched to accommodate load-bearing metal or wood studs at 16- or 24-inch centers. Walls and ceilings are delivered to the job site in partially or fully assembled sections. Tied together with wood top and bottom plates, the wall components are either tilted by hand or craned into place. The largest producers of unfaced panels are Canadian, though there are at least two U.S. manufacturers.

The Advantages

Builders and buyers cite four main advantages of structural stress-skin panels.

Speed. Sandwich panels replace three steps in standard construction: framing, sheathing, and insulation. Pop in the windows and you're warm and dry for the next trade.

The actual pace of construction will vary with panel type and size and house design. In general, the simpler the design—few ins and outs, and most or all dimensions in 4-foot increments—the faster it goes up. The more indents and popouts, the more delays.

With complex designs, you can regain some speed advantage if you order precut panels. These are more expensive, but save on-site cutting time. They work well as long as the folks in the factory don't mess up or the client doesn't want to change something between order date and delivery.

In general, the larger and fewer the panels, the faster the house goes up. Large unfaced wall panels are light enough to erect quickly even

Heave ho! Placing large roof panels or two-story verticals usually requires a crane (left). But a strong crew can handle some large panels, like the gable end, above, without a crane.

without a crane. The largest sandwich panels, however, require a crane, which, depending on the house, you might have to rent for one to three days.

In a fairly simple plan, using large panels—some as large as 8x28 feet—can speed things enough to justify the cost of the crane. Yet with large custom homes that have a number of offsets, you may not have enough long wall surfaces to justify the use of a crane. In that case, you'd be better off building even the long runs with pieces small enough for your crew to handle.

A house composed of smaller sandwich panels, typically 4x8 feet but sometimes longer, can be erected by hand by crews of two or three in three to ten days. With a somewhat larger crew, you can put up considerably larger panels by hand. Depending on labor costs, this can sometimes be the most cost-effective route.

Superior energy performance. Panels improve little upon well-built super-insulated, stick-framed homes. However, panels clearly outperform the energy systems used by most production builders today. In the few legitimate side-by-side comparisons I know of, panel homes have had significantly lower heating and cooling bills than stick-built houses of similar cost.

It's not that stick-built houses can't be as well-insulated and tight as a stress-skin house; it's just harder to make them that way, for several reasons. For starters, foam panels contain their own gap-free insulation, at high levels. Foams used in panels provide between 25 and 100 percent higher R-value per inch than fiberglass does. Fiberglass batts give between R-3 and R-3.86 per inch. Molded expanded polystyrene (EPS) gives R-4 per inch, extruded polystyrene R-5, and urethane or polyisocyanurate both supply R-6.

Foam panels also have fewer seams to seal to reduce infiltration. They do require careful caulking, however, mainly along the seams above and below band joists, and along the tops of walls where one floor meets another.

Finally, most panel systems use very little dimensional lumber, reducing conductive heat loss through framing and infiltration around it.

Strength. When tested to failure, panels exceed minimum standards by huge margins, outperforming frame walls by factors of between two and six, depending on the test. This is an underappreciated advantage, until a natural disaster strikes. There are several documented cases where panel homes have survived hurricanes, tornadoes, and even the 1989 San Francisco earthquake, in good shape, while their stick-framed immediate neighbors have suffered serious damage.

Costs

When comparing panel house-shell packages against standard frame construction costs, builders should make sure they price systems with comparable energy features. For example, a panel with a 5½-inch EPS core of 3⅝-inch urethane should be compared to 2x6 framing with 1-inch rigid foam sheathing (R-23 or better).

When comparing different panel systems, you must again make sure you're comparing apples to apples. On the surface, the simplest approach would appear to be pricing panels by the square foot. Listed prices for 5½-inch EPS panels from 16 manufacturers average $2.17 per square foot, ranging from $2.03 to $2.69. But some manufacturers may include certain items as standard—from shop drawings to an on-site factory rep—while others charge extra for them.

For example, the highest minimum price quote above ($2.69) is for panels and splines precut in the factory (gables, plumb cuts for roof panels, and window cut-outs are already made), while the low one ($2.03) is for blanks on which the builder would have to make these cuts at the site.

To level the playing field, try to price panels in one of three ways: uncut blanks, with whatever items are standard; pre-cut panels, with standard items; and erected shell packages per square foot. A final word of warning: Not every manufacturer offers precut panels, and only a few offer the option of erecting the shell.

Placing An Order

A manufacturer will usually take a few days to compile a bid. You'll help yourself on costs if you make certain that all straight wall runs and all ins and outs break down into 4-foot increments.

One other way to save money is to buy panel walls, but use a trussed or conventionally framed roof system. Denver-area builder Jim Lambert (Lambert and Son) tried this approach on a 1,200-square-foot home to save the cost of high-R-value roof panels. (Blown cellulose is a lot cheaper than foam.) The downside, as suggested earlier, is that you have to seal extremely carefully to

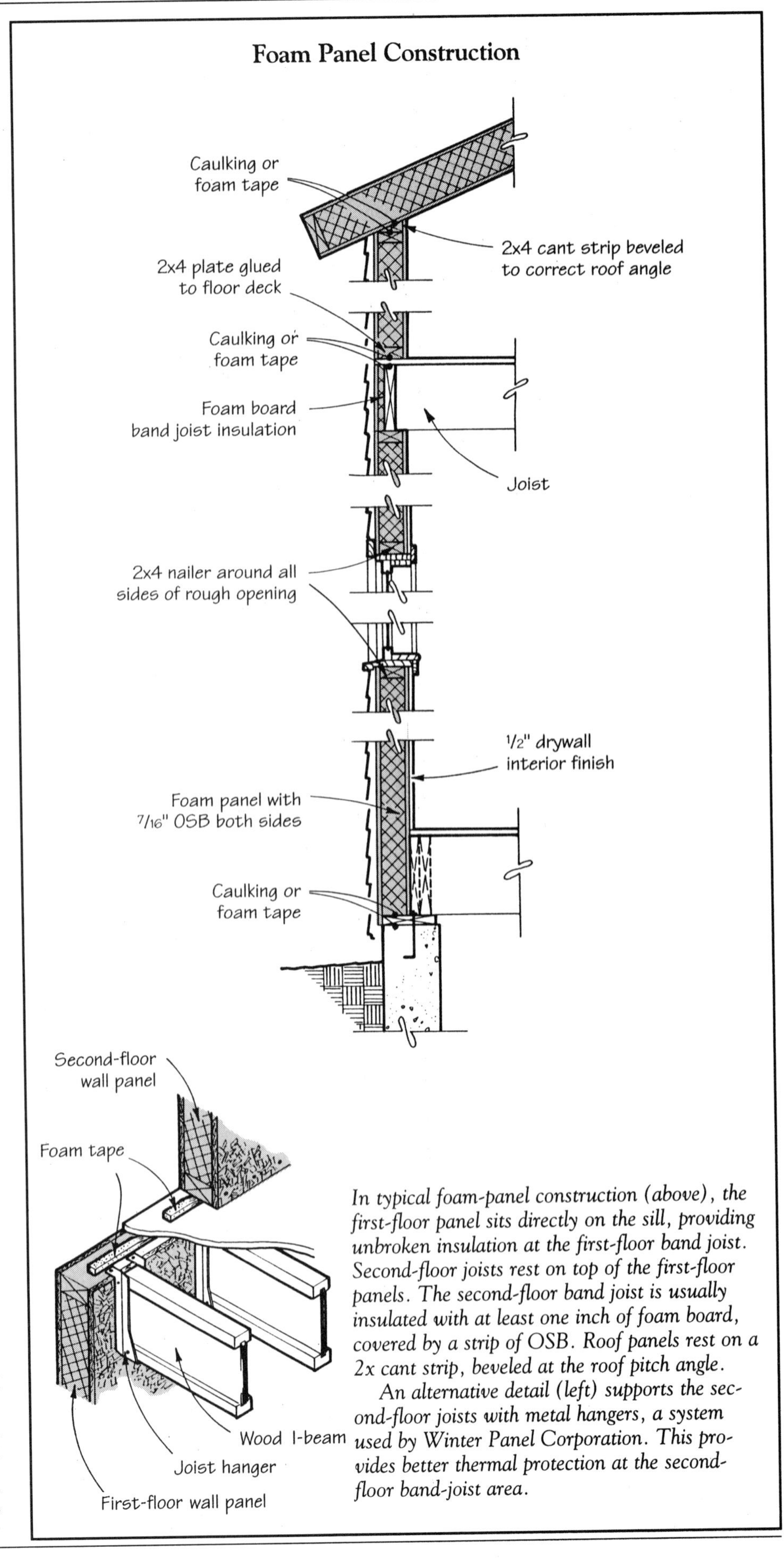

In typical foam-panel construction (above), the first-floor panel sits directly on the sill, providing unbroken insulation at the first-floor band joist. Second-floor joists rest on top of the first-floor panels. The second-floor band joist is usually insulated with at least one inch of foam board, covered by a strip of OSB. Roof panels rest on a 2x cant strip, beveled at the roof pitch angle.

An alternative detail (left) supports the second-floor joists with metal hangers, a system used by Winter Panel Corporation. This provides better thermal protection at the second-floor band-joist area.

Foam Panel Tools and Techniques

Using panels for the first time generally requires a few adjustments in the field.

Tool requirements. Working with panels requires a few special tools, such as one or more oversized circular saws, either 12-inch or 16-inch, to cut through the thick panels. You'll also need a small chainsaw to cut window and other openings if you don't have the factory do it. You'll need several caulk guns and plenty of caulk, so you can seal as you build. You'll also need lots of metal connectors, as specified (and possibly supplied) by the panel maker. And you'll want mechanical nailers or staplers, since there is a lot of nailing. The electrician may want a 4-foot drill bit for last-minute electrical changes. Finally, don't forget a sledge to snug up panels as they go in.

Putting the pieces together. Manufacturers have developed about a dozen different ways of attaching panels. Most sandwich systems use either double thin splines, a thermally broken stud, or a single 2x stud (see illustration, page 190). All work well and install easily. The stud splines drop energy performance slightly, but certainly not enough to be the deciding factor between systems.

To prevent air movement and create a monolithic wall, every panel attachment system requires use of adhesives and sealants. This takes enough time that when two people are putting up 4x8-foot panels by hand, a third can be applying the adhesives (wood-to-wood connections) and sealants (wood- or foam-to-foam) along the edges of the bottom plate as well as on both sides of the splines where panels meet. This is one more place where using large panels saves labor.

Electrical and plumbing. To make wiring easier, most panel manufacturers offer some type of internal wiring chase. This is typically a 1-inch core running horizontally 12 to 16 inches above the floor, with a vertical core to the floor, either at the center or at one end of each panel. (Long panels have these cores about every 4 feet.) To provide access to these cores, most builders predrill 1-inch holes through the bottom plate and floor deck, at spots corresponding to the vertical cores, before the panel is placed down. Unused cores can be sealed with a bit of expansive foam.

Once the wall is up, you snake cable through the hole and up the vertical chase. When building on a slab, you have to run some wire as the panels go up, working from the bottom up. This can get awkward at the corner; it may be easier to use a baseboard raceway.

Plumbing supplies are best kept out of exterior walls, especially in cold climates. If you simply can't avoid running a vent up an outside wall, order a 2x2-inch chase cut in the foam at the factory. Then mark your plans to make sure you don't forget the special panel. — S.A.

Cutting. *You'll need an oversized saw to cut any panel. For those too thick for the saw, use a small chainsaw so you don't have to flip every panel for a second cut.*

Notching. *To let in nailers for the window or door frame, you'll have to notch the foam. This is easiest using the hotwire tool shown here, supplied by most manufacturers.*

Wiring. *Most wall panels come with predrilled chases for electrical wiring. Before setting a panel in place, drill a 1-inch hole through the subfloor to align with the vertical chase in the panel. Fishing wire is then fairly simple.*

Stick Builders Size Up Panels

Contractors who try panel systems don't necessarily swear off stick building forever. But even long-time stick builders have found panels to be a great help on the right jobs.

Don Moody, a contractor since 1956, recently used panels from APC International for his retirement home on an island in Puget Sound. The home, a 2,700-square-foot two-story, went up in a tight spot, with no crane on site to help lift the 8x28-foot panels. The walls had 5½-inch EPS cores, with floors and ceilings 7¼ inches thick.

When asked what he liked best about panels, Moody cited ease of construction. "We built the shell for about $11 per square foot of floor area, and that included factory precutting on a lot of ins and outs on walls, and complex roof intersections. It took one carpenter, my wife, and me nine days to put up all the walls and ceilings. On a normal site, where you could use a crane, it would have gone up in four days."

Over on the mainland, Mike Roberts (Cascade Energy Homes), a Bellingham custom builder for 22 years, put up his first panel home in 1990. Roberts said, "The system worked fine. There were a few problems stemming from the shop drawings—we had to fix some pre-cut panels—but that was because it was our first home with panels."

Roberts has built quite a few well-insulated, airtight homes, so he was interested in tracking costs to make a comparison. He calculated that putting panels in the walls and floor of a 1,900-square-foot house cost an extra $3,000. It took his electrician about an extra hour to wire the house ("there was some messing around pulling wires through chases at the corners"), which he figures was fine for the first panel house. He had no problems with code officials.

Denver builder Gene Johnson (Johnson Homes), who has twice used panels for homes in the $200,000 to $800,000 price range, finds that panels compete well with stick framing unless a design has a lot of short offsets. "If you make minor design changes in a home to accommodate the panel, then panels are really easy to work with. It's a little slow for the first home, but after you figure out the system, it works really well. And it's a quality upgrade, because it's warmer and tighter."

Sal Vuocolo, a home builder and distributor of Heartstone timber frames in Glen Gardener, N.J., has worked with different types of housing packages for years. He feels panels offer an antidote to the falling quality of conventional framing material. That and the energy efficiency advantage impress him enough to recommend panels to clients. But he finds many reluctant to pay for energy upgrades, especially in a market where there is tremendous price cutting. He has also had trouble with skeptical building inspectors.

"Whenever you challenge the established order in the building industry, you have a real struggle," said Vuocolo. "If you have a structural panel product that doesn't have a BOCA code number, you have to work one-on-one to get code approval. It's tedious. There's simply no way to cover the cost of all the time involved."

In central New York state, Roy Hoysradt (Hoysradt Homes) recently finished his third home using sandwich panels from Winter Panel Corp. He was asked to bid on a two-story against modular construction. When the panel bid came in a little higher, he asked the client whether the modular bid included comparable insulation. It didn't, and at equivalent insulation levels the modular couldn't compete, so Hoysradt got the job.

Unlike Vuocolo, Hoysradt says he didn't have any problems with code officials. "They're kind of baffled by the system, but they accepted stamped drawings with no problem."

— S.A.

match the airtightness of the paneled system. And of course it takes longer.

Once contracted for, panels are usually ready within two to four weeks, or later, as your schedule requires. Payment schedules vary and are sometimes negotiable. Generally, however, you'll have to put up about 10 percent payment at order, 25 percent upon acceptance of a set of shop drawings, and the balance on delivery.

Lingering Questions Department

A few questions about panels persist.

Durability. How long will a panel home last, compared to a stick-built home? Though panel homes haven't been around long enough to provide a definitive answer, results from accelerated aging and field tests are quite positive. Four panels installed in 1968 in a U.S. Forest Products Lab (FPL) test house were removed a decade later and tested to failure; all exceeded the design load requirement (20 pounds per square foot) by at least a factor of ten.

When panels have failed in the field, it has usually been because quality control was lacking or the panels were used to carry loads they weren't meant to. In a few cases, panels made with the wrong adhesives failed.

If you're thinking about using panels for the first time, visit plant sites for the manufacturers you're considering and compare their quality control programs. Manufacturers who pay an independent firm to oversee quality control will charge more for their products (about 10¢ to 20¢ per square foot), but it may be worth the extra cost. You should also check warranties: Some manufacturers now offer a warranty against delamination. These vary from 15 years (Sunlight Homes) to "the life of the structure" (Insulspan).

Performance in fire. Although unsheathed foams burn readily, panel assemblies actually perform quite well in fires. The Associated Foam Manufacturers subjected EPS panels covered with ⅝-inch drywall and OSB to the standard ASTM 20-minute "full-wall" test (ASTM E-119 series), in which the panel is placed in an oven

For More Information

For information on panels or a list of suppliers, contact:

Structural Insulated
Panel Association
1090 Vermont Ave., NW
Suite 1200
Washington, DC 20005
202/408-5003

Selected panel manufacturers:

Sandwich panels
Alchem, Inc.
3617 Strawberry Rd.
Anchorage, AK 99502-7111
907/243-2177

APC International
2280 Grandview Rd.
Ferndale, WA 98248
206/366-3400

Branch River Foam Plastics, Inc.
15 Thurbers Blvd.
Smithfield, RI 02917
401/232-0270

Atlas Industries
6 Willows Rd.
Ayer, MA 01432
800/343-1437

Enercept, Inc.
3100 Ninth Ave. SE
Watertown, SD 57201
605/882-2222

Foam Laminates of Vermont
P.O. Box 100
Hinesburg, VT 05461
802/453-4438

Harmony Exchange
Rt. 2, Box 843
Boone, NC 28607
704/264-2314

Insul-Kor, Inc.
P.O. Box 116
Elkhart, IN 46514
219/262-3472

Insulspan Panels
(4 plants in Midwest)
c/o Foam Products Corp.
P.O. Box 2217
Maryland Heights, MO 63043
800/824-2211

J-Deck Building Systems
2587 Harrison Rd.
Columbus, OH 43204
614/274-7755

Korwall Industries
326 North Bowen Rd.
Arlington, TX 76012
817/277-6741

Modular Energy Systems
311 East Glen Cove
Mesa, AZ 85201
602/898-7283

Pond Hill Homes
RD 3, Box 467
Blairsville, PA 15717
412/459-5404

R-Control Panels
(18 plants nationwide)
c/o Associated Foam
Manufacturers
Box 246
Excelsior, MN 55331
612/474-0809

Remarc, Inc.
P.O. Box 174
Holderness, NH 03245
603/968-9678

Sunlight Homes
P.O. Box 1569
Bernalillo, NM 87004
800/327-5835

Vermont Stresskin Panels
RR2, Box 2794
Cambridge, VT 05444
802/644-8885

W.H. Porter
4240 North 136th Ave.
Holland, MI 49424
616/399-1963

Winter Panel Corp.
RR 5, Box 168B
Brattleboro, VT 05301
802/254-3435

Unfaced panels
Insul-Wall Ltd.
11 Mosher Dr.
Dartmouth, NS
Canada B3B 1L8
902/468-5470

NASCOR, Inc.
7803P 35th St., S.E.
Calgary, AB T2C 1V3
Canada
403/279-1966

RADVA Corp.
Drawer 2900
Radford, VA 24143
703/639-9091

Wallframe of
Southern California
11180 Penrose
Sun Valley, CA 91352
818/768-8100

and exposed to 1,450°F flames over its entire surface for 20 minutes.

When the fire was terminated, the testers stripped away the drywall and OSB and examined the panel. The OSB was still bonded so well to the foam that it had to be ripped away with a prybar. And the panels suffered no core melting, no apparent structural damage, and no deflection, despite being loaded with 21,600 pounds to simulate a three-story building.

EPS panels faced with OSB and drywall have also fared well in an Underwriters Laboratory test called the Room Corner Fire Test, in which a 1,300 to 1,600°F fire is built in the corner of an 8x12-foot room. Fire performance of other foam types is similar.

These tests were done on panels faced on both sides with OSB and on the interior side with drywall. In a test on panels that had only drywall on the inside, the drywall adhesive failed at 215°F, causing the drywall to delaminate. This exposed the foam to the fire, and the wall failed in less than 20 minutes. Unfortunately, this test wasn't as controlled as the others, so it's impossible to say whether this poor performance was due solely to the lack of OSB facing or to other factors as well. But it appears that sandwich panels with OSB beneath the drywall fare better in fires.

A frequent criticism of urethane foam is that in a fire it will release deadly hydrogen cyanide. (Other foams emit mainly carbon monoxide, just as burning wood does.) This is true, but must be balanced against the fact that many other household materials—both the building and its contents—release deadly gases when burning. Before a fire could penetrate to the foam in a urethane panel, it is likely to have consumed a lot of furniture, carpeting and drapes, all of which will be a source of deadly carbon monoxide, not to mention cyanide and other gases.

Chlorofluorocarbons. Many foam panels use extruded polystyrene, polyurethane, or polyisocyanurate foams, which have traditionally been produced with ozone-destroying CFCs and HCFCs. (EPS, or expanded polystyrene, foams, are not produced with CFCs.) In response to concern over the ozone layer, foam manufacturers are phasing out the use of these blowing agents, which will probably be completely replaced with ozone-safe substitutes in the next three years.

For the time being, however, most non-EPS foam products do use some ozone-depleting CFCs or HCFCs. HCFCs destroy only about 5 percent as much ozone as CFCs do, but they still deplete the ozone level and contribute to greenhouse gases. Until all CFCs and HCFCs are eliminated, builders and clients must make their own decisions about whether gaining foam panels' advantages (energy efficiency being one) is worth losing the ozone. It won't hurt in the meantime to press the industry for more rapid conversion to non-CFC products.

Ants and termites. Panels are as susceptible to carpenter ant and termite infestations as any other wooden structure. These beasties get nourishment from wood, not foam, but their tunneling in the foam can reduce insulation value, drive homeowners crazy (nibble, nibble in the night), and eventually compromise structural integrity. Wherever they reside, you should treat to prevent infestations.

After extensive research efforts, several manufacturers, lead by Remarc and AFM, are treating their panels with insect repellent products. Product testing indicates that manufacturers need to perform a fair amount of research to assure the success of their recipe.

The alternative to treating panels is to cut back on adjacent or overhanging vegetation and treat the ground around the structure, as you would with a stick-built house.

Shingle durability. A light-colored asphalt shingle atop a panel roof system will run up to 20°F hotter than it would over a vented attic. Shingle warranties notwithstanding, this heat will shorten the shingle's lifetime. Case in point: During a routine post-hailstorm roof inspection of a two-year-old roof, an insurance claims adjustor noted that some shingles were much more brittle than others. It turned out that the brittle shingles were above panels, while the others were over a conventional attic.

One moderately expensive solution is to make a cold roof by laying down 1x3-inch furring strips and a layer of 1/2-inch plywood. In this case you should probably provide a continuous ridge vent and screen the low edges to prevent bug residency in the 3/4-inch airspace. The added bonus: Your roof will run cooler in the summer.

The other option is to simply use a conventionally framed or trussed roof, as mentioned before. This will mean more work sealing against air infiltration, but it will save a few bucks and be easier on the shingles.

Visible roof joints. Where panels join together in roofs, you can sometimes see small ridge lines in the shingles. Frank Baker, with Midwest Panel Systems and president of the Structural Insulated Panel Association (formed in 1990), explains that OSB comes out of the manufacturing process extremely dry. If it is installed while still dry, it may expand or creep as it absorbs moisture, causing a bulge where two panels meet. To avoid this, either let the panels acclimate before installation, or leave a 1/4-inch gap along all joints. This can be sealed with an expanding foam sealant. Some panel makers actually leave the OSB slightly short of the panel edges to allow for a little expansion. Baker is hoping that use of OSB delivered at a higher moisture content (8 percent) might eliminate most of the problem. Ask the manufacturers what they would recommend in your climate. ■

Steve Andrews is a Denver-based residential energy consultant and the author of Foam Panels and Building Systems: Principles and Practice *(Cutter Information Corp., Arlington, Mass.).*

To Stick-Build or Panelize?

Panelization can ease management burdens and sometimes save money

by Joseph Carusone

For years, most builders have relied to some extent on premanufactured building components. Who doesn't use prehung doors or manufactured trusses? What builders assemble and glaze their own windows?

Until recently, only large production builders found it economical to have the entire shell built in a factory. But that is changing. Over the past few years, we have seen a surge in demand for panelized housing components in the Northeast.

A shrinking skilled-labor pool and a more competitive marketplace are making various types of prefab building more appealing to smaller builders. The options range from preassembled wall sections, customized to any design, to complete modular buildings delivered to the site—with appliances and wallpaper.

Whether a rural jack-of-all-trades or a suburban "briefcase builder," builders may find definite advantages in using panelized components. Either type of builder should ask the same question: "Can panelized housing improve my performance and my profit, and can it make my business life more manageable?"

Variety of Services

The decision to use panelized housing is really a matter of degree—to what extent should your company rely on premanufactured components?

Several types of component companies exist in the Northeast. Some produce components in conjunction with building-material suppliers, and some operate as separate businesses. The most commonly available components are, of course, floor and roof trusses, but many companies also offer wall parts—such as headers with jacks—or sectional wall panels. Wall panels are available with or without sheathing, and are typically custom-manufactured for each house. A few companies will build the walls in one piece up to about 40 feet long, with siding and insulation applied.

Another type of company offers not only the components, but complete building-material packages as well. These manufacturers, such as our own company, typically provide design and engineering services, along with marketing support and brand-name recognition.

The material packages contain panelized components such as wall sections, precut components such as joists and rafters, and the necessary siding, roofing, windows, and other accessories to complete the shell.

Finally, some companies offer modular housing, in which almost all of the building, including interior finish work, is done in the factory. The house is shipped to the site in two sections (or four, for a two-story), and clamped together.

Panelizers in the Northeast

The following is a list of companies in the Northeast and Mid-Atlantic states that panelize custom home designs. Most also offer stock designs.

A complete directory of housing and component manufacturers, the Red Book of Housing Manufacturers, *is available from LSI Systems, Inc. 11-A Village Green, Crofton, MD 21114, 301/721-3838. It cost $60 per region, or $140 for the U.S. edition.*

Acorn Structures
P.O. Box 1445
Concord, MA 01742
508/369-4111

American Standard Bldg. Systems
P.O. Box 4908
Martinsville, VA 24115
703/638-3991

Coastal Structures, Inc.
P.O. Box 2646
Gorham, ME 04038
800/341-0300

Hilton Lifetime Homes
P.O. Box 69
Leola, PA 17540
717/656-4181

K-K Homes, Inc.
Sunburst Homes,
Div. of K-K Home Mart
420 Curran Hwy.
North Adams, MA 01247
413/663-3765

Luxury Homes, Inc.
P.O. Box 277
Bomoseen, VT 05732
802/468-5045

New England Homes
P.O. Box 1138
270 Ocean Rd.
Greenland, NH 03834
603/436-8830

Nickerson Homes
Main St.
Orleans, MA 02653
508/255-0200

Northern Counties Lumber, Inc.
P.O. Box 97
Upperville, VA 22176
703/592-3232

Northern Homes
51 Glenwood Ave.
Glens Falls, NY 12801
518/798-6007

Ward Corp.
1300 Piccard Dr.
Rockville, MD 20850
301/948-0333

Assessing Your Needs

You should analyze several factors before deciding to purchase components:

- How available is labor in your area? What about the quality and cost?
- What volume of business do you realistically anticipate? Do you want to do more?
- Are you able to produce or easily obtain plans and engineering to meet local and state building codes?
- Do you need help with marketing? Do you need more prospects? A sales edge?
- What image do you wish to project?

The potential benefits from using manufactured components or complete house packages are many. Most important, a small builder can build more houses with the same crew, or the same number of houses with a *smaller* crew by using prebuilt and precut components.

Typically, all door and window openings are already framed in. This allows a builder with a three- or four-person crew to erect the shell of a 1,500-square-foot house in a single day, once the foundation is in. The house can be weathertight—with roofing, windows, and doors in place—in a work week.

Faster assembly reduces interest costs on construction financing. Quicker building also reduces exposure to weather and theft on the building site.

Good quality work can be done with less experienced help than in stick-built construction. All that's really required is one experienced carpenter—whether it's the contractor or a foreman—plus a small crew of helpers and laborers. The homes can be small and plain or large and complex.

The quality of the work is equal to that of the best builders. Because all components are built and assembled in jigs, the cuts are precise and the assemblies are square. Materials are dry, and are assembled with corrugated fasteners that hold better than nails. Individual components, such as precut joists, are preselected, so the builder does not have to reject any pieces.

The more sophisticated companies provide the small builder with fixed prices and guarantee that the materials delivered will complete the job—that you won't, for example, run out of roofing shingles. These guarantees make it much easier for the builder to bid quickly and accurately. Some companies also help the builder estimate the quantities of other materials needed to complete the house.

The larger panelizers—those offering complete house packages—can also be a big help with code compliance. Home manufacturers have the engineering and architectural staff to stay abreast of state and local codes, and to negotiate with code officials when conflicts arise. They can provide complete sets of plans stamped by engineers, along with energy audits and the required detailing.

In addition, larger companies can offer the builder marketing aids, including sales catalogues, signs, customer leads, and design services. The better companies also offer technical backup, and, if required, arbitration. Our company, for example, recently provided free arbitration to resolve a payment problem on a job that, due to a conflict, was never completed. The contractor was paid, and no one went to court.

Finally, the small builder should consider the advantages of associating with a brand-name manufacturer. Many customers are more comfortable purchasing a home from a name they recognize.

Potential Risks

Buying components or an entire packaged home does not get the contractor off the hook, because the panel manufacturer cannot replace carpentry skills. Builders are still responsible for the quality of the finished product. Even precision-cut panels and products can turn into a shoddy house in the hands of unskilled or careless tradesmen.

If you want to eliminate all skilled labor on site, modular housing might be the way to go. But that limits you to simple stock design, and leaves you with almost no control over the end product.

With some panel manufacturers, you will be limited to the stick plans they offer. Others are flexible. Builders who specialize in custom homes must look for suppliers willing and able to panelize custom designs.

How Economical?

Labor rates determine the savings a panel manufacturer can offer. If builders can find inexpensive skilled labor locally, and are buying components from an area with high labor costs, then the savings may be nonexistent.

But if they are buying from a supplier who has low labor costs, and if their own labor pool is small and expensive, the savings can be considerable. Labor rates tend to be higher in and around metropolitan areas. If an urban contractor can find a rural panelizer, he or she stands to realize the greatest savings.

The builder should also consider site accessibility. Can the supplier's truck get close to the foundation? Will the company stage deliveries to suit the contractor's schedule—or will the load be dropped off all at once?

Some panelizers use cranes, while others expect the building crew to unload the materials. Eight-to 12-foot wall panels can be handled by a couple of workers, but for large beams, full-length walls, or other large assemblies, the builder may want to rent a crane for a day.

In 1985, nearly one-third of all new housing starts used some panelized components. Small builders everywhere are learning to use the services of panel manufacturers to boost their profits and to ease the management burden.

The trend is clear: As skilled labor becomes more scarce, codes more complex, and markets more competitive, panelized components will gain wider acceptance. ■

Joseph Carusone is vice president of Northern Homes, Glens Falls, N.Y., which sells panelized custom homes.

Building Walls in the Shop

Building finished wall panels in the shop can streamline stickbuilding

by Nick Hurt

In a converted barn, walls are framed and sided flat on the bed of a house trailer. In the photo, one carpenter cuts a full bundle of cedar clapboards with a 15-inch chop saw, while another installs the siding over foam sheathing and house wrap.

A few years ago I took a hard look at how we put up houses, and thought there has to be a safer, more efficient way to build.

Consider conventional practice: After lifting up a framed and sheathed wall, and building the roof, someone sets up staging and moves it all around the house for running soffit and fascia. Then he moves it all the way around the house again for siding. Then a painter moves it all the way around to stain. Every day is spent trudging over the soft, uneven terrain of a freshly excavated site. Every day countless manhours are spent covering material, kicking aside scraps, chasing blowing trash across the neighbor's lawn, carrying tools and material up and down ladders, setting up new work stations, running to the lumberyard for another pound of nails, and securing the site. Then there's the weather: Walking icy planks or working under the drip line has never been my idea of fun. And every time a carpenter digs out a wet pile of lumber from under a snow drift, he never fails to complain that the lumber's no good.

While I wanted to get away from these hassles, I still wanted a method that wouldn't be too different from stickbuilding, because that's what I know. So I set out to eliminate my troubles one at a time.

To get away from using scaffolding, I started building completed wall panels on site, including siding and soffits, and lifting them up with wall jacks. I built three houses this way, but we got caught by the rain before the siding had been stained. As a result, the siding got soaked, and many of the clapboards curled or split at the butt joints.

This inspired us to eliminate the weather problems by moving parts of the process indoors. Having joined with a partner on a 15-lot development site, I leased a nearby barn to use as a "panel shop."

Wall System

We have completed seven houses in the shop and have a smooth system down.

We can start building the walls at the same time the foundation hole is dug. Just before the walls are completed, we send two guys to the site to frame and deck the floor.

The walls are framed just as they would be on site, except in the shop we work on the bed of an old 8-foot-wide house trailer. Because the trailer is only 8 feet wide, I don't need a special permit to travel on state roads. The walls are built in the reverse order they will go up, and stacked on the trailer with pieces of 2-inch rigid foam between each wall section to protect the siding, trim, and windows.

It takes two to three carpenters one day to complete a wall. The walls are constructed with 2x4 studs, 7/16-inch OSB sheathing, 1-inch rigid foam, and housewrap. We leave the housewrap long on three sides to lap the rim joist and the corners.

We used to use let-in metal bracing, without wood sheathing, so the walls would be light enough to lift by hand, but we never felt very good about the final product. Wood sheathing provides better resistance to racking, allows us to square the walls precisely, gives us something to kneel on when siding, and gives us a

Step 1: *Three carpenters can lift a 34-foot 2x4-framed wall without structural sheathing. Lumber strapping is nailed to the bottom plate and to the deck to keep the base of the wall from kicking out as it's raised into place.*

Expandable Floorplan

A popular house in this affordable range is a 34x26 Cape with a shed dormer. Most of our houses sell to families who have one small child and plans for more. We leave the upstairs unfinished to become bedrooms later. (The upstairs can be called an attic, if FHA financing is involved). We also build a full basement with a flush girder to extend the headroom, and an interior frame wall insulated with fiberglass, which can be finished at a later date.

To make the transition upstairs or down, no alterations need to be made to the first floor layout: The entrance opens into the living room. There is a combined kitchen/dining room at the back, and two bedrooms at each end. One bedroom opens onto the livingroom, and can be used as a future family room. The other downstairs bedroom can be converted to a formal dining room after the upstairs is completed.

34'
26'
Bath
Kitchen
Bedroom (future Family Room)
Bedroom (future Dining Room)
Living Room
Entry

We keep the systems simple—poured concrete foundations, direct bearing framing, 2x10 floor joists, 11/12 pitch roof, 150-amp service. To satisfy energy requirements, the house has R-30 ceilings, R-18 walls, low-e thermal-pane windows, and a Heatmaker boiler for heat and hot water. — *N.H.*

▲

Step 3: *A carpenter scabs the top plates to lap at the corners, while another straightens and nails the bottom plate to the deck.*

Step 2: *After lifting a wall into place, the crew plumbs and braces the corners.*

substrate on which to install exterior lighting and trim.

Before installing the windows, we snap layout lines for the siding on the housewrap, and then install the soffit and fascia using a 2x4 ladder for lookouts. We use a rafter tail pattern and a framing square to position the ladder and snap a line. The soffit, vent, and 5/4 clear cedar soffit go in very smoothly, compared with working over your head on scaffolding. The soffits come out dead straight. There's never a need to reassure yourself by asking how things look from the ground.

While the walls are flat, we install the windows, cedar trim, and cedar siding nailed with stainless-steel, ring-shank nails. We leave the corner boards off until the walls are erected, but we screw on a temporary straightedge to butt the siding to.

An enormous amount of labor is saved by siding the walls flat. We used to need two people on staging to hold up long clapboards, and one on the ground cutting. Inevitably, when a piece was miscut, it would have to be passed back down to the cut man, while the two guys waited. A single clapboard could easily take three people five minutes to install. This kind of time adds up quickly.

In the shop, we can square the walls precisely so the corner boards and casing are perfectly parallel to each other. As a result, each piece of siding in a run is identical. With a 15-inch chop saw, one guy can cut a bundle of clapboards at once, while one guy installs.

As a final step, we stain the siding and trim. Working in the barn, we don't have dust blowing around, and we don't worry about rain. Because the walls are flat, we don't get drips. With a sprayer on hand, staining becomes one more part of a natural progression that gets done at the end of each day before going home.

Advantages of a Shop

Working in the shop, we always work on level ground. We don't spend much time breaking down workstations or setting them up. We can store materials easily, and keep track of our scraps. As a result we generate very little waste.

We rip down scrap OSB and foam sheathing on the table saw, and take these to the site in a bundle, to sheath the rim joists. On a two-story house, we will reach over the edge of the second floor deck before the walls go up, and bring up a few courses of clapboards. Then all we have to add later is one course of clapboard, which we can do quickly from a ladder. We also chop up the foam scraps to fit between the joists to further insulate the rim joist. We use scrap lumber for blocking, cripples, and soffit lookouts. The only pieces we don't use are under 8 inches.

I can now buy building materials in larger quantities since I have a place to store them without rehandling them. This saves considerably. I not only get a better price, but I get material that hasn't been pawed through at the yard.

When to Use a Crane

At the site, the walls can be stood up by hand, with wall jacks, or with a crane. Without sheathing, 2x4 walls are light. A 34-foot wall can be rolled across the deck on 4-inch PVC drain pipe, and lifted with three people. It takes at least three hours to stand up the walls. With labor rates between $12.50 and $15 per hour for a carpenter, the base cost to lift the walls by hand is between $112 and $135.

With sheathing, however, the walls are heavy, and we have switched over to using a crane. Fortunately, the sheathing provides enough support that the walls can be picked up by the top plates.

With a crane, raising four walls off a trailer takes us one hour. I can hire a 100-foot tree crane for $100 to complete the job. I still need one guy to work with the crane operator, however, and I keep the other two busy until we need to tie the walls together and frame the roof. So I actually spend somewhat more money with the crane. But we're ready to frame the roof much sooner, the job runs a lot smoother, and I end up with a better product because of the structural sheathing. We also have the crane lift ceiling joists and rafters onto the top plates.

Recently, we've started building the gable-end walls in the shop, and lifting these into place with the crane, too. We have to build the gable frame in one section and then cut it in 9-foot pieces to fit on the trailer. Working on the flat allows us to complete the rake eaves in place. On one project, we installed decorative shingles with minimal effort. This took a little prefiguring to get everything to line up, but went very smoothly.

From this point on, we frame the roof in conventional fashion. I've considered using trusses, but I'd rather frame the roof myself when I can. I'm a stickbuilder by trade, I'm not interested in just assembling panels and trusses.

After building several houses with rigid foam and let-in bracing, the author went back to using structural sheathing. To lift these heavier walls, a crane is needed. Here, a crane lifts a gable-end wall into place. The walls were built in 9-foot sections to fit on the trailer.

Affordable Design

Panelizing walls doesn't work well on complex house designs, so I've stuck to a simple design. Building houses this way, I've been able to break into a more affordable housing market. This has worked out well since current house sales in New England are very slow, while there continues to be a need for "affordable" homes in the $90,000 to $120,000 range (see "Expandable Floorplan," page 202).

This house design sells. I'm now producing and selling houses in a slow economy faster than I did in the boom years. I've had almost no requests for anything different, so I can streamline production in the shop. But each house is still built one at a time, so I can offer my clients quality and flexibility. Even in the "affordable" range, the housing market still favors the custom builder. ■

Nick Hurt is a builder in Milton, Vt.

Buying Walls From The Factory

Panelizers promise premium shells, salable designs, and few hassles. Can they deliver?

by Gary Mayk

Factory-made panels for the second floor of this home in Maiden Creek, Pa., come right off the truck with the help of a scissors lift—and a willing crew.

Schedule for Framing Crew

Monday:

7:00 a.m.	Frame the first floor
10:30 a.m.	Frame the second-floor deck
2:00 p.m	Frame the second floor
5:00 p.m.	Go home

Tuesday:

7:00 a.m	Erect roof trusses
1:00 p.m	Sheathe roof

Wednesday:

7:00 a.m.	Complete interior framing: kitchen soffits, etc.
1:00 p.m.	Start on second

What sounds like an ambitious schedule for any four-man framing crew becomes routine with one simple ingredient: factory-made wall panels. With these panels—already squared and sheathed—and with factory-made roof trusses, Lot No. 187 in Maiden Creek Estates in Maiden Creek, Pa., will go from a completed foundation to a completely framed 2,200-square-foot house in 2½ days.

Project foreman Hank Mengel rolls his eyes when asked if he'll ever go back to stick-building. "Oh, no," he says. "Never." It's no wonder. At Maiden Creek Estates, crews are framing a house in half the time of stick-building crews down the street. In a competitive area like this, speed is important. So are quality and cost containment, and Mengel says the factory-made panels give him an edge in several ways:

- A truer product. Factory-squared panels combine with laser-leveled foundations to almost guarantee plumb walls.
- Uniform openings. Windows and doors always fit, and openings are almost always in the precise location.
- Dry materials. Panels and materials don't sit in the yard or on site, exposed to the elements, for days or even weeks on end.
- Better materials. Panel manufacturer H.M. Staufer and Sons of Leola, Pa., culls out wood that doesn't grade. Usable cut-offs go into panels as cripples below windows or in T-braces, not as wall studs.

A Small Builder Takes A Big Step With Panels

Bricktown, N.J., custom builder John Carnesi trades successfully under a name that seems bigger than his company: Atlantic Cedar Development Corp. Typically, Carnesi has one or two people on his payroll. He subs out most of his work, and to keep his overhead, paperwork, and headaches down, he buys panelized homes.

Carnesi is a builder/dealer for Hilton Lifetime Homes of Leola, Pa. For his franchise and initial advertising material, he paid $3,000. When he sold his first house, the fee was refunded. With some customers, Carnesi negotiates directly. With others, he uses Hilton's staff to negotiate contents and price. In all cases, the customer pays him for his services, and in most cases, the customer pays Hilton directly for materials. As a builder/dealer, he earns a commission on the materials customers buy. He also will earn a commission on materials for Hilton homes bought and erected by other builders in his area, although so far the idea has yet to catch on.

After buying the home packages—from the basic framing to complete package with kitchen and bath— he says he won't return to stick building except when it's demanded. Here's what he had to say about panelized construction.

Completeness. "I've been a builder for ten years. I've been building panelized homes for about four years; I've done quite a few. Hilton can manage anything from a ranch to a 50-unit condo development. I'm just doing single family. The biggest I've done is a 4,000-square-foot, two-story contemporary. Hilton can supply everything—tubs, ceramic tile, cabinets for the kitchen and bath. But buyers don't have to take the whole package."

Ease of assembly. "Even a do-it-yourselfer who's half a carpenter can do it. It's as easy as ABC. We frame a house in about four days.

"With the trim package, we save installation time. Trim sets are already mitered and assembled. They come to us picture framed—that's how we nail them up."

Quality. "These panels are perfect. I know with a 2,000-square-foot floor plan, if these frames are out of square 1/8 of an inch, it's a lot. The openings—they're right on the money.

"These are stronger and better. We can have panels built from any material we want. They sheathe ours in 1/2-inch fir or CDX fir, and then 1/2-inch Ultra-R on top of that. I can get the same R-value in 3 1/2 inches as most builders are getting in 2x6 studs. I'm getting R-20, and actually most of them are only getting R-19 with 2x6s."

Buyer resistance. "The only problem is that people confuse it with modular, and it's not. At first, when you talk about panelized, they say they are not interested in it. They think the windows come installed, and the plumbing and electrical are in. But the panels just come sheathed."

Customer-factory contract. "It cuts my paperwork. People can get a 2 percent discount for paying cash for materials. They can get a 1 percent discount for paying cash in two stages.

"As a builder/dealer, I get a commission for all the materials the customer buys. It equals the discount I would get at the lumberyard."

Returning to stick-built. "No way. But if buyers say they still want a stick-built, we'll build it for them."—G.M.

- Lower overhead. Maiden Creek developer Jim Saunders is building 70 units a year with a skeleton crew. The time needed from subcontracting crews is also reduced because the panels are only erected, not constructed, on site.
- Less waste. "We save $500 a house on waste disposal," Mengel says. A building lot with panelized construction has a fraction of the scrap generated by on-site framing.
- Easy customizing. The panel factory's design department uses computers to add custom features to otherwise stock plans. Computerized stress analysis assures that even unusual vaulted and cathedral designs will stand up.

What If I'm a Small Builder?

Like many panelizers, Stauffer primarily serves big developments due to the economy of scale. Its newly acquired division, Hilton Lifetime Homes, however, is good to the small guys. The small builder gets the same service and same technology from the same facilities. "We have an advisory program to walk the inexperienced builder through our process," says Paul Sheak, general manager of Hilton Lifetime Homes.

In buying Hilton, Stauffer was acknowledging a dilemma facing builders in the Pennsylvania market and nationwide: The pool of skilled construction workers was shrinking, and the workers who remained became more expensive to hire. This was especially true in the cyclical home-building business.

The solution: Sell as complete a product as possible, using lower-cost factory labor in place of higher-priced field labor. The key was to maintain—or improve quality. Now, in effect, Hilton sells smaller builders not just materials, but labor, too.

Can You Build This for Me?

Builders often mail in nothing more than a basic floor plan and a picture from a magazine. "Can you build this?" they'll ask. "How much?" At no charge, Hilton supplies a "not-more-than" cost. The factory's fees start with design. For $850, Hilton will spec out the design from even the crudest of drawings.

Builders can buy just the framing and truss-roof system; a complete package that includes bath, kitchen, and interior finish trim; or any combination. If the house is ordered in any of several levels, the $850 fee is refunded at delivery.

The assembly line: *Interior and exterior walls can be either nailed by hand with pneumatics or squared and nailed on a Morgan nailer (top). After sheathing is laid on, a router is used to cut window and door openings (middle), and a "gang nailer" attaches the sheathing with staples (bottom).*

"Hilton's niche," Sheak says, "is the single-family builder who's inexperienced or needs a lot of technical help or the builder who's done it all and says he's not interested in the little incidentals of ordering and managing loads. The minimum we sell is a shell package of rough-framed wall panels and a roof system. We don't do foundations, rough wiring or plumbing, Sheetrock, or interior finish such as paint, carpet, or sheet goods. Everything else, we are able to supply."

Builders' preferences on materials vary by region, so Sheak supplies according to a builder's specs. Materials will vary even on the standard Hilton designs, depending on where the house is going.

Hilton's contract is designed to reduce builder paperwork. "We contract with the customer," Sheak explains, even if the builder comes directly to Hilton to request a package. Hilton will even negotiate with the customer on appearance and content. Payments for materials are made by the customer to Hilton on a cash-in-advance, cash-on-delivery, or bank-assignment basis. Buyers can earn cash discounts. Buyers also must pay the builder for work performed. "That helps the builder's cash flow," Sheak says. (See "A Small Builder Takes a Big Step With Panels," previous page.)

Step No. 1: Computer-Aided Design

Hilton works out the design in the Stauffer design center. The computer-aided design includes stress analysis of all components—something a small builder might not otherwise be able to afford. "Then we send the plans to inspectors," Sheak says, to make sure they meet code and get a state seal. The staff deals with code officials from Massachusetts to Virginia. He says panel manufacturers in other areas also get to know what's expected by regional and even local officials.

Jim Estakhrian of the design staff explains that for panels to fit, quality control is paramount. "By the time we get a panel out the door, about ten people are involved, from the architect to the construction people. What you see on the architectural drawing may not be the final product." The computer may say the original design has a weakness. "We will make recommendations, and if they want to follow them, fine; if not, that's their decision."

At the site: *The framing crew places panels as close as possible to their final location (top), then raises and nails them in place (bottom).*

Sheak says, "Even on prints that we don't originate, we take the responsibility of verifying basic bearing conditions, such as on the headers."

For the roof system, a computer program comes up with truss shape and quantity. Trusses can be standard models or custom designs.

Step No. 2: The Cutting Room

"We do not cut, drill, or screw a piece of lumber unless it's sold," Sheak says. "We want to fabricate before delivery. That means a nice, fresh product. It turns gray if left to sit in the yard. Code inspectors don't want to see that."

Cutting orders are grouped by project. The orders are then moved to production of trusses in the Leola plant, or into panels at the Myerstown plant, about an hour to the north. Radial arm saws, over-sized tables, and jigs assure fast and accurate cuts. Sorting of off-grade lumber takes place here—something a stick builder has to do on site.

Paperwork is grouped by floor and then into interior and exterior walls. A pull list tells cutting room employees which lumber to cut. The cut list is drawn to match the pull list. Bottom plate drawings show the location of studs, cripples, jacks, and T-4 braces (2x4 construction) or T-6 braces (2x6 construction) where walls will intersect. For clarity, some panels get a drawing with full-face view. The factory notes any odd-sized panels or members.

A small plywood label identifies each bundle after cutting and before staging for the interior and exterior assembly lines. A letter code allows quick identification in the factory of interior and exterior wall bundles.

Step No. 3: The Assembly Line

The plant has two exterior lines, an interior line, and a sub-assembly section for door and window openings. On the less mechanized line, workers on each side of a roller-equipped table line up studs, cripples, jacks, T-4s, and pre-fabricated openings in a jig, according to markings. They tighten the jig and nail with hand-held pneumatics.

On the other exterior line, a Morgan frame nailer enables one man to build one wall in one minute. The labeled parts are laid overhead, where they're easily read and reached. Studs are placed between top and bottom plates, then hydraulically grasped and nailed. The press of a button advances the panel to the proper spacing—say, a standard 16 inches—until the nailers line up with the stud markings on the plates. The process is repeated until all studs are nailed.

Next comes a crucial point in panelization: the squaring table. A jig makes sure the panel is square and grasps it. A metal diagonal T-brace is nailed to maintain squareness. Sheathing of the builder's choice goes on top, with corner panels and front panels getting waferboard at the builder's request. Other panels typically get foam board. Some panels get a corner of waferboard when openings prevent the use of a metal brace for stability.

Hand-held pneumatic nailers tack the sheathing, including over openings. Workers rout out openings in far less time than they would need to cut sheathing to size, as site builders do. A pneumatic gang nailer delivers 10,000 pounds of pressure to complete the stapling.

Interior panel assembly is done in jigs with hand-held pneumatics, much as on the less mechanized of the two exterior lines. Interior panels get bracing only at the request of the builder. All interior panels are finished open-faced.

Completed panels are grouped and labeled according to the assembly plan. Panels remain in the yard until shipment.

Step No. 4: At the Site

Panels arrive by pre-arranged schedule. When the truck arrives, the framing crew should have the first-floor deck in place, and lines should be struck for aligning bottom plates. Panels can be removed from the truck in bundles by forklift, or crews can unload panels separately, Because they're no longer than 8 feet, most panels can be handled comfortably by two framers.

A floor plan identifies each panel by dimension and number. Crews lay the panels around the perimeter as close as possible to assembly position. After tilting up each panel, crews align it and nail. Interior panels are raised and nailed last. The framing crew must also nail on a second top plate. One floor takes about 3 1/2 hours for a house of 2,000 square feet.

With the first floor complete, carpenters frame the second-floor deck. This lumber is usually bought from the panel maker. Some builders prefer traditional dimensional lumber, while others order manufactured joists. The manufacturer delivers tongue-and-groove subflooring as ordered.

When second-floor panels arrive, typically in the afternoon, crews have enough time to unload and erect them before quitting time. A truck delivers the second-floor panels by scissors lift. Panels are arranged as first-floor panels were, and they are erected in the same order.

Roof trusses and sheathing arrive the next day for installation in the traditional manner. All that remains is finish framing: kitchen valances, for instance, or special framing to accommodate tubs.

The home—2 1/2 days after setting the sill—is now ready for roofing and mechanical subs. In addition, it is square and true. The process deviates little from conventional stick-building, but the benefits can be worthwhile. ■

Gary Mayk is a former editor for The Journal of Light Construction.

Section 10. **Pole & Timber**

Photos by Sarah Knock

Hybrid Timber Frames

It's fine to mix timber and stud, but you must understand the structural systems to make them work together

by Steven Chappell

Over the years, I've been approached by many individuals who thought it would be a great idea to have exposed timbers in one or two rooms in their new house. For one reason or another they did not want a complete timber-framed home. Their reasons for wanting a partial or "hybrid" timber frame are usually sensible. But the results I've seen in the field are often less so. Some of the hybrid frames I've seen rely far too much on hocus pocus for my peace of mind. This does not have to be the case. Integrating some of the finer points of timber framing into otherwise conventionally built custom homes can be accomplished very easily and economically if one takes a *whole-system* approach to building. Although we have yet to build a project, I feel there is a place for it in the custom-building market.

Whole Systems

Before analyzing the various forms that a so-called hybrid frame may take, it will be helpful to look at the two building forms that are being combined. By a hybrid frame we mean one in which the general principles and methods of traditional joined-timber frames are combined with conventional platform or stud-frame construction. It is important to keep in mind that the structural principles inherent in each system must be present in the hybrid offspring if the combined frame is to be structurally sound. Just as conventional framing and timber framing are whole systems, so must the hybrid frame be when complete. Magic just doesn't work. To accomplish this successfully, it is essential to have a working

This timber-framed living room was almost completely enclosed by stick framing in the completed house. Many builders and clients like the timber effect, but see no need to be a purist about it.

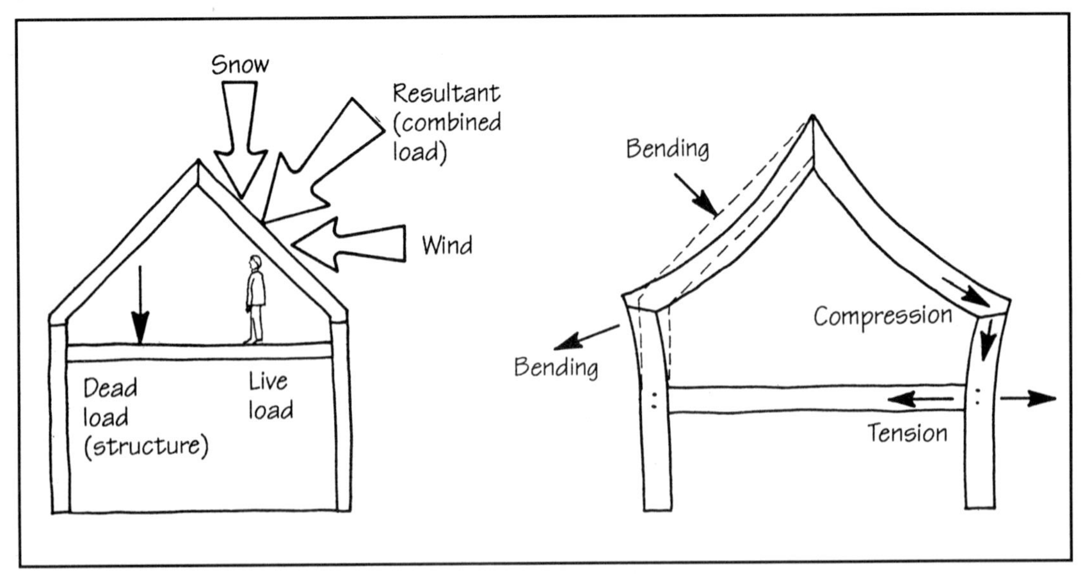

Figure 1. *In timber frames, the primary loads (far left), are concentrated on a few key members (left). The great demands on each timber and joint leave small margins for error compared with conventional framing.*

knowledge of each system's basic engineering.

The Structural Frame

Once they left their caves, it became clear very quickly to our ancient ancestors that a rigid structural frame was the essential ingredient of a sturdy shelter. The frame, therefore, is a necessary evil that we must have in order to hang and support the sheathing, roofing, floors, walls, and windows. It must support all of its own weight, as well as the additional weight that will be loaded upon and in it, both now and in the future.

If this sounds a little too simplistic, let me say that the most horrendous cases of bad building I have ever seen were the results of misunderstanding the simplest concepts of structure. It's fairly common for non-timber builders to expect more out of timbers than is physically possible. Perhaps after building with 2x4s all day, a stick builder views a section of 6x8 as a much greater timber than it really is.

Despite obvious differences in appearance, both stud frames and joined-timber frames serve one and the same function. Each is a skeleton over which the enclosure materials are attached. The differences lie in the ways the imposed loads and external forces are distributed throughout the frame. Whether the fastening method in a frame is nails or pegs does have an effect. However, for practical purposes we can assume for now that all connections are fixed and solid.

The primary forces acting upon any frame (see Figure 1) are:

- Dead load—the weight of the structure itself.
- Live load—the forces acting on the frame as a result of its intended use, that is, occupants and furnishings.
- Snow load—based on the maximum accumulated snow that can be expected to fall and remain on the roof.
- Wind load—the forces exerted on walls and roof by the wind (either a positive pushing force or a negative lifting).
- Resultant or combined load—due to combined wind and snow loads.

As these loads are imposed on the frame, each member is subjected to a number of forces. Primarily each member is subjected to one of two distinct loading stresses—compression or tension. Often, as in the case of a horizontal carrying beam, a timber is in both compression and tension simultaneously. In the example of the carrying beam, the wood fibers on the top half of the timber are under compression loading, while the fibers on the bottom are subjected to tension loading.

Of course, not every member nor every connection has to be dissected and analyzed in every frame that you build. A good example for this is a conventional stud frame. Most experienced framers learn quickly to recognize proper construction methods. There are standards used throughout the industry that remain pretty much the same from coast to coast (ask your local building inspector for a copy of your local code and suggestions as to other reference materials). It is likely that many stick framers go through their careers without ever having to acknowledge the thrust of a rafter or the compression on a stud. However, a prerequisite to timber framing is to gain a general working knowledge of these facts. In some ways, it becomes even more important when working on hybrid frames.

Distribution of Loading

In a conventional frame, the primary goal is to distribute the imposed loads so evenly throughout the frame that each member ends up carrying only a minute portion of the total load. The demands on each member and each connection are so dramatically reduced that removal or destruction of one member or connection has little or no effect on the stability of the frame as a whole. In contrast, a timber frame is designed to concentrate loads on a few primary members, significantly increasing the demand on each member. What this means for timber framing is that the margins for error commonly accepted in conventional framing cannot be used. Fewer structural members and greater forces call for much stricter tolerances in workmanship. Dead loads also become a bigger issue. Whereas the actual dead weight of each common rafter in a conventional roof is insignificant, the weight of the wood becomes an important factor when building with timbers.

Timber-Framed Walls and Ceiling

A common form of hybrid frame is one in which only the first-floor walls and ceiling are timber-framed. On top of this platform, trusses are placed or a site-built roof is framed, which will allow second-floor living space. In both cases, the rafters are placed conventionally at 16 or 24 inches on-center. In this design the posts are all too often 6x6s spaced at 4 to 12 feet on-center with a 6x6 top plate. It seems common in this design for the builders to overlook two very important concerns: (1) the outward thrust of the rafters and (2) the distribution of loading on the top plate by the roof rafters.

As previously mentioned, joined-timber frames are a whole system. The design of the *bent* is nothing more than a very heavy truss. The same principles used to design conventional 2x4 trusses are employed. The tie or bottom chord prevents outward thrust of the rafters. The queen posts and struts prevent rafter sag and simultaneously stiffen the tie beam. If one or more elements of the truss were to be left out, there would be cause for great concern.

In a traditional four-bent principle-rafter-and-purlin frame (see "Parts of a Frame," page 218), the bents are commonly spaced 8 to 16 feet apart. Because purlins span between bents, the roof places outward thrust on the top plate. The roof load is transferred across the purlins, in equal proportion, to each bent. In turn, the sum of this load is transferred down along each rafter. If not secured at their feet, the rafters would give way to the accumulated thrust (horizontal outward force) and collapse. Understanding this possibility beforehand, the timber framer can easily design the bent to resist the load comfortably. In this type of traditional design, all of the outward thrust remains within the bent. All of the roof load is directed vertically onto the posts.

Let's imagine that our four-bent frame was 28x36-feet with the bents spaced at 12 feet on-center. Let's also assume that the maximum total roof load for this house is 45,000 pounds. Now let's see what happens if we project these figures into our hybrid design.

If we were to cut off the top portion of the frame leaving only the tie beams and posts, and then build a conventional rafter roof on top, what will we see happening? Two very important changes have occurred. First we have completely altered the way in which the forces are distributed to the first-floor wall framing. Instead of concentrated loading on the few posts, there is now an evenly distributed load along the length of the top plate of approximately 625 pounds per foot or 6,250 pounds along each 10-foot span between posts (see Figure 2).

Second, the outward thrust of the rafters is now an evenly distributed force along the top plate. No matter how comforting the term "evenly distributed" sounds, this is not a desired

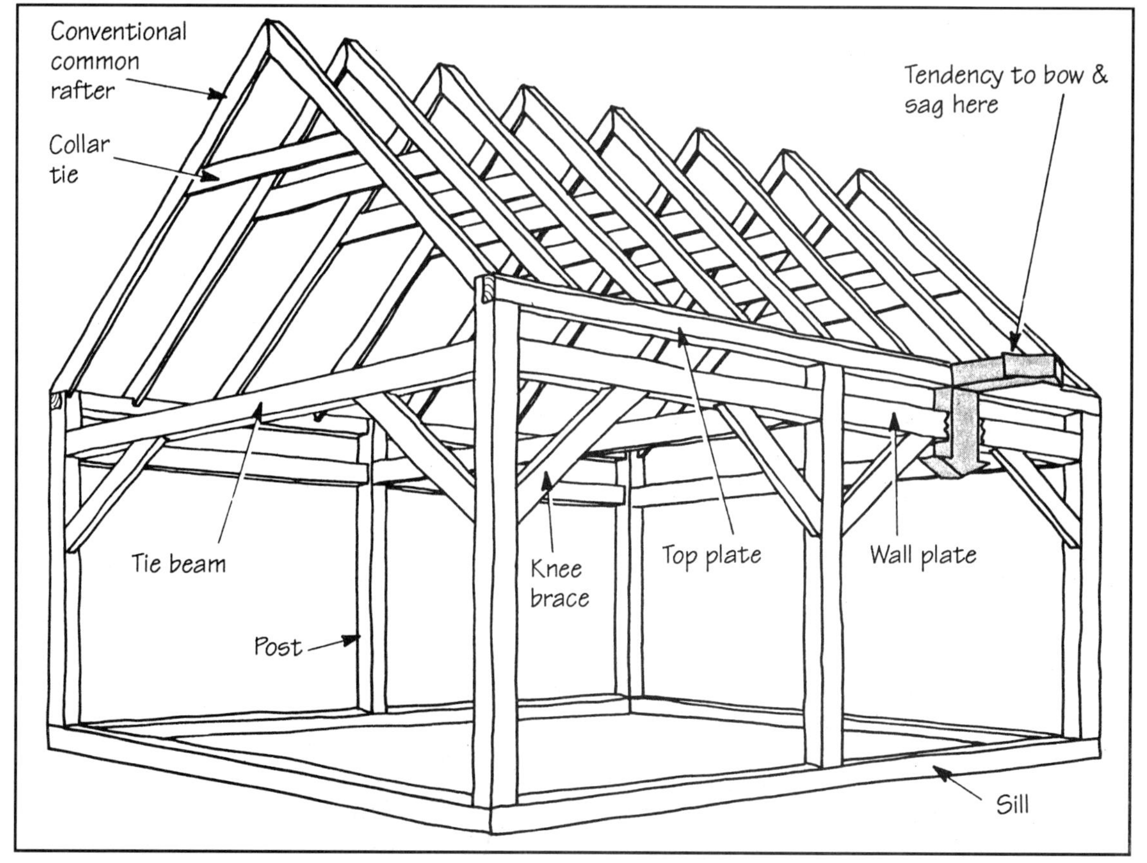

Figure 2. *Conventional roof with timber-frame walls and ceiling*. *A common-rafter roof imposes evenly distributed loads downward and outward along the top plate (shown by arrows). The timber frame, however, is designed to carry concentrated loads, not distributed loads.*

condition for timber frames. We must find a way to counteract the outward thrust and support the mid span of the top plate to prevent sagging.

The sagging top plate may easily be remedied by placing intermediate posts or by infilling between posts with 2x4 or 2x6 wall framing. The outward thrust, however, can be a little more difficult to restrain.

Two things are happening to the top plate because of the outward push of the rafters. First, the plate will tend to bow out, placing a twisting force on the post and second, the top plate will want to roll out at the top. If living area is desired on the second floor, the easy option of using trusses will not do. And collar ties will not work either. They will not alleviate the thrust adequately — unless they are placed so low you would hit your head on them.

A very direct way to reduce the thrust imposed on the top plate is to redistribute the load with two stud walls (kneewalls) framed the entire length of the building (see Figure 3). These will pick up the roof load and transfer most of it to the tie beams, which must then be supported by posts or stud walls on the first floor.

But what will support these second-floor stud walls between the points where they rest on the primary tie beams, which may be 12 feet on-center? Often a summer beam is placed directly under the wall, as shown in Figure 3, and is strong enough to carry the load. If the tie beams are placed more than 12 feet apart, however, pay very close attention to this detail to determine accurate loading information and to assure safety.

Timber-Framed Second Floor and Roof

Another common approach to *hybrid framing* is having the first-floor walls conventionally framed and the roof and second floor timber framed. Although this may require more timber-framing talent, it is really a more complete system. In this way the first-floor wall framing is nothing more than a foundation for the roof fram-

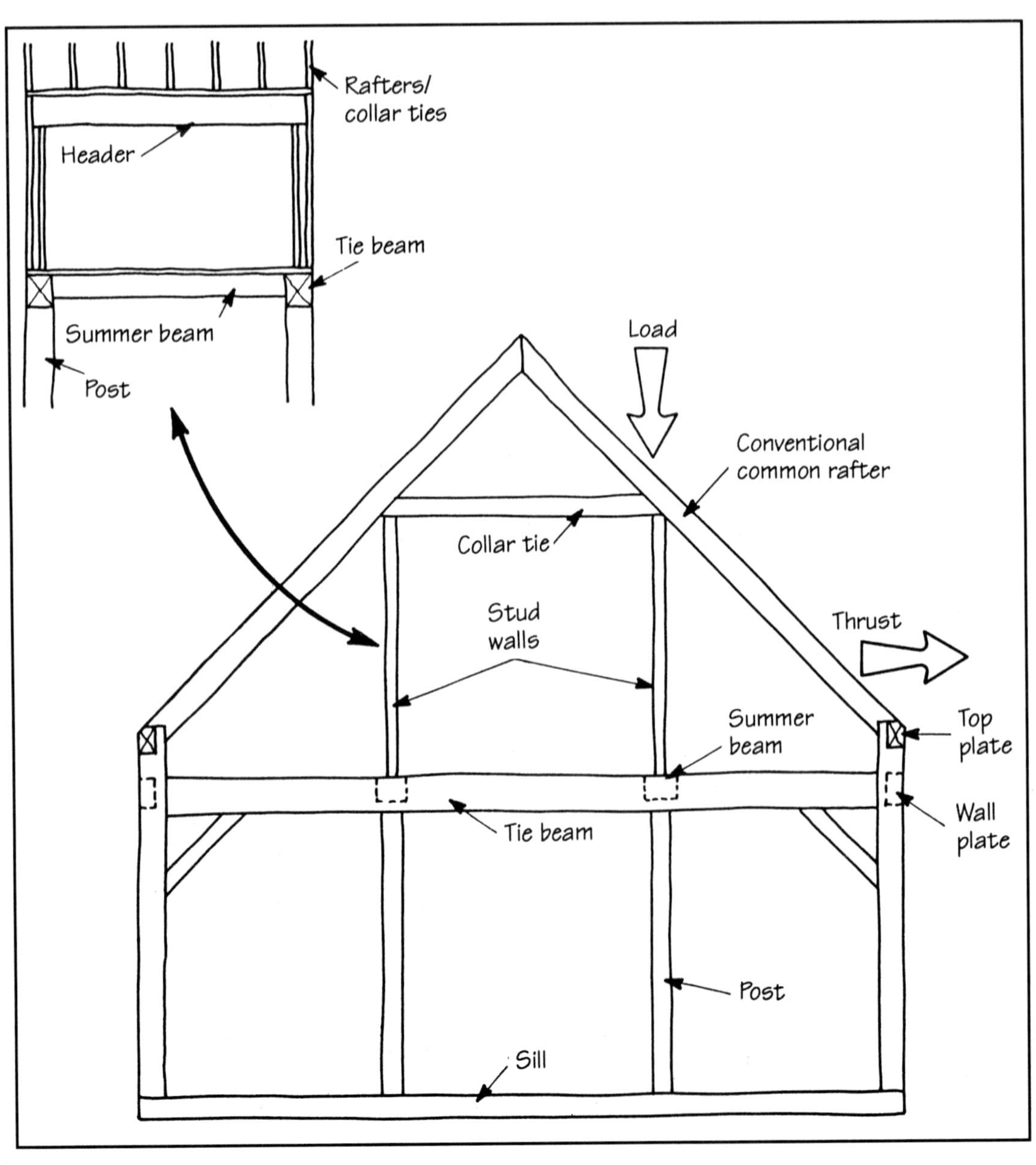

Figure 3. *Collar ties are not adequate to reduce rafter thrust. One solution is to add stud walls (kneewalls) the length of the building, as shown. Openings are allowable in the stud walls with appropriate headers (see inset). To support the added walls, first-floor posts are needed as well as hefty summer beams placed under the walls.*

Figure 4. ***Stick-framed walls and timber-framed ceiling and roof.*** *If the second floor and roof are timber-framed properly, they should subject the first-floor walls to vertical loading only. Use minimum 2x6 stud walls in this type of hybrid, and use multiple studs under rafter ends.*

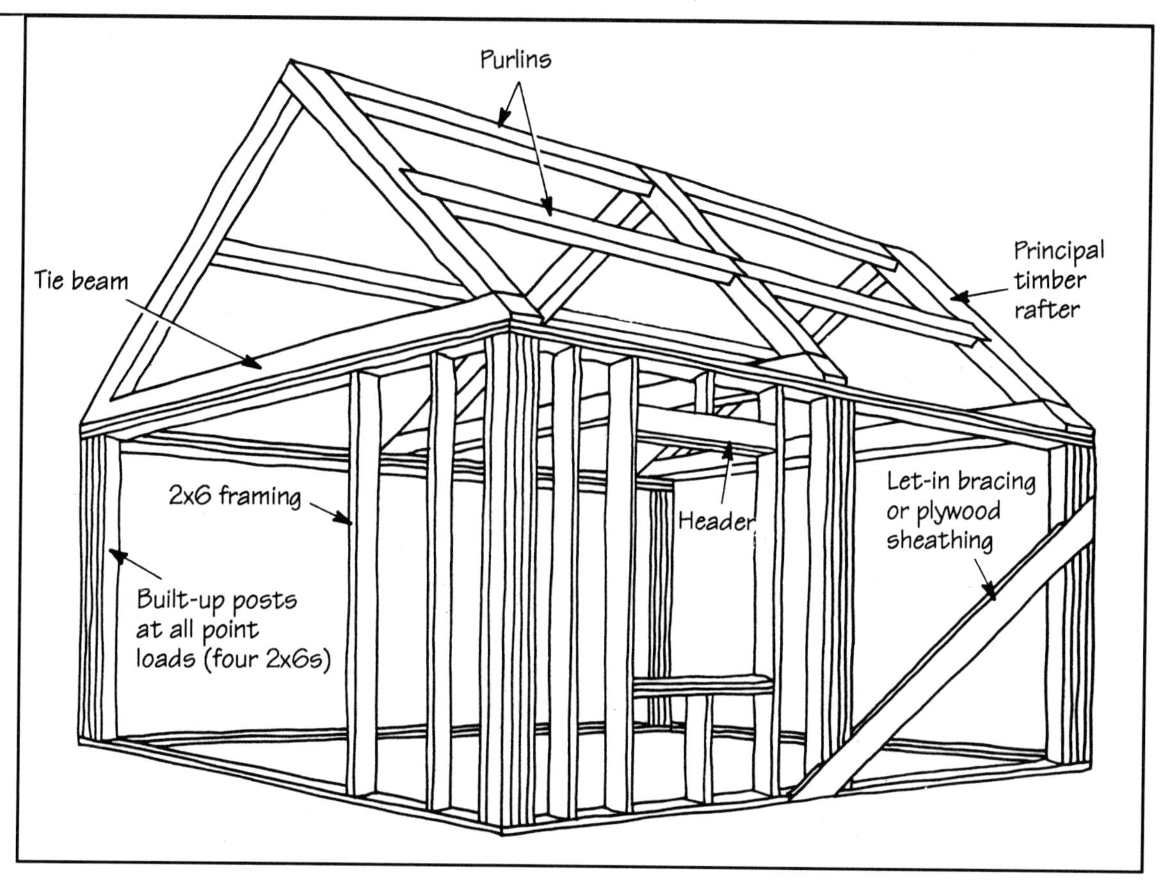

ing and the roof framing is nothing more than a timber frame with no posts. The major structural forces, such as the thrust of the rafters, that usually dictate the design criteria can be completely handled within the roof framing system. Therefore, if the roof frame is properly designed, the first-floor walls will only be subjected to vertical loading (see Figure 4).

Racking of the walls due to wind loading can be prevented by letting in corner bracing or solid plywood sheathing (I do not recommend oriented-strand board). If the roof frame is a principle rafter system with purlins, you must also consider the concentrated loading on the walls where the rafters bear on the top plate. Under each rafter end, use solid studding the width of the rafter as shown in Figure 4. This will be sufficient to support the load if the wall is built with 2x6s or greater. I would not recommend building 2x4 walls in this system.

Since the roof system is the major structural element in this type of hybrid frame, the quality of the joinery and the type of timber-framing system are extremely important. Although the specifics of the joinery and framing system are beyond the scope of this article, some mention of timber-framing concepts is in order.

Joinery

We have assumed up to this point that all the joinery and connections have been adequately strong. Perhaps the hardest thing to decide when building a hybrid frame would be where to stop the timber joinery and begin the nailing. Steel plates and straps work wonders, but they do detract from the appearance of the timbers. They also do not work well at resisting twisting forces, which are important when working with timbers. Outside of glulam, there is no such thing as a dry timber. Mortice-and-tenon joinery then, is still the best approach to connecting timbers. Since several articles would be necessary to do that topic justice, I'll just offer a few suggestions.

Tension. Tension is the force that tries to pull things apart. With any timbers joined under tension, it is best to use full *through-mortice-and-tenon joints*. In a typical timber frame this might mean that a 7x12 tie beam would be joined to a 7x9 post with a tenon of 7 inches high by 9 inches long (see Figure 5). The width of the tenon will vary according to the situation at hand, the average being 1½ to 2 inches. Two of the most important factors influencing the strength of a mortice-and-tenon joint are the width of the tenon and the amount of wood behind the peg resisting in shear. The wider and longer the tenon, the greater will be the resistance to shearing and the stronger

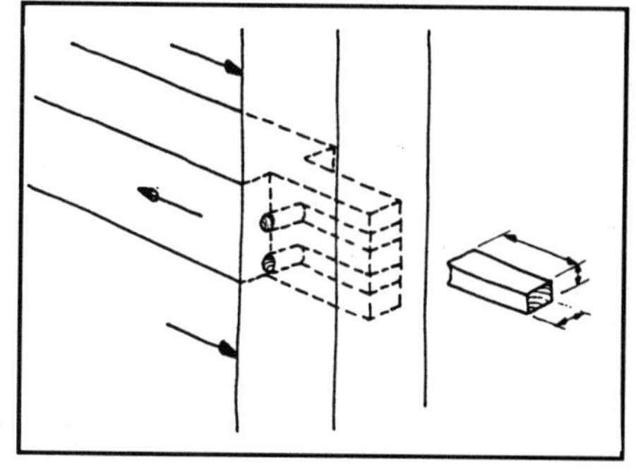

Figure 5. *A through-mortise-and-tenon is used to join main beams loaded in tension. The strength of the joint is determined largely by the amount of wood behind the peg resisting in shear.*

the joint will be.

Pegs are, of course, an important and required element in most joinery details. It should be the goal, however, to design joinery and framing details that rely less on the pegging for strength and more on proper structural design. If adding another framing member would greatly reduce the load to a given timber, thereby reducing the demands on the joinery holding it together, then it should be added. The desire to have an attractive open span of timbers should not be the justification for doing so. The justification should be calculated and proven structural ability.

Compression. Compression, as opposed to tension, is the force that wants to crush timbers together. The more the surface area in contact under a given load, the stronger the connection. Wood is much stronger in compression than in tension. For this reason it is desirable to use smaller compression members, such as struts and queen posts within a timber bent or truss, to distribute the roof and floor loads in a balanced way. This will reduce much of the stress placed on the key tension joints.

Contrary to what some believe, timber frames are not capable of spanning long distances on the merits of single timbers. The greatest reasonable span for load-bearing beams is 16 feet. When spans beyond this are necessary, more elaborate truss designs must be used. Attempting long free spans must be approached wisely and sensibly.

While almost anything may be possible, the goal is to be practical. The decision to combine different building systems in the construction of a frame should be made only after acquiring a thorough understanding of all systems involved, and only if the end result will be a better product. ■

Steve Chappell owns and operates Fox Maple Post and Beam in West Brownfield, Maine, which builds timber-frame homes and conducts workshops on traditional timber joinery. He is also editor and publisher of Joiner's Quarterly.

Parts of the Frame

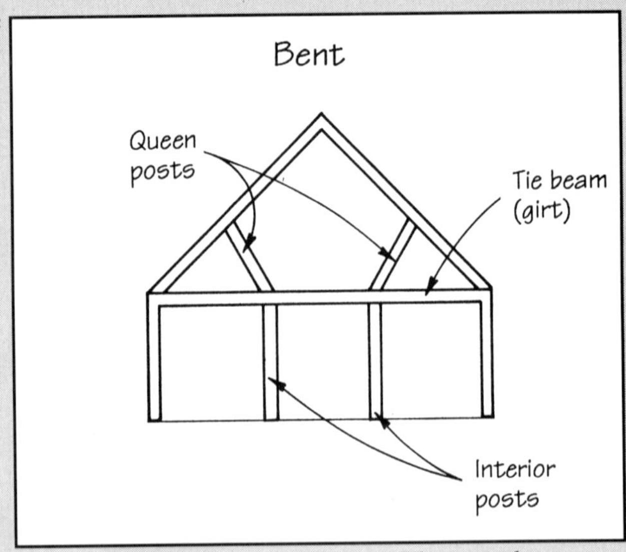

The traditional frame shown in the photo is called a "four-bent principle-rafter-and-purlin frame."

Why the name? Simply because the frame uses four bents, and the roof system is comprised of *principle rafters* (each part of a bent) and connecting pieces called *purlins*. Most frames are built from pre-assembled bents, which are essentially timber trusses that extend from sill to ridge. Bents come in many shapes and sizes. The one shown in detail (above) uses angled *queen posts* to help support the rafters, and interior posts to support the additional load on the tie beams (girts).

Other parts of the timber frame are identified in the large illustration below at right. Note that it is a complete system in which loads are concentrated and transferred vertically to sturdy posts. Problems arise when stick framing is mixed in, since it distributes loads evenly across beams (top plates, summer beams, etc.), which sag or twist if not supported.

Also note that the second-floor framing, shown in detail below, is not designed to resist the outward thrust of the rafters, except at tie-beams. This can be a problem if standard roof rafters are used on 16- or 24-inch centers.

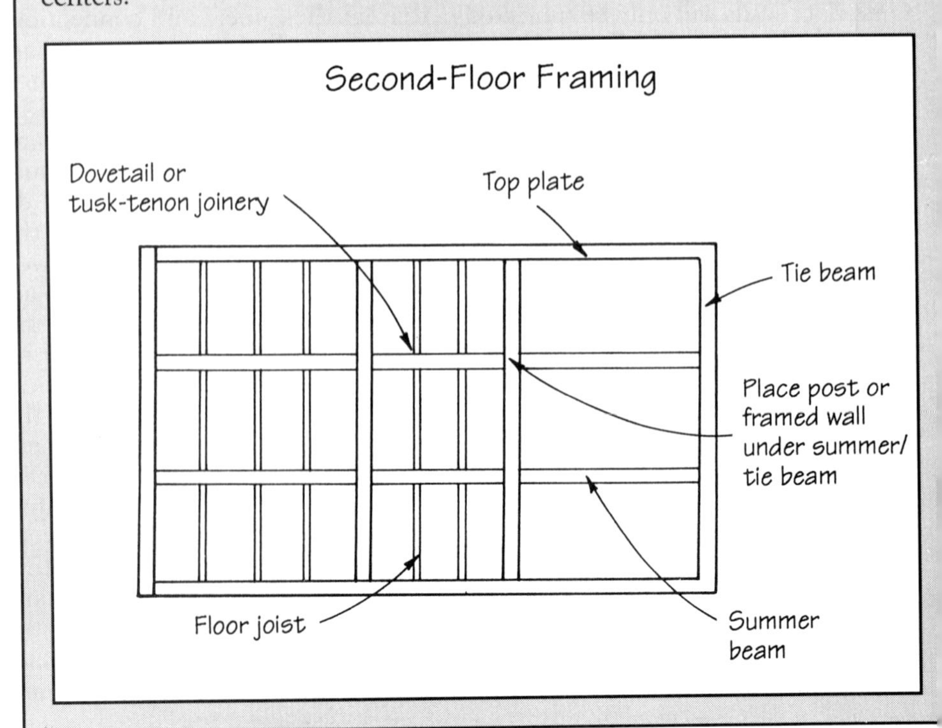

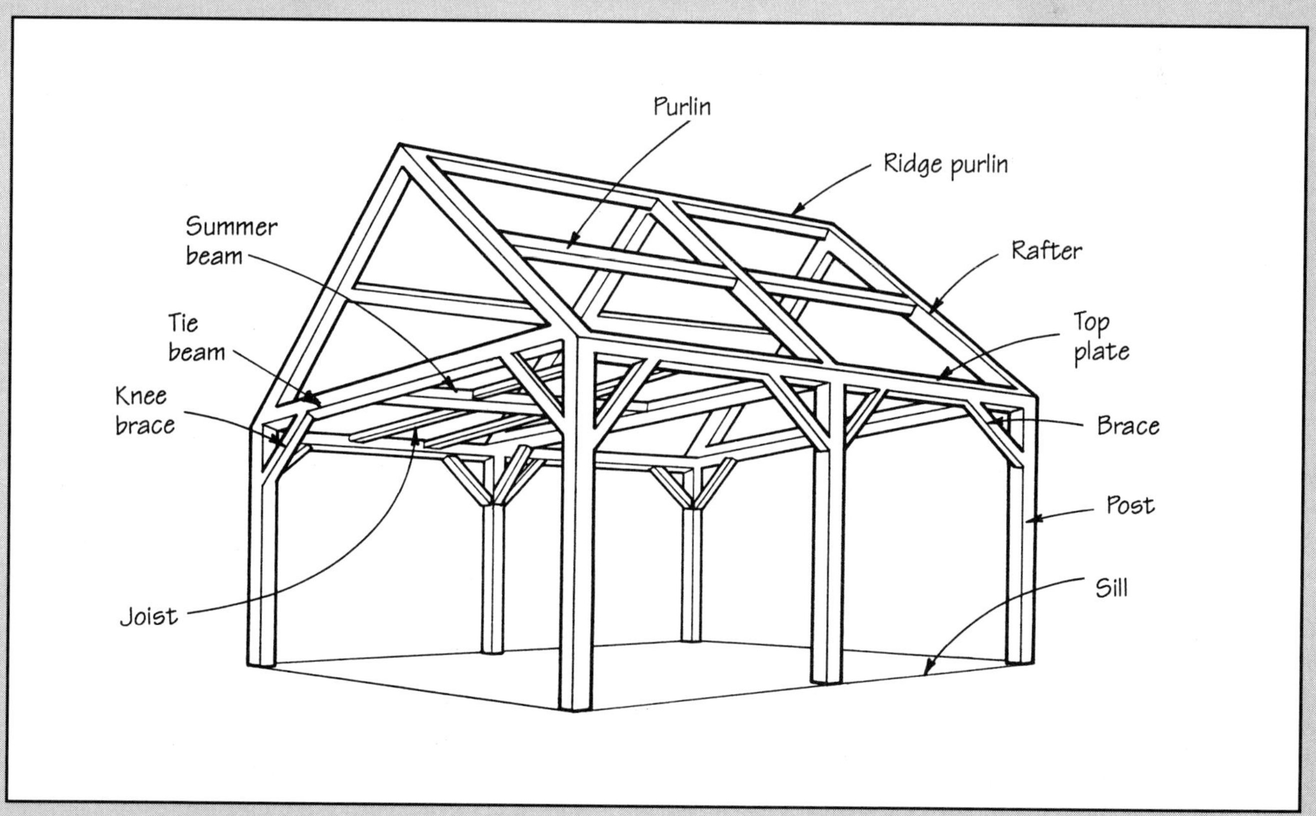
Purlin
Ridge purlin
Summer beam
Rafter
Tie beam
Top plate
Knee brace
Brace
Post
Sill
Joist

Wrapping Frames With Foam Panels

Recipe for a one-step shell: Plan for minimal waste, measure and cut with precision, and be sure to seal 'em tight

by Tedd Benson

Photos by Sarah Knock

A little ingenuity can replace a crane when installing roof panels. This 4x16-foot panel is 4½ inches thick and weighs in at about 250 pounds.

In 1976, after building my first few timber-frame houses, I began to search in earnest for a better method of insulation. The built-up system of horizontal 2x4 nailers and foam board we were then using was labor-intensive, suffered from infiltration and thermal breaks, and did not take advantage of the structural characteristics of the timber frame. A decade later, after much experimentation, the stress-skin panel has become our preferred insulation system.

A stress-skin panel is nothing more than a sandwich of a thick core material bonded to two thin outer skins. The structural principle is similar to that of an I-beam, while the insulation core performs the role of the I-beam web.

The panels we now use have a polyurethane core with an outer face of oriented-strand board (OSB) and an inner face of plaster-base drywall (blue-board). We use OSB because it is stronger than chipboard, has better nail-holding capacity, and costs about the same as regular chipboard. We prefer blueboard rather than regular drywall because it is more durable, and holds up well if it rains before we get the roof on. Blueboard gives our clients the option of either plastering or painting.

If we need extra strength or a nail base on the interior, we sometimes use "doublechips"—panels faced with oriented-strand board on both sides. When we use doublechips, we specify that the interior skin have low formaldehyde-glue content. Otherwise, outgassing into the household air could be a problem.

Making a Plan

In theory, installing stress-skin panels is similar to installing other types of sheathing, except that the panels are thicker, heavier, and a lot more expensive. To keep waste down and make installation efficient, it's important to map out a strategy before getting started.

The locations of the timbers determine the final lengths, widths, and cuts for the panels. Therefore, use a blueprint of the framing plan as the basis for the panel plan.

You'll need a plan for each wall of the house and each roof surface. Before you draw in any panels, locate the rough openings for windows and doors. All openings should be dimensioned from the edges of the

Making Connections

Panel-to-panel connection is just as important as the way a panel is made, because it's the connection at the seams that is really the weak link in a panel system. There are a number of manufacturer-suggested connecting systems, but we've found they don't offer the thermal performance we're looking for. Instead, we use the double-spline system shown in the drawing.

The panels come with square edges, which we then rout out to accept two 5/8x3-inch plywood splines. Our spline router (actually a grinder) is designed to take out 1/8 inch of insulation at each edge between the splines to create a 1/4-inch cavity into which we spray expandable foam.

The foaming is done after the panels have been installed by drilling a 1/4-inch hole every 14 inches through the splines to accept the nozzle of the foam can. We fill all roof joints as well as the wall joints, and work our way up the joints, not down, because the foam tends to rise as it expands.

Experiment on some scraps before you start, because climatic conditions affect foam expansion. We have found a two-second count generally fills to the next level quite nicely. After the foam has completely expanded and hardened, scrape the excess from the surface.

At this point we also foam the gaps left between the extended sill and the panels, at all outside corners, between wall and roof at the eaves and rake seams, and at the ridge.

We treat the roof joints a little differently. On a couple of houses, despite a thoroughly sealed roof joint (foam in the joint and roofing cement on the exterior surface), we had trouble with the shingles wrinkling over the rafters. After poking around, we discovered that the wrinkling wasn't caused by moisture getting through the panel joints, but by these panels being pulled together as the frame shrank. For that reason, we now use a single 5/8x3-inch spline in vertical roof joints.

All this foaming may seem like a lot of bother, but if you want to achieve the best thermal value, these extra steps are the cat's pajamas. —*T.B.*

Plywood splines are either pre-installed in one side of the joint before the panels are assembled (above), or driven in from above (right) after the panels are in place.

The spline cutter is designed to leave a small gap at the foam-to-foam joint between the panels. After the panels are up, this gap is foamed through 1/4-inch holes drilled through the splines (below).

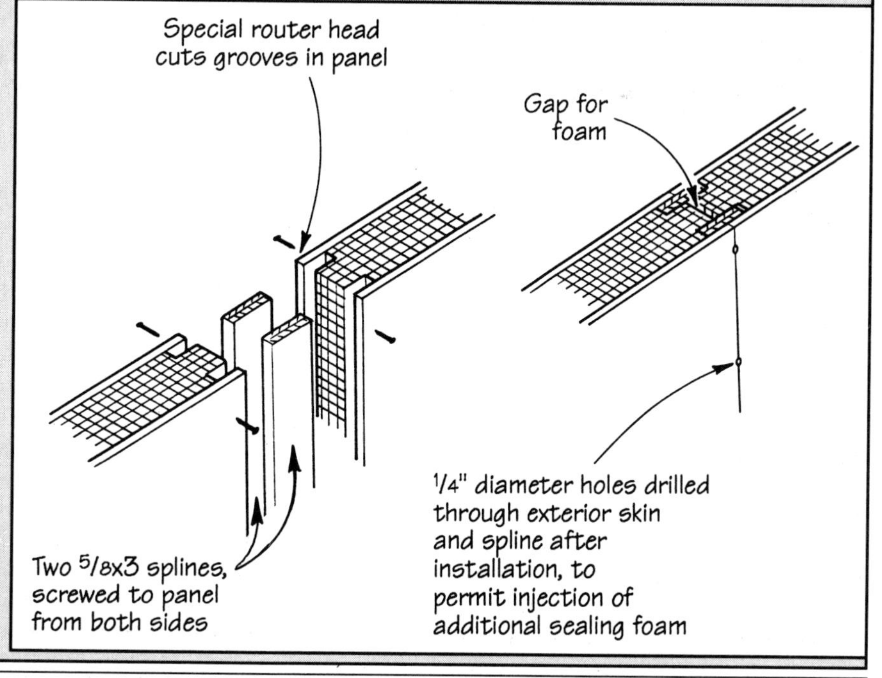

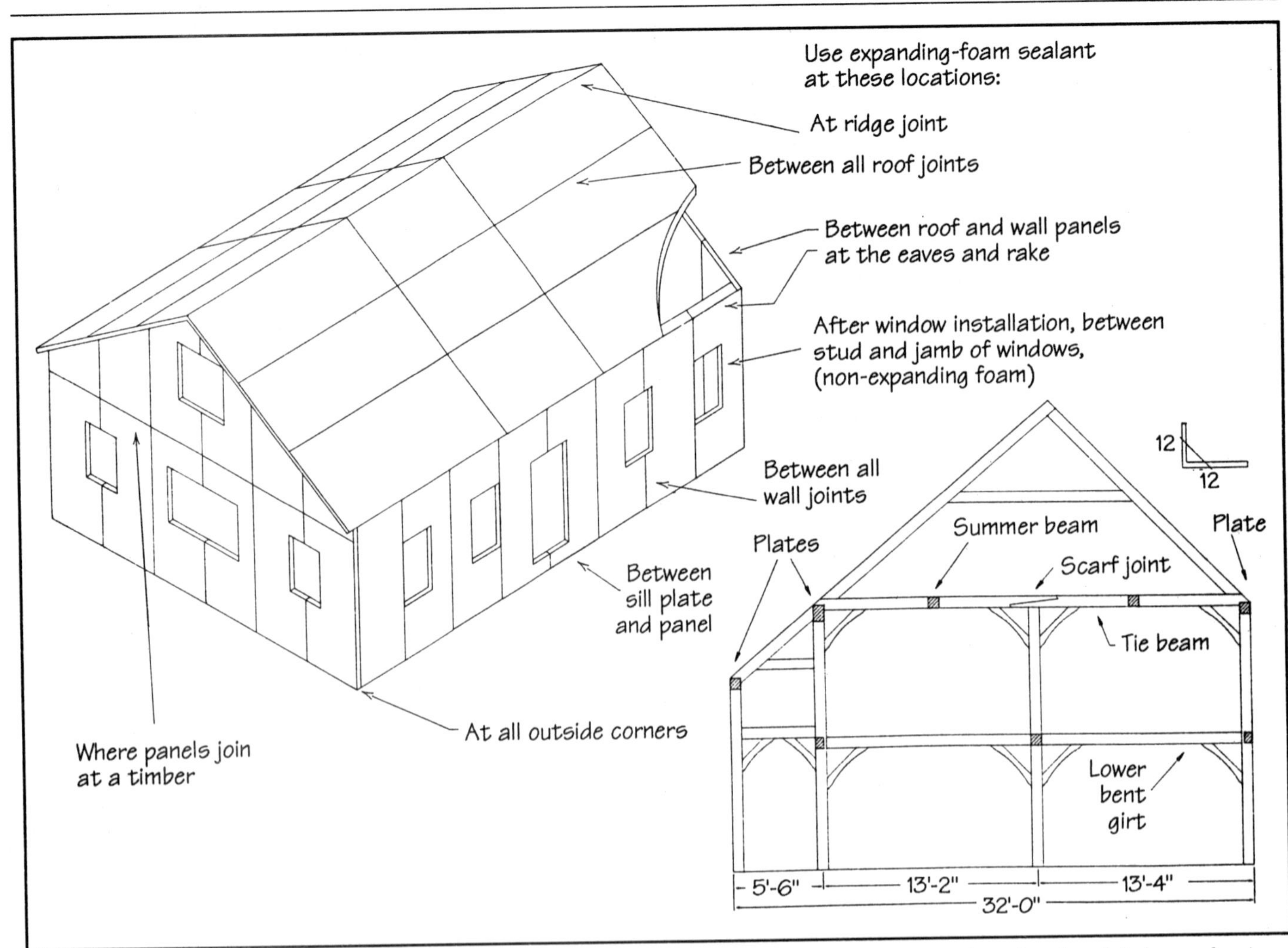

Figure 1. *The panels should be laid out over the framing plan (at right). Dimension all rough openings from the edge of timbers—there's nothing else to measure from. When installing panels, leave a small gap between non-splined panels to receive expanding-foam sealant. The critical areas are shown in the panel plan (above). Pay particular attention to roof seams, or risk moisture damage and higher heating bills.*

timbers, because the panel installers will have nothing else to measure from. Use the same rough openings as specified for standard frame construction (see Figure 1).

After you've located the rough openings, you're ready to draw in the panels. Laying out the panels is a matter of juggling several considerations: strength, panel waste, rough openings, and roof overhang. Here are a few guidelines:

First, check the requirements of the local building code and the loading characteristics of your panels. Most manufactured panels are engineered to span 8 to 9 feet on the walls and 4 feet on the roof. Be wary of a manufacturer who cannot provide documentation.

Second, don't waste costly panel material just to make the spans work out perfectly. For example, suppose the distance from the extended sill to the centerline of the girt is larger than your 4x8 panels. You could cut the 4x16 panels to fit, but you'd be creating a lot of waste.

A better solution would be to position the 4x8 panels above the extended sill (see Figure 2) to span from box sill to girt, keeping in mind that 2 inches is the minimum for good nailing. To do this, you would snap a line along the face of the box sill and nail a temporary ledger to the line to support the panels during nailing. Then you would fill in below the panels to the extended sill with offcuts.

Third, windows wider than 3½ feet must be reinforced with 2x4s on either side of the span from beam to beam. Door openings must be similarly reinforced. Smaller windows can be cut directly into the panels without this reinforcement.

If the window is small enough, try to locate it toward the middle of the panel, and leave 6 inches of panel on either side. But if the window has to be closer than that to a panel edge, try to move the panel edge to the edge of the rough opening. This puts the joint in a place where it can be strengthened with 2x4s and splines.

Fourth, non-supported roof overhangs should be limited to 1½ feet beyond the wall panel at the rakes, and one foot beyond at the eaves. Using doublechip panels here allows you to increase overhangs to 2½ feet on the rakes and 2 feet at the eaves. You could run doublechips just along the roof edges, and use standard pan-

Figure 2. *The lower sill, shown in detail below, is extended out to support the wall panels. The upper sill and deck can be made of timbers, or it can be conventionally framed, as in the photo above, where the box sill is being drilled for wiring.*

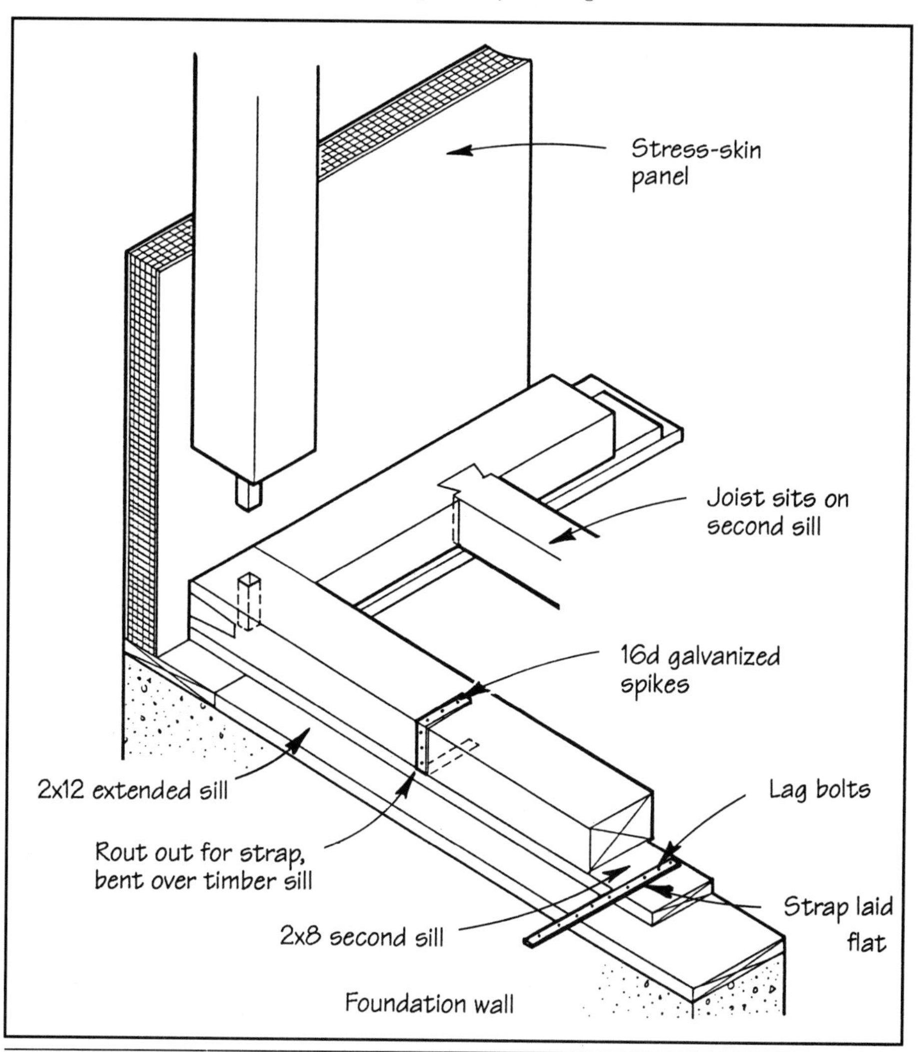

els for the rest of the roof. Some trim details support the eaves panels, making it possible to use greater overhangs with the standard panel.

Some other things to consider when laying out the panels on the plan: Panels should run vertically on the wall, unless the posts are spaced closer together than the horizontal girts and plates. After laying out the full panels on either side of door and window rough openings, plan to fill in above and below the openings with panel pieces. Try to have tapered factory edges of the drywall meet at all seams.

Use the same procedure on the roof, laying out full panel sections on either side of openings. You will have to juggle all these factors with an attempt to use offcuts efficiently. Dot the panels in over the framing plan, making sure each panel and offcut is clearly marked.

Cutting the Panels

Once the panel plan is drawn up, you can lay out, cut, and install the panels. We try to install the wall panels before the roof panels, particularly if an overhead winch or crane is used.

In some situations, it is difficult to fit the wall panels under the eaves panels and over the extended sill simultaneously. If the eaves panels extend over the upper plate or girt, nailing the top edge of the wall panels is difficult, if not impossible. The obvious disadvantage is that the wall has no protection until the roof is on.

Begin by marking panel joints and the rough openings for windows and doors on the outside face of the sill and beams. Then string a line along the eaves to represent accurately the eaves intersection. Next, using a chalk line or a large T-square, draw the rough openings and all cut lines on the panels. Panel edges that will fall on posts are sawn to allow a 1/4-inch gap at the seam for expanding-foam sealant. Panels that will span the full height from sill to eaves should be sawn about 3/8 inch short at the bottom for the foam.

We've had some problems at the eaves joint with ants getting into the

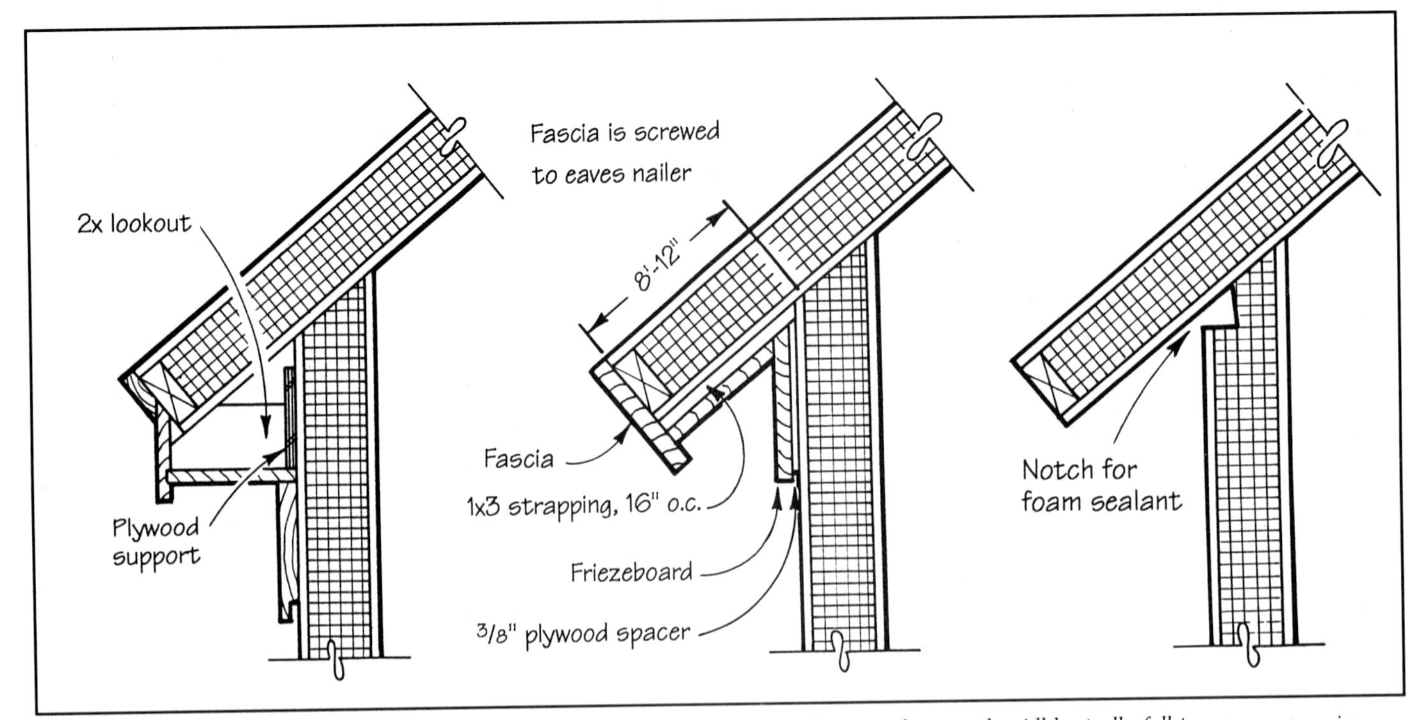

Figure 3. *Trimming the eaves.* *A wide variety of roof details can be used with stress-skin panels. All basically fall into two categories—the flat soffit (left) and the open soffit (center). Trim is generally connected to the roof by using a nailer let-in to the edge of the roof panel. Generally, the overhang should be held to 8 to 12 inches for a regular panel (drywall to one side) and 16 to 24 inches for a doublechip. In the open soffit, because there is no nailing on the bottom of the panel, 1x3 strapping is wedged in place by the frieze board. For either detail, cut a notch in the top outside edge of the wall panel (right) about 1 inch high and 1 1/2 inches deep to receive expanding-foam sealant.*

foam. So now we spread a healthy layer of plastic roofing cement on the extended sill before installing the panels. The roofing cement coats the edge of the panel, and so far appears to keep out the ants.

We usually lay out and cut the wall panels one at a time, but precut all the roof panels at the same time in preparation for crane assistance. The advantage of this system is that you can check your work on the wall as you go. The possibility of making mistakes on the roof is pretty slim, as there are usually few openings.

When you cut, always wear a mask and goggles. Clean up panel dust and chips often. The dust from urethane panels is inorganic and ugly. My most disturbing experience with it was the time a client's beautiful cranberry bog was fouled with a frost of panel dust. The house needed insulating not the bog.

The panels are heavy, so it's best to make all the cuts in a panel at the same time. If your saw cannot cut through the thickness of the panel, cut from each side and finish with a hand saw, if necessary. Tilt the saw base to cut the roof angle on the upper edge of the wall panels. Then cut a notch in the top outside edge about one inch high and 1 1/2 inches deep to receive an expanding-foam sealant (see Figure 3).

Routing for Splines

With a specially adapted grinder, rout all the seams that do not fall on posts to accept double splines (see Figure 4). The perimeter of all door and window openings, and the outside edge on each corner, are routed to accept 2x4 nailers. Where the panel edges need additional reinforcement, such as in a vertical joint over a wide window, we rout the panels to receive a solid 2x4 nailer rather than a double spline.

After the panels have been cut, and while they're still on the sawhorses, install and secure the splines to one edge of a pair of mating panel. Attach the splines first on the nail-base side with a 1 1/4-inch galvanized drywall-type screw on the drywall side. Nails and screws should be on 8- or 10-inch centers. Don't install any nailers in door or window openings until the panels are installed, because you want the nailers to span the panel seams for reinforcement.

Before bringing the panels into position, start the nails through the exterior sheathing. The nails you use to attach the panels to the frame will depend on the material of the frame. For softwood frames, use an annular ring nail or spiral-shank nail (pole-barn nail) that will penetrate the timbers at least 2 inches. To resist a tannic-acid reaction with metal, use galvanized nails for oak frames. The nails should be long enough to penetrate the wood 1 1/2 inches. These should be on 8- to 10-inch centers around the perimeter.

Raising the Panels

After starting the nails, apply a bead of construction adhesive to the faces of the posts, beams, and box sill where the panels will attach. Starting with an outside corner, lift each panel into place, resting its base

Figure 4. *A specially adapted grinder (left) cuts both spline slots quickly—but with an accompanying spray of noxious urethane dust. The plywood splines are screwed in place (right) with drywall screws. Vertical joints over wide windows get solid 2x4 splines.*

on the extended sill (or on a ledger on the face of the box sill). Slide the panels together horizontally. When installing the top panels in the wall, lift them to the eaves string by prying with a flat bar. Drive in the nails.

After the first panel on each wall is in position, you will need to drive subsequent panels together to engage the splines (you can use a mallet or sledge). Protect the edge of the panel from the mallet with a block of wood. If the panel is no longer than 10 feet, it is sometimes easier to drive the splines from above than to pre-install the splines and mate them edgewise. This is another reason to put the wall panels on first.

In places where panel edges meet at a timber and there's no reason for a spline, use 1/4-inch-thick plywood pieces temporarily to space the panels apart while nailing. The gap is later filled with foam. Make sure the bottoms of the stacked panels have at least 2 inches of nailing in the timbers. Gable-end panels should also be shy of the roof panels by 1/4 inch for foaming.

Before installing the roof panels, snap a line across the roof where the first panel seam will be, to ensure that the eaves and rake will be straight for the rim. Put a bead of construction adhesive on the roof timbers and on the beveled edge of the eaves panel prior to placing the roof panels.

With the nails already in position, attach the panel when it is aligned with the chalk line. Most roofs require few spline joints because the timbers are on 4-foot centers. Space the panels 1/4 inch from each other with plywood spacers. This creates a cavity for foam.

Finishing Up

If you choose to ignore my instructions about foaming the wall seams, don't carry the illogical thinking up to the roof. You'd be disappointed with the results. Not only is more heat lost through air leaks in the roof, but there is also a greater likelihood of moisture migration through the roof seams, which can cause rot in the timbers or damage the panels. I have even seen unfoamed roof joints that built up so much moisture that ice formed in the wintertime in the joint and lifted the shingles.

When all the panels are completely nailed to the frame, install the nailers in all the rough openings. Nail them with 8d galvanized nails from the outside skin, and screw with drywall screws from the inside skin. Also make sure all the spline joints have been properly screwed and nailed. Now you are ready to fill the roof and wall seams with expanding-foam sealant.

You've just completed the exterior sheathing, interior wallboard, and insulation in one step—saving labor and time. Using this system, the panels take the beating from the elements. If necessary, they could be replaced in time, but the frame should survive the ages intact. ■

Tedd Benson is a timber framer in Alstead, N.H. He is the author of Building the Timber Frame House, *published by MacMillan, and* The Timber-Frame Home *published by The Taunton Press.*

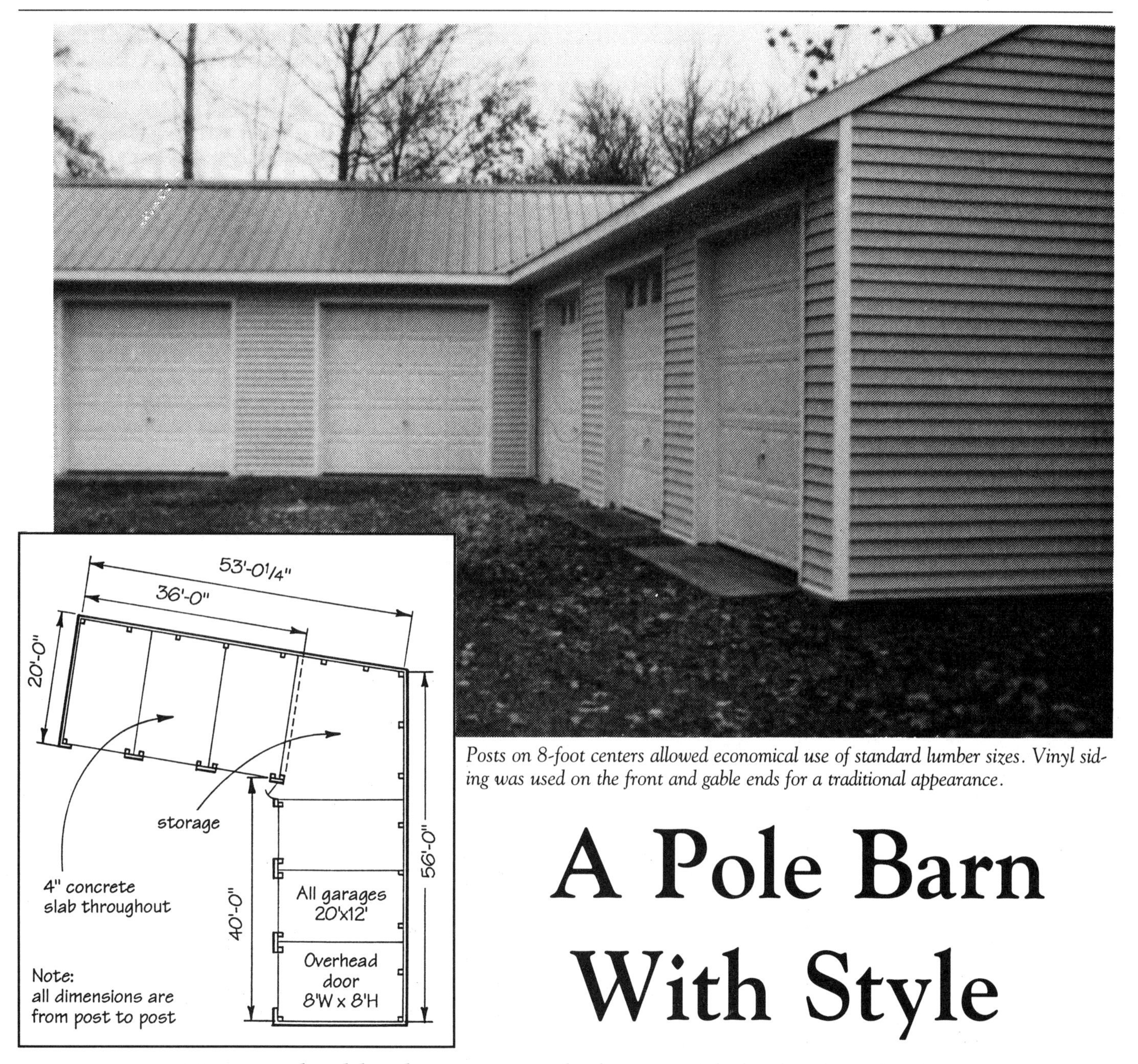

Posts on 8-foot centers allowed economical use of standard lumber sizes. Vinyl siding was used on the front and gable ends for a traditional appearance.

A Pole Barn With Style

How to build a low-cost pole barn with high-class looks

by Richard Cooley

I decided once and for all that this year I would build a storage building for myself. The decision became final when I crawled into my pickup bed looking for a drywall gun. My truck has a cap, and the front two feet of the bed collects the debris that gets plowed up from the area near the tailgate.

I didn't find my drywall gun, but under two canvases and a tangle of leads and ropes, I did discover many other long lost goodies. I found my Makita palm sander, my favorite Marshal-Town curved taping trowel, a shingle sampler, a top bracket to a pump jack, a come-a-long, a bottle of air-gun lube, and many other less notable items. "We're building a barn this year," I called out to my crew.

We had built several pole barns over the years. I was impressed with their versatility, speed of construction, and relatively low material cost. I had also found that many municipalities, including mine, assessed them for taxes at a lower rate than conventional buildings. Also, the building site I planned to use had conditions that made a conventional foundation almost impossible.

Site Considerations

On any job, careful consideration of the site is time well invested. At one time, this site had been a transfer

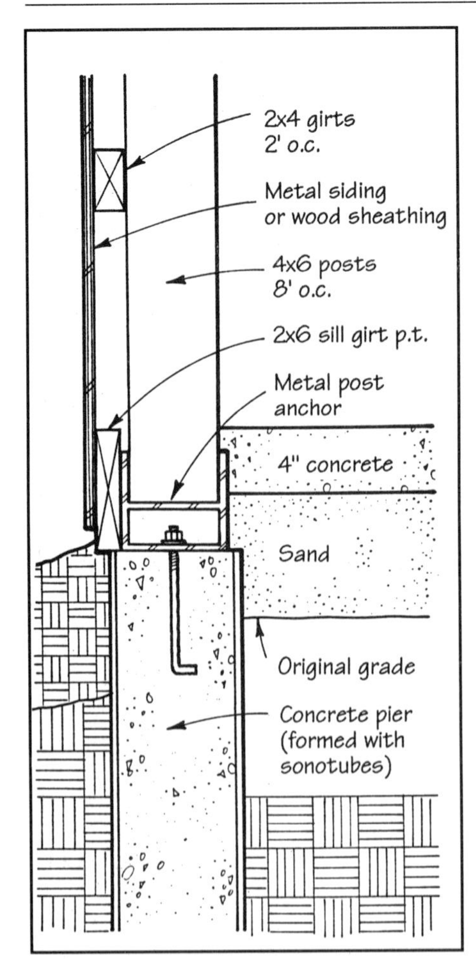

Figure 1. *On this site, the author found concrete piers more convenient to use than poles. The piers could be laid out with less precision, and a backhoe could grade a layer of sand right over them.*

Cost Breakdown Excavation and Backfill

Eight hours backhoe at $45/hour	$375
Fill sand: 70 yards at $8/yard	560
Sonotubes: Ten 12-foot tubes	150
Concrete for piers: Four yards 2,000-pound mix	250
Poles: 32 pressure-treated 4x6s	450
Headers: 320 lineal feet 2x12s	300
320 lineal feet 2x6s	125
Trusses: 28 Fink trusses	1,200
Purlins: 200, 16-foot 2x4s	700
13 pressure-treated 2x4s	75
Ribbed steel: 3,600 square feet total (see article for breakdown)	1,800
Valley and ridge material:	250
Fascia: Two rolls, 24 inch x 50 feet, 019 alum.	130
Soffit: 170 square feet vinyl, 340 lineal feet "j" channel	225
Sheathing: Ten sheets ½-inch aspenite	75
Vinyl siding: 320 square feet, including trims	350
Concrete slab: 24 yards, 3,000-pound mix	1,500
Overhead doors: Six installed	2,400
Exterior steel door with glass	180
Total material cost for completed exterior and poured concrete floor	$11,095

yard for heating coal. What was now a bike trail had at one time been railroad tracks. There were various piers for coal chutes, remains of building foundations, and deep down we even discovered a granite cobblestone road base. The cobblestone proved to be impenetrable even with a backhoe, but fortunately it was 4 feet below our final grade. Otherwise, when the backhoe hit something, we would simply trench next to it and slide it over to allow for our piers.

Installing the Piers

I decided to use concrete piers at this site instead of sinking the poles themselves for three reasons. First, the site had poor-draining soil, and we planned to pour a concrete floor right around the piers. This makes hostile conditions for wood, treated or not, so it was best to use concrete piers below grade.

Second, I needed to bring in some clean fill sand. A backhoe could grade the sand right over the piers. Poles would have been hard to work around. The sand not only brought the area up to a level grade, but it also provided a good drainage base for the slab floor. Good drainage is the key to a stable slab poured on grade.

Finally, piers don't have to be set exactly in a straight line like poles. We could make the final adjustment when we set the anchor bolts in the concrete. The piers could actually be off a few inches one way or another, as long as the bolts were set true to the string lines.

We used 12-inch cardboard Sonotubes for forms. The advantage of using cardboard forms is that you can leave them sticking out to rough elevations and cut them off later. After we backfilled, we marked them using a level and cut them off to the desired level before we poured.

Attaching the Poles to the Piers

We set the piers 8 feet on-center. This made for a header span that could be handled with two 2x12s. The trusses were designed for a 50-pound-per-square-foot load, and a double 2x12 header on the 8-foot span was rated higher than that. Also 8-foot centers make good use of standard 16-foot lumber for wall purlins.

I used 4x6 treated poles. I wanted the 6-inch dimension on one side so that I could notch in my headers. (The 4x6 poles are better than 6x6s, because they're easier to handle and cheaper.) Since the poles weren't in contact with the concrete and were not exposed to the weather, we could have gotten away with pine or something else untreated. But untreated poles are harder to come by. It's not a stock item in most lumber yards, so you won't save much unless you can find them at a small mill.

To attach the poles to the piers we used conventional galvanized-metal post-anchor boxes. They are standard items at lumberyards. These anchor boxes bolt to the pier and nail into the bottom of the pole. They also hold the pole an inch or so off the concrete, which further discourages rotting (see Figure 1).

We stood and temporarily braced

Figure 2. *Double 2x12 headers are boxed in with 2x6s on the top and bottom. The 2x12s are notched into the poles for a solid joint with no chance of splitting.*

Figure 3. *Three trusses are ganged to make a bearing beam at the inside corner of the building. Trusses with their tails cut off hang from the beam with joist hangers.*

the poles. We marked them with the aid of the dumpy level. I didn't assume the bases of the posts were level with each other. After marking the posts with a level line, we could measure off them to lay out our header notches and form our concrete floor.

Framing and Trusses

We built the headers out of two 2x12s and boxed them top and bottom with 2x6s (see Figure 2). The 2x12s were notched into the poles. This is more dependable than any mechanical fastening system. It doesn't rely on the strength of the fastener or the fastener's ability to hold in the wood. Also, there's no chance of the 2x12 splitting at a fastener.

It is best to run the top 2x6 continuously across the tops of the poles. It adds rigidity and makes for a good top plate to nail trusses to.

Trusses designed for 4-foot centers cost about $40 each. We needed 28 of them. Trusses designed for 2-foot centers were about $28 each, but we would have needed almost twice as many. So we used 4-foot-on-center trusses and saved about $500.

We nailed three trusses together to make a bearing beam at the inside corner of the building where the building turns 90 degrees, making an ell. The header allows this area to stay open underneath. We cut the tails off the trusses here and fastened them with joist hangers (see Figure 3).

We used Fink trusses with an open triangle down the middle so that we could store aluminum ladders and what-not in this space. We installed a hatch door in the gable to gain access to this area.

For purlins (roof supports nailed across the posts), we used 2x4s every 2 feet on-center. On the bottom of the walls, we used a treated 2x4 that doubled as a concrete form. This bottom 2x, called the "sill girt," is partially buried to seal the building at grade (detail in Figure 1).

Watch For Wind Racking

The biggest enemy of a pole barn is the wind, because it has a large surface area and a relatively small mass. This mass-to-surface ratio may be good if you're designing an airplane, but it's bad for a building that you want to keep on the ground. Everyone has seen the remains of a lawn shed after it has blown across a yard. For poles sunk in the ground or fastened to piers, this isn't the immediate problem. The problem there is with racking, the constant shifting action from wind gusts.

Wind can cause the building to move enough to loosen the sheet metal, cause leaks, and eventually loosen the panels. Wind can also stretch nail holes in the sheet-metal roof and siding. (The nails come

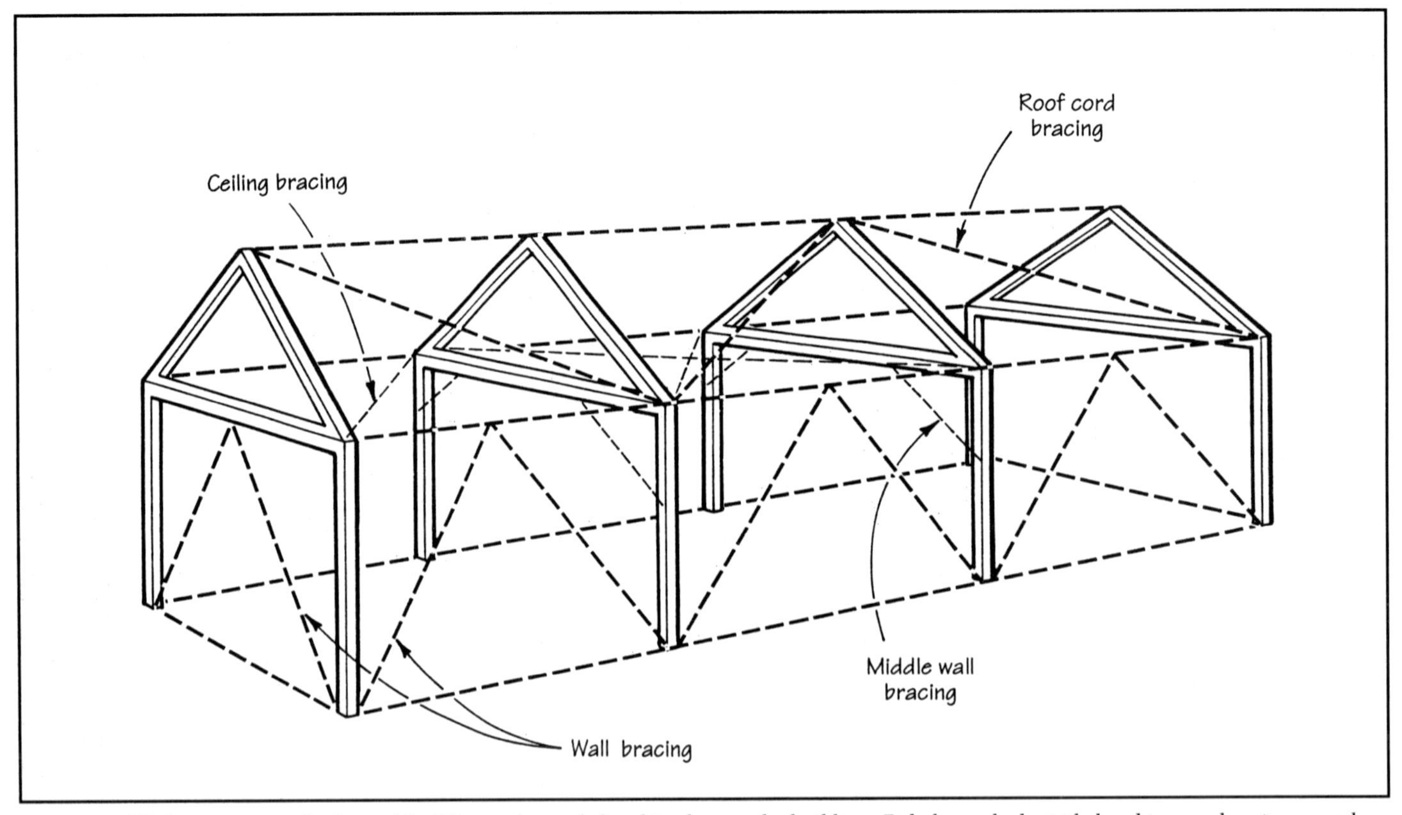

Figure 4. *With conventionally framed buildings, plywood sheathing braces the building. Pole barns lack rigid sheathing, so bracing must be added throughout the structure.*

loose with time anyway, and the gaskets on the nails can deteriorate.) Our site was well protected from the wind, but we still took all precautions.

Bracing For Wind

With conventionally framed buildings, plywood sheathing braces the building. With pole barns, which lack rigid sheathing, this bracing must be added. Two areas of our building—the bottom storage area and the trussed loft space—needed to be braced in three different planes or "dimensions." For the bottom part of the building, this meant bracing the length of the storage area, the depth of the storage area, and the ceiling or top of the storage area.

Along the length of the building, it's easy to install diagonal bracing on the exterior walls. That's the first dimension. Across the ceiling or top of the storage area, you can easily brace diagonally along the bottom of the trusses. That's the second dimension. Bracing for the third dimension stabilizes the depth of the building, so this bracing must go across the building. The end walls stiffen the building's ends. But the middle can be a problem, because the bracing must go through or into the usable space. Since we divided the building into several stalls, however, that wasn't difficult for us. If you want the building open in the middle, which is common, you have to put knee bracing at each pole. If it's kept high enough, it won't cut into the usable space of the building too much.

You need to brace three dimensions of the trussed area as well. The bottom of the trusses (the storage area ceiling) was taken care of when we braced the top of the storage area. That's the first dimension. The other two dimensions—length and depth—can be braced at the same time since the angle of the roof crosses through both. Simply put diagonal bracing inside the truss and fasten it against the underside of the rafters (see Figure 4).

Galvanized Roofing and Side Panels

Twenty-nine-gauge bold-ribbed galvanized metal roofing seems to be the standard. It comes either painted a variety of colors or "mill finished," which is plain-galvanized metal. Both are generally warranted twenty years maintenance-free. They may be repainted after that.

The painted metal sells for about sixty cents a square foot, and the mill finish sells for about forty cents a square foot. Because the front of the building was in view, we used the red metal on the roof there. On the back of the roof and the back sides of the building we used the mill. We needed about 3,600 square feet of metal. We used 1,100 square feet of the red and 2,500 square feet of the mill. We saved $500 doing it this way, and it gave me a very attractive facade.

The panels are stocked in most lumber yards in 2-foot increments, from 10 to 16 feet. Longer panels—up to 40 feet—can be ordered, and custom cut to length. Panels cover 3-foot widths. When designing a pole barn, it's wise to work around stock lengths, and plan on little or no cutting. The metal can be cut with a fiber blade.

Figure 5. *Metal roofing is installed with ring-shanked, gasketed nails driven in the ridges, not the troughs. The valley was custom-fashioned from 26-gauge galvanized steel.*

Figure 6. *One-by-six rough pine, blocked out with a 2x4 spacer, makes a simple, attractive rake board. The siding tucks easily underneath.*

Use standard 2-inch ring-shank gasketed nails, available at most lumberyards. They come galvanized, or colored to match the steel panels. Nailing must be done along the ridges (see Figure 5), not in the troughs. This may seem elementary, but I was called to look at one leaky roof and found that the installer had nailed it in the troughs.

Siding

This building had about 1,800 feet of wall surface to be sided. It would have taken two men about eight hours to cover this area with sheet metal. If I had conventionally sheathed and sided it, even with vinyl, it would have taken three times as long. We ended up doing about 20 percent with sheathing and vinyl—the section you see from the street.

There's a gasket for the back side of the metal to make it airtight here too. The gaskets are available with adhesive on them for ease of installation. I used conventional vented vinyl soffit and aluminum fascia. We bent up the fascia out of roll stock.

On gable ends, when I'm not using an overhang, I always place the fascia board out $1^1/_2$ inches. In this case, I used a 2x4 spacer, and a 1x6 rough-pine board covered with aluminum. It's quick and easy to run siding up behind the fascia with this system (see Figure 6). It works well for cedar siding and fascia as well. It adds a rich looking shadow to an otherwise inexpensive trim.

A Pole Barn's Uses and Advantages

I love the versatility of a pole-barn structure. I did a horse barn with bevel-cedar siding that looks like a classic colonial structure, another storage barn I did with rough pine board and batten. I've done conventional steel buildings that were open on one side, and for myself I made a hybrid.

The greatest advantage is in the foundation costs, but also important is the speed of construction. We took about three days for two men to get the piers in, and another ten days for three men to finish the exterior and floor, including the metal bending. (The overhead doors I subbed out.) Not bad for an 1,800-square-foot building.

There's also material savings, especially on the roof, a maintenance-free exterior for at least 20 years, and often a property tax savings. Also, by simply studding out the exterior walls and strapping the ceilings, you could insulate and put up drywall. It would make a great space for heated storage or shop space too.

For More Information

Two useful books are available on pole building construction. *Pole and Post Buildings: Design and Construction Handbook*, is available from The Cooperative Extension at the University of Massachusetts. Write them at the Bulletin Center, Cottage A, Thatcher Way, U. Mass., Amherst, MA. Also, *Practical Pole Barn Construction*, by Leigh Seddon, from the Williamson Publishing Co., Charlotte, VT 05445. ■

Richard Cooley is the principal of R. Cooley Construction Co., in Schenectady, N.Y.

Section 11. **Building With Steel**

Unimast Inc.

Steelwork in Wood Frames

Once you know the rules, there's no great mystery in working with this useful material

by Eric Bauer

Using structural steel in wood frame construction is not as hard as it looks. It's mostly a matter of recognizing that steel, like wood, has unique characteristics that creates its own set of rules. A good craftsman can translate many existing woodworking skills over to steelwork.

Builders who haven't worked much with structural steel assume it is hard to find, expensive, and difficult to handle, but nothing could be further from the truth. In many instances, in fact, steel is the easiest and cheapest material for the job. And it's often no more difficult to handle on site than wood components of the same loading capacity.

The best reason to use steel in residential construction is its high strength-to-size ratio. In other words, you get a large loading capacity from realtively small cross-sectional dimensions (see Figure 1, next page). This means that comparatively small steel beams can span much longer distances and leave more headroom than their wood counterparts. And steel columns can carry enormous loads and still fit within standard wall dimensions.

As with any major structural element, it is foolish to guess at the size of a steel beam required to accept even residential loads. You should always have a professional engineer review your plans whenever you consider using structural steel. This is less expensive than needlessly oversizing a steel beam just to be safe, and the documentation the engineer provides will reduce your liability.

Buying Steel

For structural work, the kind of steel you need is called "hot-rolled" because it is shaped while still red hot by a series of rollers (see "Recipes For Steel," page 240). Hot-rolled steel comes in many different grades, shapes, and sizes (see chart, page 238).

Steel supply yards, which are fairly numerous, especially near metropolitan areas, usually have the most common shapes and lengths in stock. A small company will be cheaper to deal with because their overhead is lower. Be sure to ask about delivery. Most steel suppliers will deliver to your site, but they may add transportation costs to the price of the material, particularly on special orders or oversize pieces.

Occasionally, you can find usable structural steel at salvage yards. This is a hit-or-miss affair, and you should know exactly what you're looking for. You will rarely find stock lengths, and some of the beams you find may be bent. They can be straightened, but this is work for experienced steelworkers. Almost everything you uncover will have had the flanges cut back (coped ends) or steel angles (clip connections) welded to them. And you usually have to arrange transportation yourself. But, if these kinds of problems don't cost you too much to deal with, the savings can be dramatic.

Almost every piece of steel you use will have to be cut to length. Steel suppliers will do what most lumberyards would do if you walked in and

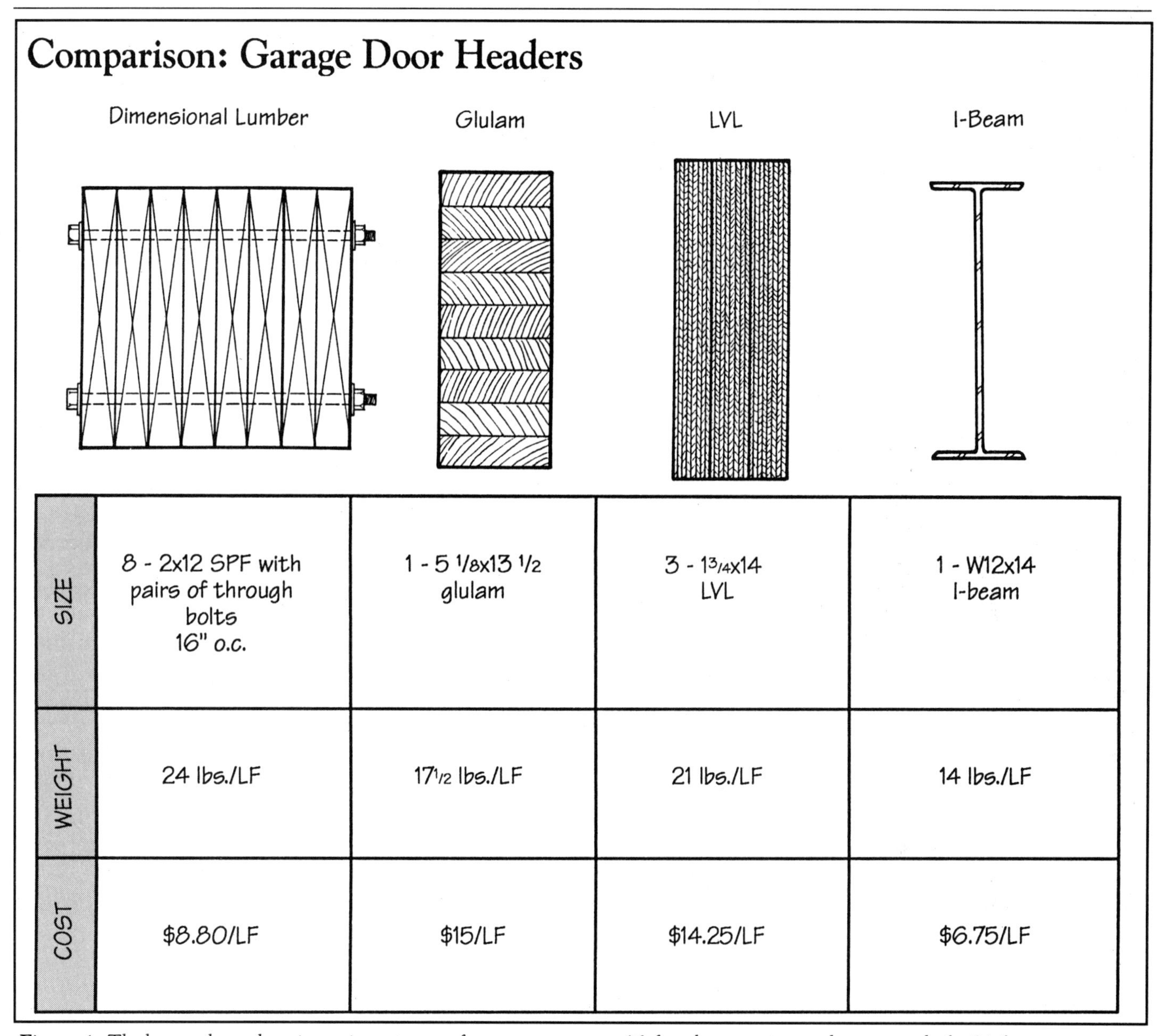

	Dimensional Lumber	Glulam	LVL	I-Beam
SIZE	8 - 2x12 SPF with pairs of through bolts 16" o.c.	1 - 5 1/8x13 1/2 glulam	3 - 13/4x14 LVL	1 - W12x14 I-beam
WEIGHT	24 lbs./LF	17½ lbs./LF	21 lbs./LF	14 lbs./LF
COST	$8.80/LF	$15/LF	$14.25/LF	$6.75/LF

Figure 1. *The beams shown here in section represent four ways to span a 16-foot door opening on the eaves end of a 24-foot-square garage. (Beam supports 780 pounds/LF.) Not only is the steel beam less expensive than the others, it is among the lightest.*

asked for a spruce 2x12, 13 feet long. They either refuse to sell cut lengths or charge so much per cut as to effectively discourage you from asking for that service. And just like the lumberyard, when they do agree to cut the material to length, you pay for the stock length, whether or not you take the "drops" (leftover pieces) with you.

Nevertheless, if you only use steel occasionally and in small quantities, the cheapest, most convenient thing to do is to have the supplier cut the steel to length before delivery. In this case, make sure you price the steel as stock lengths, not cut pieces, and include the cutting fee. If you are going to use steel in quantity on a regular basis, however, you should consider tooling up to cut it yourself.

Don't attempt to cut steel with your circular saw. To begin with, it's only practical with thin materials, such as steel studs. Eventually, the very fine particles of the abrasive thrown into the tool's innards will destroy the bearings. A reciprocating saw isn't much of an improvement. It takes forever, and will probably wear out both the operator and the tool, not to mention a lot of expensive blades.

Making Holes

Another operation performed on steel is cutting holes. If you only use steel occasionally, this is another operation a supplier can perform for you at extra cost. If you want to do the drilling yourself, there are three tools for the job: a torch set, a magnetic-base drill, or a hand drill.

A torch is best for piercing large holes in thick (1 inch or larger) steel. Holes less than ½ inch in diameter may have ragged edges, especially when cut by an inexperienced operator, but a small grinder will take care of them.

A magnetic-base drill is a portable version of a drill press. It will drill small (up to 1 inch) holes cleanly and reasonably fast, depending on the thickness of the steel you're working with. But it's hard to justify investing in this expensive tool when a torch set is more versatile and about the same price. If you have a lot of small holes to drill, consider a rental.

You can also drill small holes with a hand-held half-inch drill, but it takes longer and it's hard work. Use high-speed steel bits for this. Exotic bits, like those with tungsten or carbide cutting edges, will last longer, but should only be used in a stationary tool. These bits are so hard that even slight misalignment, such as you are likely to get with a hand-held drill, will cause them to chip. You can increase the life of your high-speed steel bits by using a liquid or a stick lubricant, such as Accu-Lube (Lubricating Systems, 22261 68th Avenue S., P.O. Box 805, Kent, WA 98035; 800/999-5823) or Fisk Darl Cutting Oil #2 (Fisk Bros. Refining Co., 129 Lockwood St., Newark, NJ 07105; 201/589-9150), while drilling.

Don't waste time drilling more holes than you need to bolt on lumber nailers. Most nailers need no more than one fastener every 24 to 36 inches, staggered from side to side along the length of the board. You can also fasten wood nailers to steel with a powder-actuated fastening tool. For a good connection, use a knurled shank pin, and a charge strong enough to drive the point fully through the steel. (Before using powder-actuated fasteners, consult the Powder Actuated Fastener Tool Manufacturers' Institute, 1000 Fairgrounds, Suite 200, St. Charles, MO 63301; 314/947-6610, for proper procedures.)

You can cut holes in steel structural members to allow pipes or wiring to pass through them, but avoid guesswork. Try to anticipate this ahead of time and have an engineer review the size and location of the holes you need. Large holes may require plates welded in place to reinforce the web or flanges.

Making Connections

Most steel connections are either bolted or welded together. Welding is useful when the surface area to be joined is small, such as the edge of a steel column. It produces a strong joint that is almost invisible in the finished product. But welding is by far the most difficult and specialized skill used by steelworkers. If you plan to do a lot of steelwork, it's a skill worth learning (see "Cutting and Welding Steel," page 241), but in most cases you'll be better off hiring an experienced welder to do this work.

Bolted connections, on the other hand, don't require special skills. The most common bolted connections are in-line beam-to-beam splices (Figure 2) and where columns join footings

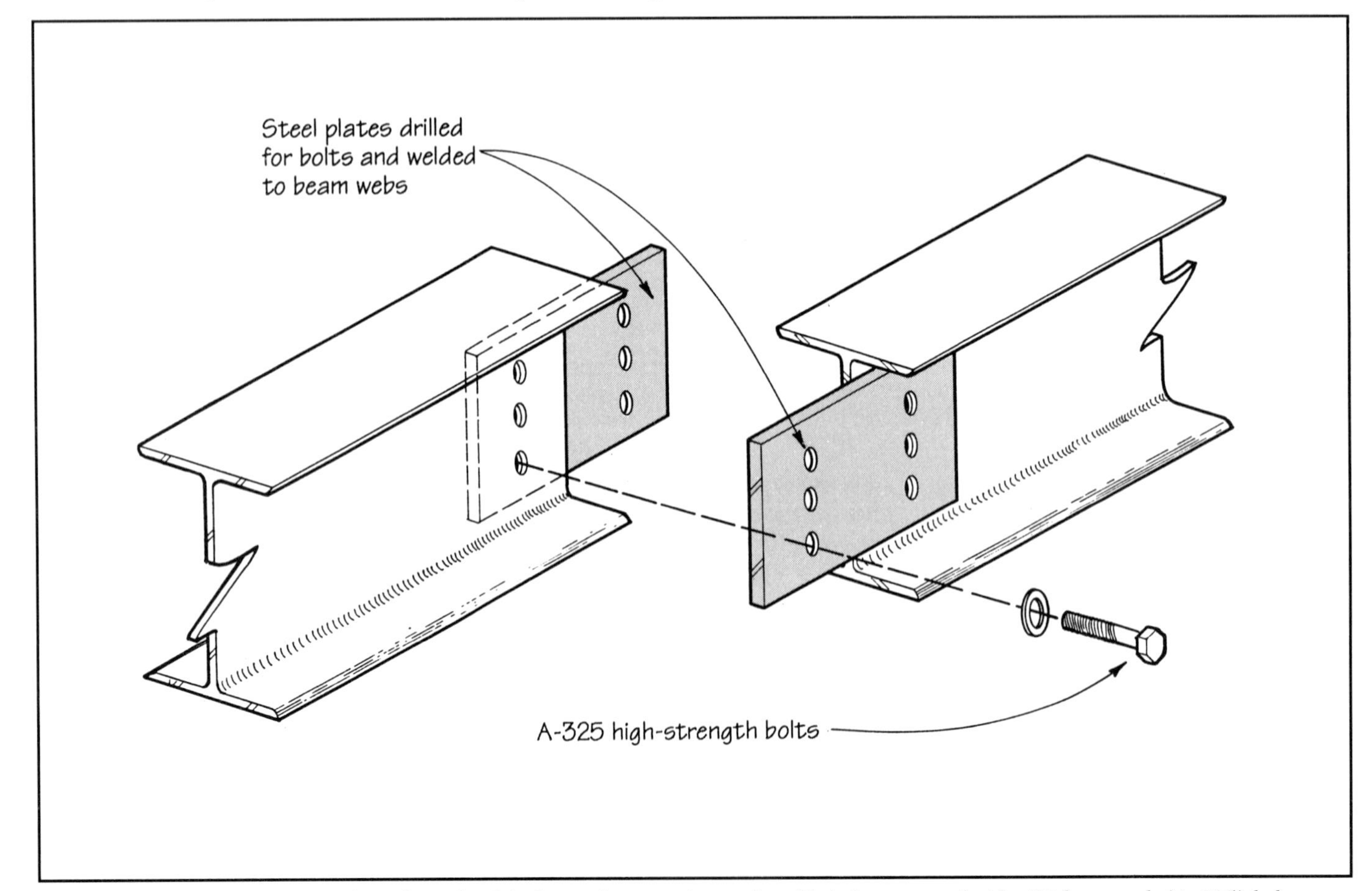

Figure 2. *In an in-line beam splice, the ends of the beams butt together with welded plates on each side. High-strength (A-325) bolts placed in pre-drilled holes complete the connection on site.*

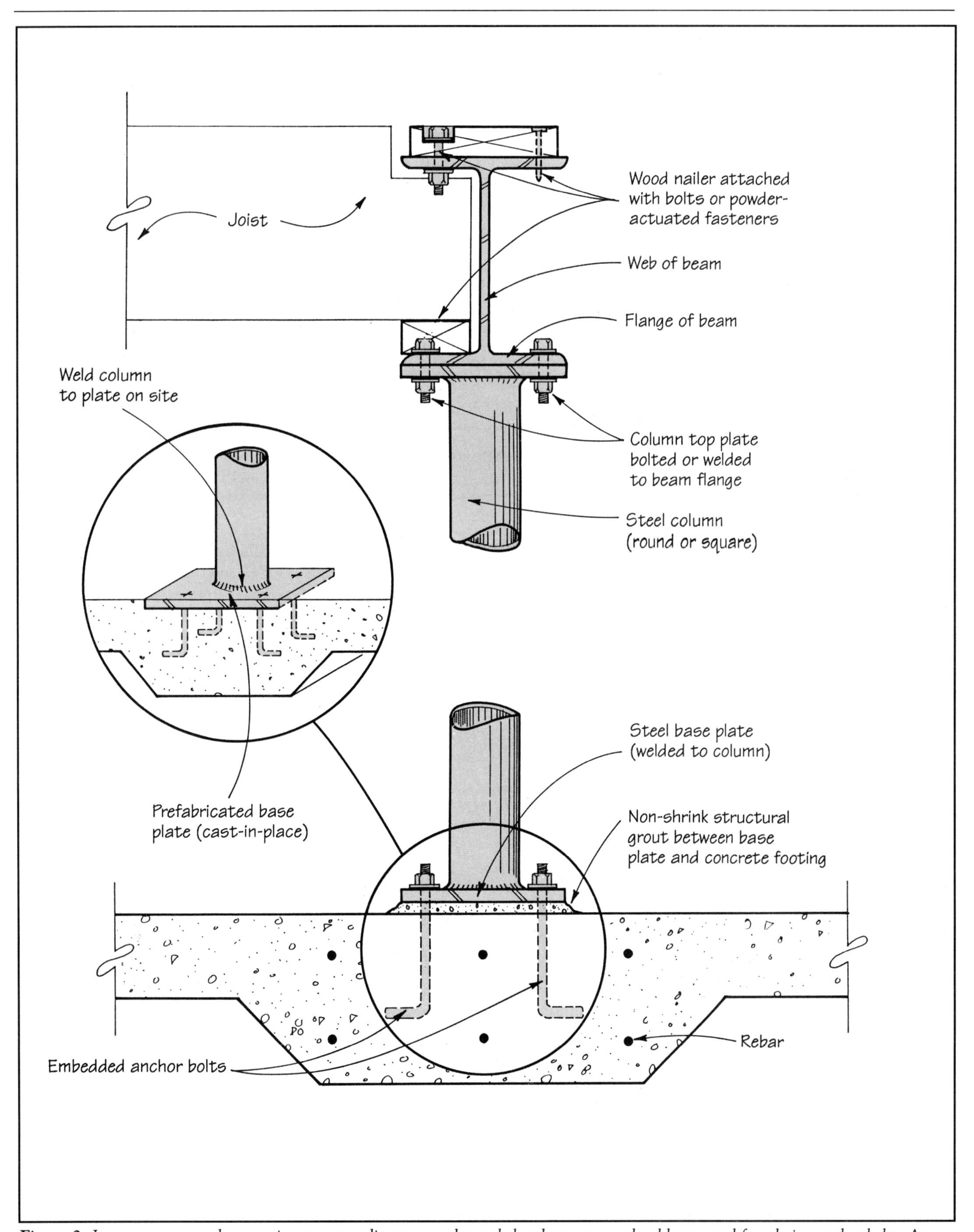

Figure 3. *In most cases, workers on site can use ordinary wrenches to bolt columns to overhead beams and foundation anchor bolts. An alternative is to weld column plates to beam flanges, and weld the column bases to steel plates cast into the slab or footings (inset).*

Steel Types and Sizes

	SHAPE & TYPE		SYMBOL	SIZE SPECS	STOCK LENGTHS	FINISH	USES
I-BEAMS	I	Junior *	M	Height in inches x pounds/LF (Ex: W8x24)	20, 40, 50, and 60 feet	Hot-rolled, mill finish	Carrying beams, headers, ridge beams, cantilevered beams
		Standard	S				
		Wide Flange	W or WF				
CHANNELS	[	Stringer *	MC	Height in inches x pounds/LF (Ex: C5x9)	20 and 40 feet	Hot-rolled, mill finish	Carrying beams, flitch plates, headers, ridge beams, columns
		Standard	C				
		Ship & Car **	MC				
TEES	T	Tee †	T	Height in inches x width in inches (Ex: T6x8)	20 feet	Hot-rolled, mill finish	Lintels, ledgers, light-load columns
ANGLES	L	Equal Legs	Angle in degrees or L	Leg x leg x thickness (Ex: 3x6x1/4")	20 and 40 feet	Hot-rolled, mill finish	Lintels, ledgers, web and flange reinforcements, joint clips
	L	Unequal Legs					
BAR STOCK	—	Flats	N/A	Thickness x width (Ex: 1/2x8)	20 feet	Hot-rolled, mill finish	Column plates, splice plates, machinist parts, tools
	●	Rounds		Diameter (Ex: 2")	12 feet and random	Cold-rolled, pickled and oiled	
	■	Squares		Width of one side (Ex: 1")			
PIPE	○	Sch. 10 *	BPPE (black pipe plain end) or BPTC (black pipe threaded coupling)	Inside diameter x schedule weight (3" sch. 40 BPPE)	21 to 24 feet	Hot-rolled, mill finish, painted, hot galv.	Columns
		Sch. 40					
		Sch. 80 **					
TUBING	○	Round	ERW (elec.-res. welded) DOM (drawn over mandrel)	Outside dim. or diameter x wall thickness (Ex: 2x1/8" round; 2x4x1/4" rectangle)	20 and 40 feet	Hot- or cold-rolled, pickled and oiled	Handrails, balusters, specialties
	□	Square					
	▭	Rectangular					

* Also called "Lightweight"   ** Also called "Heavyweight"   † Made by splitting I-beams in half

Note: *Each grade of structural steel has a specific quality as described by the American Society for Testing & Materials (ASTM) standards. ASTM A-326 is the predominant grade in the structural steel market. It has a carbon content of .26%, which gives it relatively high strength (60,000 psi tensile), yet it is easy to weld and fabricate. It is produced in several different shapes, each of which has its own descriptive nomenclature, typical finishes, and stock lengths.*

and beams (Figure 3). Make plywood templates for the bolt hole locations, and use these to lay out the steel for drilling or to locate anchor bolts in concrete. If your steel is being fabricated by a steelworker in the shop, templates are the best way to ensure that everything fits when it's time to put it together on site.

An engineer should specify the diameter and length of the bolts needed for the type of connection and the load. These will probably be high-strength, structural steel (A-325) bolts, which are available from steel supply houses or steelwork shops. Most bolts can be tightened with an ordinary box-end or open-end wrench. If you have a large number of connections, you may want to use an impact wrench. A torque wrench is almost never needed unless the engineer specifies it.

Don't count on concrete to hold embedded steel in place. The concrete will shrink away over time, so you should attach the steel to anchor bolts. Or you can attach hooks to a portion of the steel that will be surrounded by concrete. These will "grab" and prevent movement of the steel once the concrete has shrunk.

Finishing Structural Steel

Most hot-rolled structural steel is covered with a smooth, hard, black scale of iron oxide that forms as the hot material comes off the rolling line and reacts with airborne oxygen. Left untreated, this "mill finish" steel will rust immediately if exposed directly to the weather.

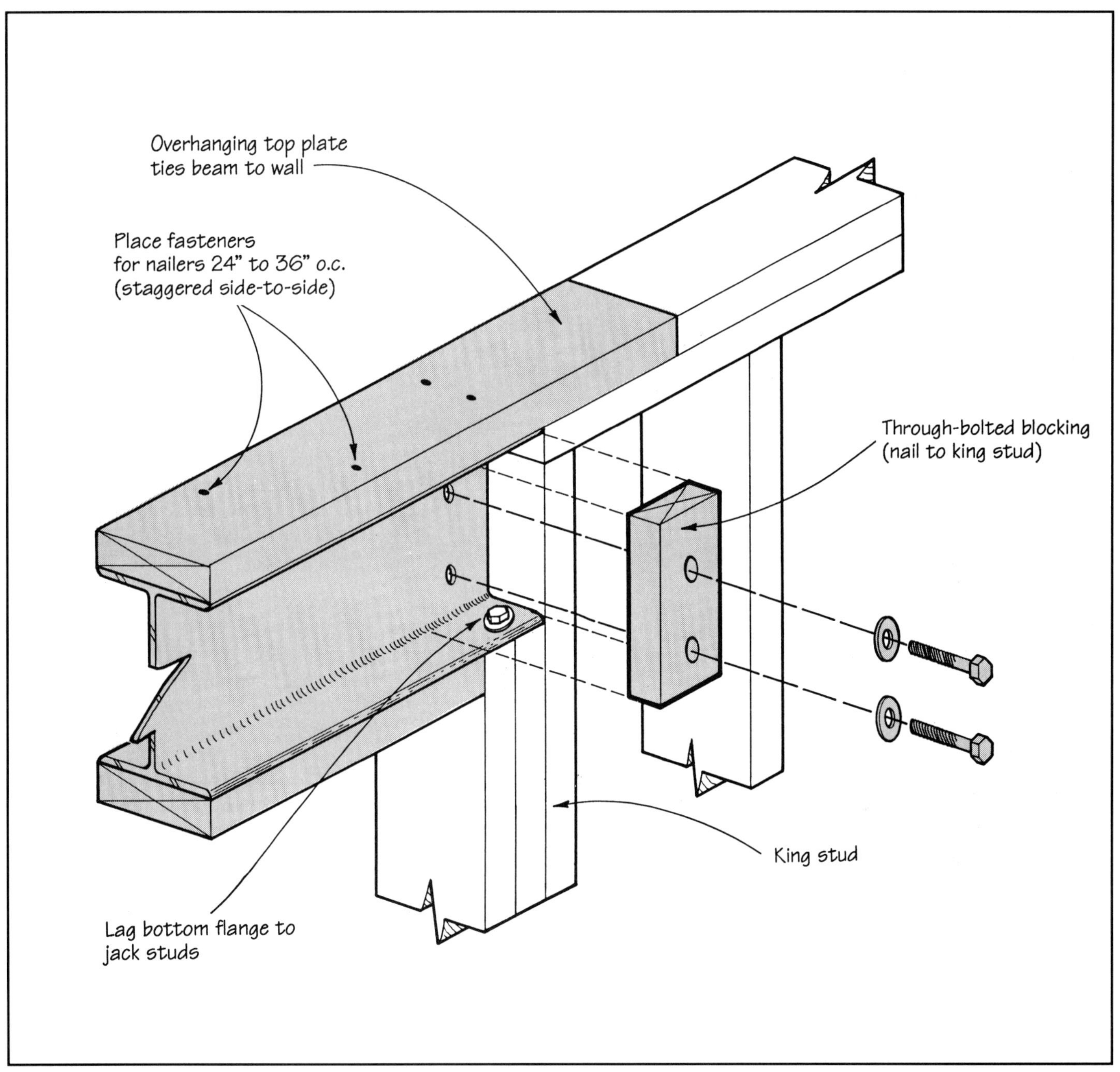

Figure 4. *Attach wood nailers to the top and bottom flanges of a steel beam with small-diameter bolts or powder-actuated fasteners. Space the fasteners 24 to 36 inches on-center, staggered side-to-side along the length of the beam. The overhanging nailer on top splices into the top plate of the adjacent stud wall. The bottom flange should be lag bolted into the jack studs, and vertical wood blocking bolted to the web and toe-nailed to the king stud.*

The best time to apply the finish is after all cutting and drilling have been done, but before you lift the steel into place. If you need to weld the steel after it's in place, you can either mask the weld area before applying the finish, or grind the finish off before welding. In either case, be sure to go back and finish these spots later.

For the simplest finish, remove the loose scale with a wire brush; and spray or brush-coat the steel with a coat of red or grey metal primer. This is sufficient for most situations where the steel will be protected from the weather, and will delay the inevitable growth of rust by a year or two, longer if the steel is buried in framing.

But even when steel isn't exposed directly to the weather, it can deteriorate from condensation. In cold climates, steel used as headers in exterior walls, for example, gets cold faster than the surrounding materials. This increases conduction heat loss through the wall and causes moisture to condense on the steel. Eventually, this moisture can damage interior finish materials and exterior siding and trim, as well as the beam itself. To prevent this, provide a thermal break between the steel and exterior sur-

Recipes For Steel

The word "steel" covers a large group of materials with many different characteristics. The basic recipe for steel consists of iron combined with varying amounts of carbon, which directly affects malleability and hardness. Steel with 1.5% carbon is extremely hard and brittle, as in cast iron. Carbon content of between .75% and 1% is considered "high" carbon steel and is the material commonly used for cutting tools. "Medium" carbon steel has less than 0.5% carbon and accounts for most structural steel which is hard but also flexibile and malleable. This oversimplified description doesn't account for a variety of trace elements added to structural steel to give it specific characteristics. Adding nickel, for example, produces stainless steel, and adding manganese increases resistance to wear.

The steel shapes you are most likely to use in residential construction are made by a process called "hot rolling." An I-beam, for example, is produced by passing a red-hot steel ingot repeatedly through a series of rollers. Hot-rolled steel varies slightly in dimension, but not enough to matter for most uses. It also results in a "mill finish"—a dirty, flaking scale that must be removed before finishing.

Another method creates common structural steel shapes from hot rolled coil stock. In this case, an I-beam is constructed from three flat pieces—two flanges and one web. The joints are welded together simultaneously in a process called "electro-resistance welding." This is a common method today among "mini-mills," which fabricate steel shapes, but do not actually manufacture the raw material.

When hot-rolled steel is put through another set of rollers after it has cooled somewhat, it is called "cold rolling." This process is used to size steel tubing and flat bars for precision machining. Consequently, cold-rolled steel has crisper edges and its dimensions are very accurate. Also, the scale is removed in a "pickling" bath and the steel is either sprayed with an oil preservative ("pickled and oiled") or galvanized. Because of the additional steps required in its manufacture, cold-rolled steel costs about twice as much as hot-rolled steel.

faces (rigid or expanding foam work well), and apply a good finish to the steel. An inside vapor barrier also helps protect the steel from interior moisture migration.

Outdoors, steel needs a much more durable finish, for example, on pipe handrails for exterior porches or stairs. Prefabricated pieces should be sandblasted and immediately coated with a steel primer. After installation, follow up with at least one top coat of enamel paint formulated for metal. This finish should last four to six years in direct weather before signs of rust appear.

Masonry lintels and other items which are to be embedded in concrete should have a hot-dipped galvanized finish, but this must be done by the supplier and is often too expensive for small builders. The alternative is to sandblast the raw steel and then use a special zinc-rich epoxy paint, such as that made by TNEMEC (6800 Corporate Dr., North Kansas City, MO 64120; 816/483-3400), on the surfaces that will be in contact with concrete.

For steel that has rusted but doesn't warrant a costly sandblast treatment, I have found a finish that seems to work. First, use a wire brush to remove the loose rust. Then brush, dip, or spray on several coats of Watco Danish Oil Finish (Minwax Co., 15 Mercedes Dr., Montvale, NJ 07645; 800/526-0495) or Penetrol (The Flood Co., P. O. Box 399, Hudson, OH 44236; 800/321-3444). The surface will be gummy unless the finish cures completely, which means you should apply this finish at temperatures over 70°F. The result is a polymer coating that encapsulates the rust and holds it to the steel. An additional advantage is that paint adheres well to this base finish. It is far from permanent, but is a good, quick alternative to more labor-intensive methods.

Lifting Steel Into Place

You can install many common steel components as easily as their wood counterparts because the major obstacles are the same for both materials: site access, length and weight of the pieces, and their height off the ground. The method you use depends on the conditions you face.

Preparation. Do as much as you can while the beam is still at ground level. If you ordered a stock length, first cut the piece to length. Even though steel beams are supposed to be very straight, check for a crown or bow in the steel by blocking it at each end and sighting along the edge of the flanges. As with wood beams, you will usually install steel with the crown up.

Then attach the wood nailers you need to the beam flanges and web. Prepare beam pockets or jack studs just as you would for wood framing (see Figure 4). Theoretically, steel doesn't require a larger bearing area than wood. But because of the greater loads it often carries, it's a good idea to increase the bearing area to avoid

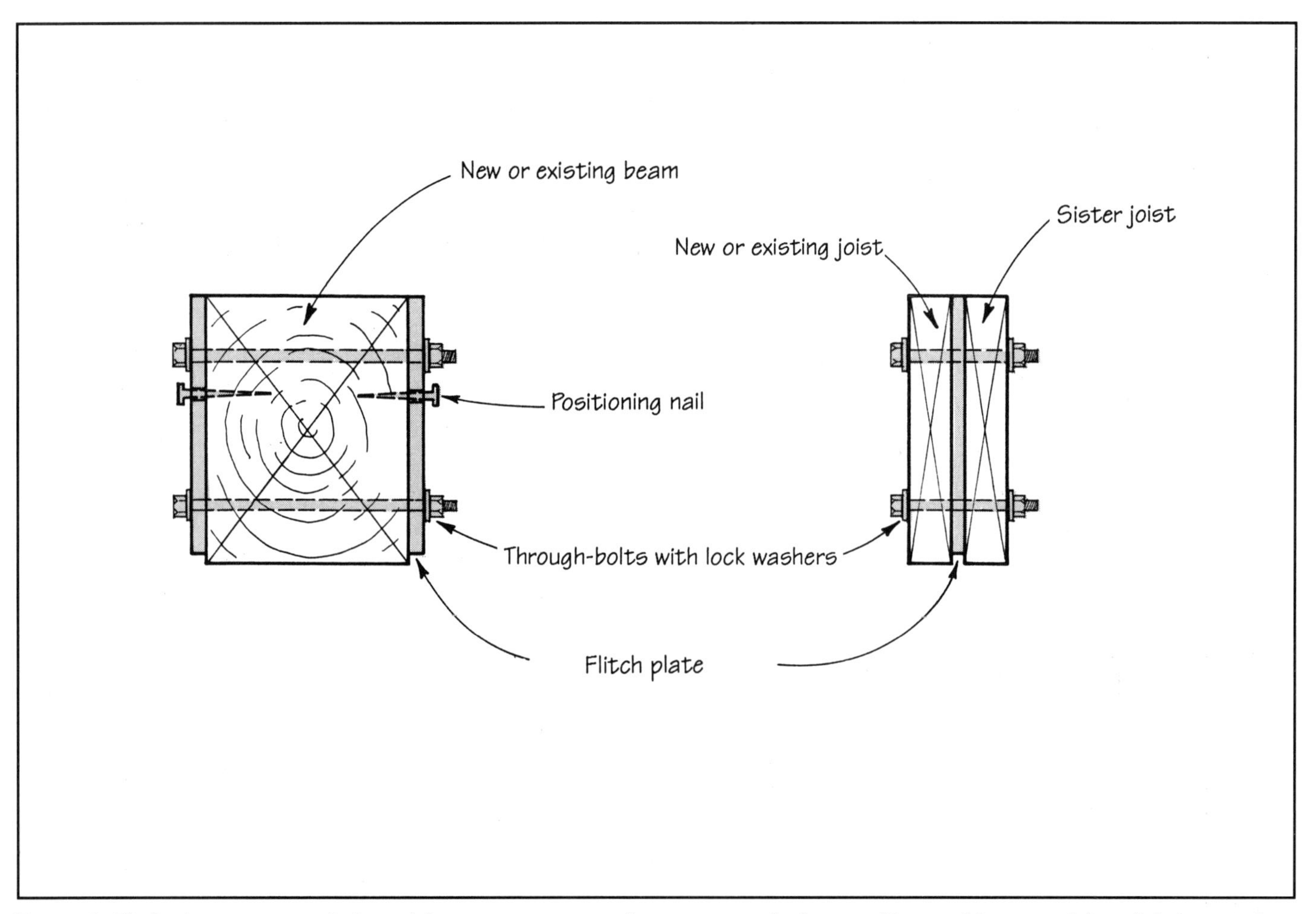

Figure 5. *Flitch-plates are particularly useful in renovation to reinforce sagging or broken wood beams. Use pairs of through-bolts spaced 16 to 24 inches on-center to make the connection.*

Cutting and Welding Steel

The most versatile tool for cutting steel is an oxy-fuel gas torch. It uses oxygen in combination with one of several gas fuels to produce a high-temperature flame that burns through the steel. Each oxygen-fuel combination has unique characteristics. Oxy-acetylene, for example, can only be used at low pressure, but burns at high temperature, so it is excellent for cutting steel. Any welding supply house can help you decide which fuel best suits your needs.

A medium-duty torch set costs about $300, and extra tips start at about $15 each. You can't buy gas tanks, but you can lease them for about $40 each per year. The cost for fuel will vary depending on the kind and volume you use. There are also minor costs for related items, such as eye protection and gloves.

A set of torches is a valuable addition to your toolbox. They are portable, require no power source, and, when used properly, can cut through up to 3-inch-thick steel plate with ease. At first your cuts will be rough, so you'll need a small portable grinder to smooth out the bumps. A grinder will also remove welding slag, which intereferes with finishing and detracts from the appearance of the final product.

There are other ways to cut steel, such as a portable band saw or a gas-powered saw with an abrasive cut-off wheel. These will produce clean, square cuts in small-section (up to 3/4-inch) materials, and require little operator experience. But they cost almost as much as a torch set and are far less versatile.

Welding. Welding is one of the most common methods of joining

Continued on next page

crushing the ends of wood columns. Always support steel on the end grain of wood posts, since wood resists crushing much better in this direction than across the grain. Where large point loads are concentrated, you may also need to add welded stiffeners to the web to keep the beam from deforming. This should be clearly called out in the engineer's specification.

Using a crane. A crane is sometimes the only practical way to lift steel into place. It is most useful when overland access to the building is poor (a steep slope, for example) or when the steel needs to be placed in the second story or higher, as with a ridge beam. Since a crane with operator can be expensive—about $60 to $90 per hour, including travel time to and from the site—make sure everything is ready before the crane arrives. If you can, make good use of the crane for other work, such as setting trusses or stress-skin panels.

Brute strength. If access to the site is good, however, steel can often be placed by hand. One extremely practical use for steel is in flitch-plate beams (see Figure 5). These can be used in renovation to shore up overstressed or broken wood beams, or in new construction to reduce the size of wood beams.

Another common use of steel in residential construction is a garage door header. Two men using brute strength can coax an 18- or 20-foot-long steel beam weighing up to 350 pounds into place without special equipment, as long as they know their lifting limits and have help at hand.

Even smaller carrying beams are heavy, so you should lift them in stages. Build a crib or step scaffold so you'll have a series of "platforms" to lift to. You can also use steel pipe scaffolding. If it doesn't have horizontal rungs at appropriate intervals, use solid planks or small wood beams that you can reset as you go up.

Using a backhoe. It may be tempting to use a backhoe to do the lifting. This is certainly possible, and may be convenient if you have a machine on your site, but I don't recommend it. Backhoes are not designed to do this kind of work. They can lift, to be sure, but backhoe controls often produce jerky motions that can be dangerous. If you try this, observe a few simple rules that apply to all situations where you are lifting overhead:

- Always stand to the side—not underneath—when anything is being lifted into place by machinery.
- Use a choker hitch with chain or a sling at the balance point of the beam. Chains or slings should be in good shape and rated to the required capacity.
- Attach a light rope "tag line" to one end of the beam. This allows someone at ground level to control the swing of the beam in midair.
- Wear a hard hat. Steel is much harder than wood and has sharper edges. You can get a nasty bruise if you bang your head against it.

Mechanical advantage. With a block and tackle, you can do the same job with fewer people and less physical strain. Make sure the equipment is rated for the weight you are trying to lift. A chainfall will do the job, too, but it lifts more slowly. Either method is inexpensive and far safer than lifting by hand.

You can also lift a steel beam using one or two hydraulic "bottle" jacks (available at any auto parts store) and a generous supply of cribbing. This is often the only practical method of placing steel inside existing structures where a crane or backhoe won't work, and where there isn't much room overhead for a hoist. The cribbing should be as large as possible in section, but easy to move by hand. Roughcut 6x6s are ideal, and 4x4s or sections of smaller dimension lumber are handy for shimming during the

---

pieces of steel. You can weld with a gas torch, but most welding in residential construction is shielded-metal-arc or "stick" welding. Arc welding uses electricity to fuse steel at a joint with a small-diameter, specially-coated welding rod. With the proper rod and welding technique, an arc weld is as strong as the steel itself.

Welding equipment is fairly expensive. And, like torch work, welding requires a great deal of job experience to produce good, reliable welds, especially in out-of-position field situations, such as overhead work. Nevertheless, I would encourage anyone interested in steelwork to learn how to weld. Be sure to give yourself plenty of time, and learn from qualified teachers. To do otherwise is to cheat yourself and invite trouble.

Safety precautions. Even if you subcontract all your steelwork and never use these tools yourself, you should understand the potential hazards they create. Oxy-fuels are volatile and must be properly stored and transported. And the fumes produced during oxy-fuel cutting or arc welding are extremely toxic. Be sure the work area is well-ventilated and avoid direct exposure to the fumes for extended periods of time.

The shower of sparks generated while using these tools is actually molten steel. It can very quickly cause severe burns, so you should always wear protective clothing (usually leather), including gloves, apron, heavy boots, and a face mask. Protect combustible materials near the work area from the heat and flying sparks. Be sure to check for smoldering materials, like sawdust or insulation, and have a fire extinguisher handy.

Finally, irresistable as the urge may be, never look directly at the flame of a cutting torch or the burning electrode of an arc welder. The light they generate is so intense it can permanently damage your eyes. Always wear welding goggles (sunglasses won't work), and be sure to have extra eye protection on hand for others working nearby.—*E.B.*

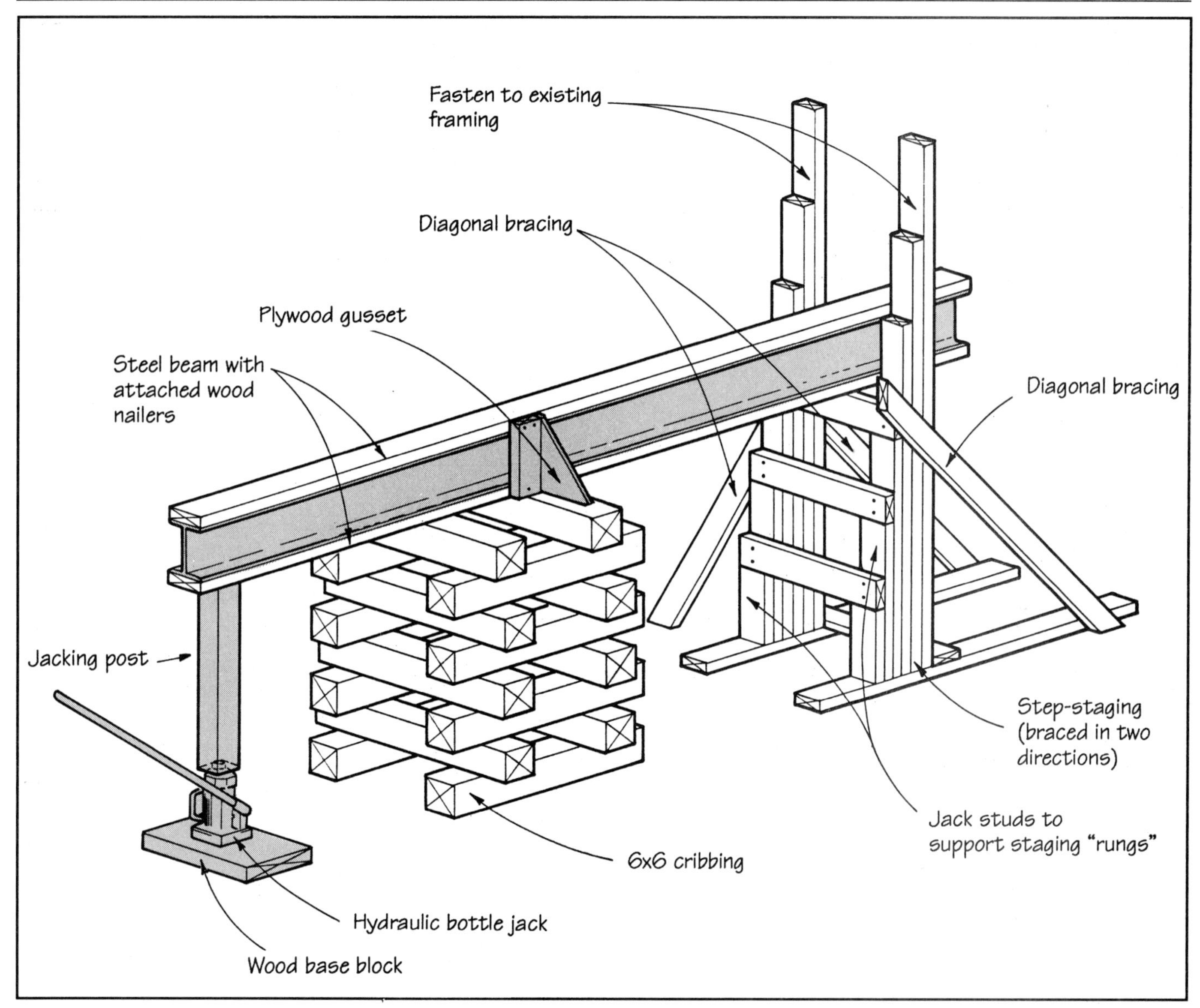

Figure 6. *When a hoist or crane is impractical, you can lift a steel beam into place with one or two hydraulic bottle jacks. Build a crib or step-staging to temporarily support the beam while you reset the jacks. Work the beam from the ends, but don't let one end get too much higher than the other or the beam may slide. Plywood gussets attached to wood nailers will help keep a tall beam from tipping.*

last foot of the lift. Keep the base of the crib as wide as possible (3 feet minimum), and narrow it slightly as you go up. If you can't use cribbing, build a step-staging with 2x4s or 2x6s. Use jack studs to support horizontal scaffolding pieces that will carry any weight (see Figure 6). Never rely solely on nails to do this; jack studs are not only safer, but they inspire confidence.

Work the beam close to the ends, lifting and blocking first one end, then the other. Don't lift one end too much higher than the other or the beam might start to slide. A long beam, especially one whose section is taller than it is wide, can tip on its side if you're not careful. Plywood gussets screwed to the wood nailers will keep the beam stable. If you go slowly and are sure of your moves, you can lift many thousands of pounds safely into place this way.

Securing the steel. Don't put the jack studs or columns in place until the beam is at its final elevation. Keep some thin plates of steel in 1/8-inch-thick increments for shims. These should have holes cut to match the bolting pattern of the beam and column plate. If you can't use steel shims, at least use hardwood or plywood, which will compress less than softwood. And make sure the shims spread the load fully over the bearing surface.

Once the beam is in place, use lag bolts to fasten the bottom flange to wood jack studs, or bolt the flange to the top plate of a prefabricated column. You can further secure it by nailing into the adjacent framing through the wood nailers you bolted into place while the beam was still on the ground. Triangulate the assembly with braces, which you can fasten to these nailers as well. If the beam is bowed, string a line on it and straighten it as you frame it in. It won't give as much as wood does, but it's usually straighter to begin with. The overhanging top plate shown in Figure 4 is strong enough to both hold the wall together and keep the beam from rolling.

When To Hire A Sub

Many times the best approach to steelwork is to hire a subcontractor to handle all or most of it. They will have the proper tools for the job, and their experience will ensure a safe, high-quality installation. Another thing to consider is the effect that doing the work yourself will have on your workers-compensation rate. Insurance companies classify steelwork as hazardous and will probably raise your premium to extend coverage to your employees. If your employees are not experienced steel workers, limit your involvement to routine installations.

If the job calls for two or three good-sized beams or requires crane work or a lot of fabrication, you ought to consider subbing out the materials and installation as a package. You can price the steel and decide for yourself if the subcontractor's quote is fair. Unless you've had some experience with steelwork, you'll have to guess at the labor costs. But you can assume that most experienced steel fabricators will get the job done faster than you can.

This doesn't mean you can't work together with a subcontractor. In most cases, you can do almost all of the prep work at the site and some of the installation. This includes taking dimensions and establishing elevations, preparing templates, and making sketches of joints, bolt locations, and other details. If your information is accurate and complete, the steel subcontractor can fabricate the steel in the shop, and needs to come to the site only to weld the columns to the beams.

You can save more time and money by planning ahead a little. Instead of making templates to locate holes in column plates, you can have your subcontractor fabricate steel column base plates with anchor bolts welded to the underside. In this case you need to accurately place the base plates in the concrete footing or slab before it cures, as shown in the Figure 3 inset. Then the fabricator can come to the job with over-length columns with no base plates, shoot the elevations on the spot, and cut and weld the columns directly to the plates embedded in the slab. ■

Eric Bauer is the owner of Fayston Iron and Steel in Fayston, Vt., which fabricates, installs, and repairs steelwork of all kinds.

Metal Building Basics

Assembled like a big erector set, metal buildings offer residential builders an easy inroad to commercial construction

by Donna Milner

Layout and assembly is best done on the ground with this post-and-beam endwall.

The thought of cold-rolled steel construction may not warm your heart, but it might pay to consider it as an alternative to wood-frame, residential building. After all, how many really *good* clients have you had in the last two years? And how many *bad* ones, where that last payment was held up because of a nail pop? With metal buildings, you won't have nail pops; you will have clients who want a lot of space in a hurry and are willing to pay for it.

Pre-Engineered Metal Building Types

There are two basic types of pre-engineered metal building: the rigid-frame building, and the metal post-and-beam building.

Rigid-frame buildings. Rigid-frame buildings consist of modular frame sections. These frame sections are designed to provide a clear span without interior columns. In a rigid-frame system, the rafters and columns are rigidly bolted together so that the connections transfer all forces down to the foundation. In this way, they function like a continuous arch.

These heavy and rigid "moment" connections transfer "bending moment" from the rafters to the columns, allowing for a more efficient use of the material. It is important that the contractor and crew take great care to assemble the building properly so connections function as the designer intended.

The rafters and columns in a rigid-frame system are usually tapered, providing the thickest cross section of steel at the places where the load is the greatest (see Figure 1). Because rigid-frame systems are designed for standard clear spans of up to 120 feet (with spans of 300 feet available on special request), they span large, uninterrupted spaces that can be used however the clients wish.

Post-and-beam metal frame buildings. The post-and-beam metal frame works the same way a post-and-beam wood frame does. Roof rafters are supported on columns that transfer the vertical load to the foundation (see Figure 2). Some fabricators use horizontal members similar to collar ties, to stiffen the frame sections. Otherwise they must be attached to an adjoining rigid frame.

Post-and-beam frames are less costly than rigid frames, because they use less steel in the rafters and columns. However, they require the use of interior columns, which interfere with the interior space. Consequently, they are used primarily for end walls only (in conjunction with rigid frames).

Unlike rigid frames that are modular and can be expanded at any time by removing the wall panels and adding more bays, post-and-beam end walls are not expandable.

Erecting a Rigid Frame

Since the rigid frame is the most common structural system for pre-engineered metal buildings, we'll focus on that for the rest of the article. First, we will outline the con-

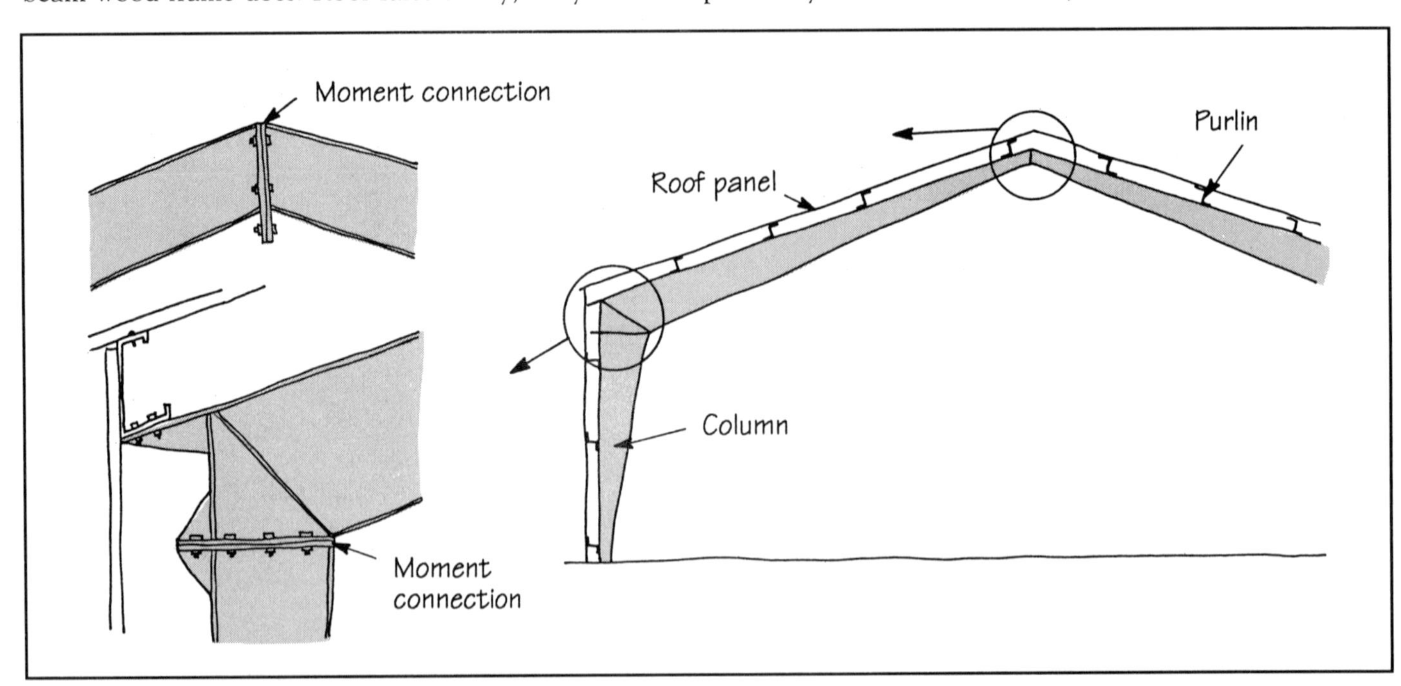

Figure 1. *Rigid-frame metal buildings provide uninterrupted space. The rigid "moment" connections (details enlarged) make the frame function like a continuous arch. Rafters and columns have the largest cross sections where loads are greatest.*

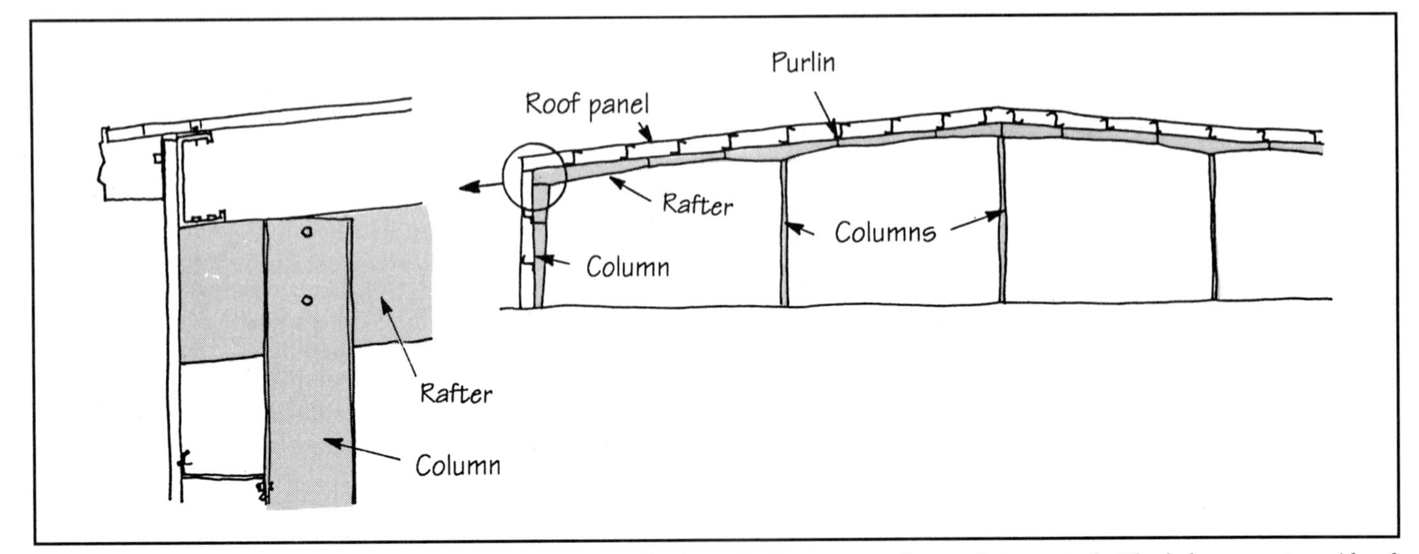

Figure 2. *In post-and-beam buildings, columns transfer the loads to the footings—so less steel is required. The bolt connections (detail enlarged) are simple pinned joints—so bracing is needed for rigidity.*

struction sequence.

The foundation. Rigid-frame buildings start with a foundation. Before assembling the building, the contractor must make sure the concrete is sufficiently cured. If not, there is a chance that the anchor bolts may pull loose, the concrete itself may chip and crack, or heavy equipment used to set up the frame may crack the foundation.

Choosing the right concrete can speed up the completion time. Plain portland-cement concrete takes fourteen days to reach half its final strength, but seven days is often long enough to wait before beginning erection. High early strength concrete can be used in only three days.

Because the frame will require specific tolerances, the contractor should double-check the foundation measurements and the location of anchor bolts. The foundation should be level, square, and the right size.

The braced bay. The need to erect a braced bay is not a concept familiar to stick builders. But it is important in this kind of construction. The braced bay serves some of the same functions as diagonal let-ins or as plywood sheathing at the corners of houses. These stiffen the structure against wind and seismic loads.

When erecting a rigid frame for a metal building, erect the braced bay first. Wind load from the end walls is transmitted through the structural members into the braced bay. The braced bay contains the brace rods and other diagonal bracing required to withstand these loads (see Figure 3). It is critical that this bay be constructed level and square as all other bays of the building are built off of this one.

There are various ways to erect a braced bay, but one of the most common is to begin with the columns. Column flange braces or clips are attached to the columns first. Then set the columns flat against the foundation and at the proper elevation. After the columns are secured, tie together with horizontal girts and eaves struts to increase their stability (see Figure 4).

Once the columns for the braced

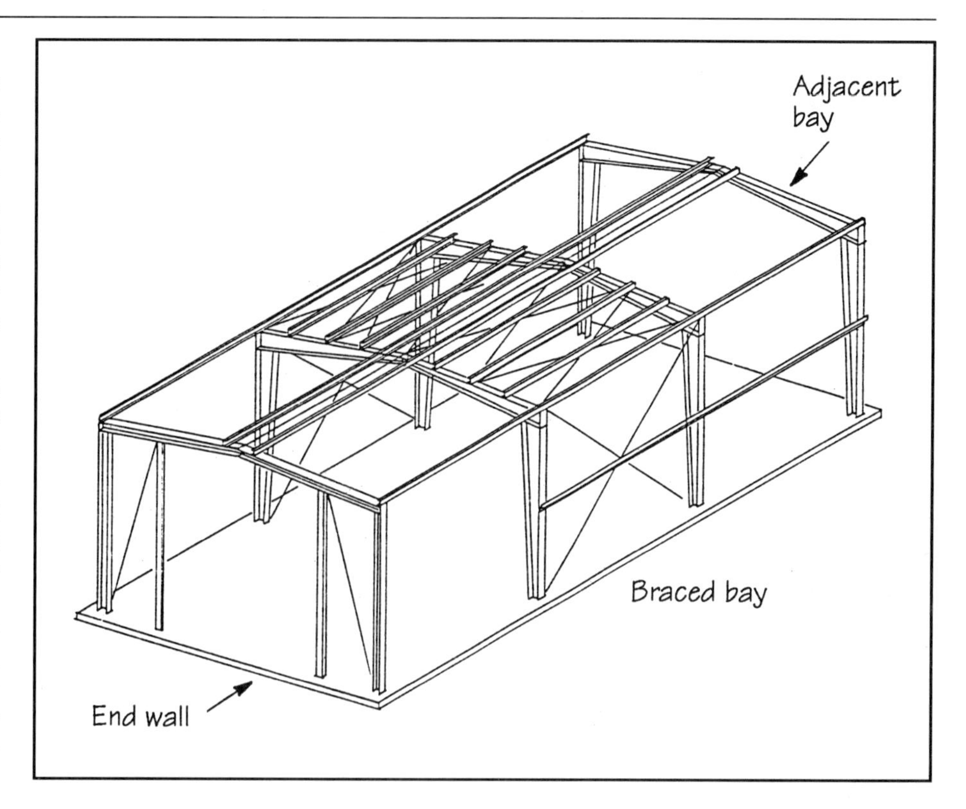

Figure 3. *With no sheathing to provide racking resistance, a braced bay is needed to prevent wind loads from collapsing the building.*

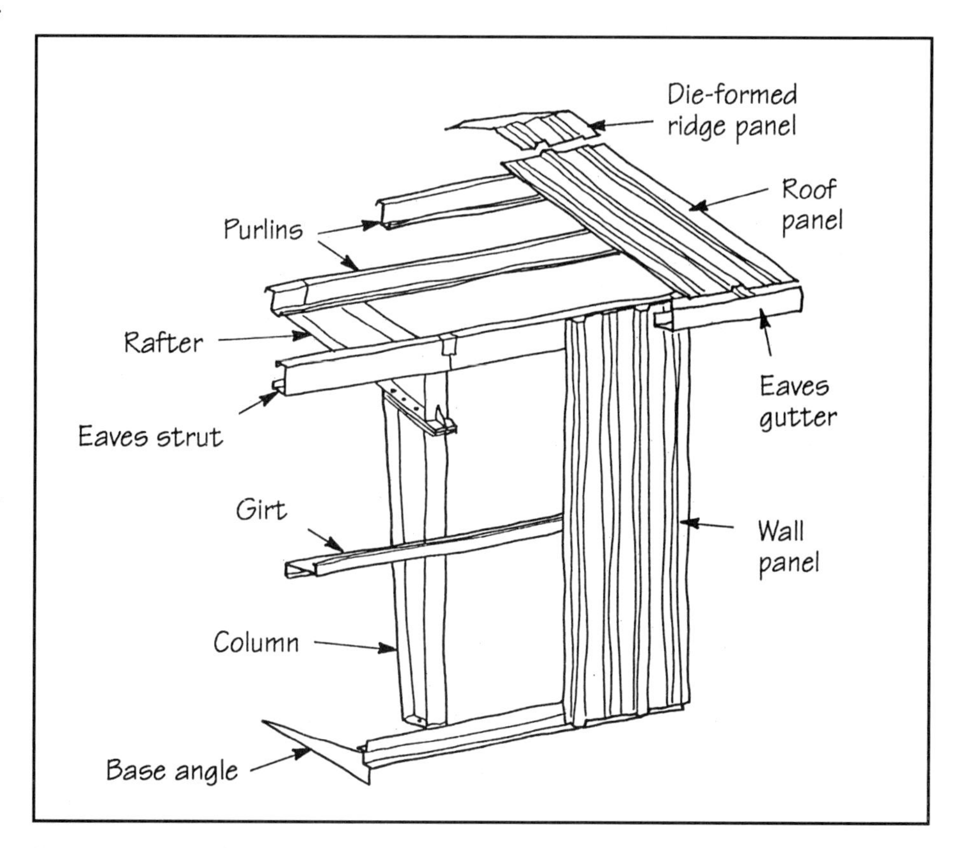

Figure 4. *On a rigid-frame building, girts, eaves struts, and purlins tie the individual frames together and provide support for roofing and siding.*

Marketing Metal Buildings

Manufacturer support makes the contractor's job a lot easier

by Chuck Stockinger

Because of their low cost, speed of erection, and open space, metal buildings are well suited to the space demands of small businesses, a sector of the economy expected to grow. In fact, four out of ten metal buildings constructed today are built for small clients: stores, banks, car dealerships, and other service businesses. And some 80 percent of metal building contracts are for structures 15,000 square feet and under.

As their use has increased, so has the range of design. Brick, wood, glass, and concrete are routinely combined with metal building systems. Improvements in metal roof coatings and finishes have helped make them more attractive as well.

Sales and Turnover

The transition from house building to metal buildings is not difficult. Erecting metal buildings requires a different business strategy, however. The metal building industry is heavily oriented toward negotiated design/build contracts. And the builder must be committed to staffing the company with an effective sales team since so many projects are negotiated directly with a building prospect.

Selling metal building system products has been made easier through the years, thanks to marketing support from manufacturers and to technical improvements in the systems themselves. Also, the prospects for improved cash flow can be attractive. Clayton Richardson, chairman of the Metal Builders Manufacturers Association (MBMA) says, "You will routinely find metal building contractors completing 10,000-square-foot projects within 90 days of their groundbreaking. Many ordinary buildings are just coming out of the ground when a metal building contractor is banking his final payment."

Builders Are Trained

The relationship between manufacturer and builder begins with a manufacturer's training school. Typical training schools cover the marketing, selling, pricing, and erection of the manufacturer's product line.

The manufacturer's engineering department is available for consultation should a builder or architect require special design features. However, most metal-building components are off-the-shelf items. The manufacturers provide catalogs with information about the span and load capabilities of various components.

Today, almost all metal building projects have some features custom-designed for the client. To allow such flexibility, many manufacturers provide customized, computerized design services.

Design/Build Popular

Many metal building dealers classify themselves as design/build contractors. They coordinate the entire job for the owner, including the design. So an owner has to deal with only one party.

The design/build contractor will plan the building with the owner, order the building, prepare the site, construct the foundation, erect the building, and install utilities. Any engineering or architectural work is done by either in-house staff engineers or architects, or by arrangement with outside consultants. The design/build contractor hires all subcontractors.

The size of an erection crew will vary with the job size. As a general rule of thumb, the work will start with a small crew to lay out structural members. More workers are added as required. The structural steel crew is usually separate from the crew installing the sheeting and flashing.

A small 7,500- to 10,000-square-foot building can be erected with one crane and its operator and a five-man crew, consisting of a working supervisor, two lead crew members, and two assistants. Buildings over 10,000 square feet typically will require two more crew members.

Marketing Support

Manufacturers also provide a wide range of marketing and promotional support. Many use co-op advertising programs, such as newspapers, television, the *Yellow Pages*, and *Sweet's Catalog*.

MBMA offers sales support to manufacturer members as well. The Association has published an extensive library of insurance literature, including a slide presentation. The presentation is designed to help MBMA manufacturers assist their builder networks in selling metal building systems against competing forms of construction.

Interest in metal building systems also has grown among allied construction organizations. Associated Builders and Contractors, Inc., for example, has established a National Metal Building Council for members wishing to learn more about this form of construction.

All forms of construction are vulnerable to economic downturns. But metal building systems can add a profitable new dimension to a home builder's operations. For more information, write to Metal Building Manufacturers Association, 1300 Sumner St., Cleveland, OH 44115. ■

Mr. Chuck Stockinger is general manager of the Metal Building Manufacturers Association.

bay and adjacent bays have been set, plumb them and tighten all permanent bracing rods. Bracing allows the columns to form a rigid support for the frames when they are raised and set into position.

Assembling rafters. As much assembly as possible is done with the members on the ground. This is because it is easier, safer, and faster to align and tighten bolts on the ground than it is in the air. For bolting structural members, high strength A325 steel bolts are usually used.

First, lay out the members on blocks in their approximate positions and fasten them together. Bolts are tightened, either with a torque wrench or by the turn-of-the-nut method, to the tightness required by the manufacturer.

The turn-of-the-nut method is probably unfamiliar to most stickbuilders; however, it is commonly used to tighten high-strength steel bolts. First, align the holes with spud wrenches and drift pins. Bolts are placed in any holes that do not have drift pins in them, then brought to "snug." ("Snug" is the point at which an impact wrench begins to impact. If a regular wrench is used, snug is as tight as can be turned by a man exerting full effort.)

When all the bolts are snug, all surfaces are drawn into full and complete contact. The drift pins are knocked out and remaining bolts installed. Then tighten all the bolts to a fraction of a turn as specified by the manufacturer.

To maintain the integrity of the moment connections, use caution when reusing A325 bolts. (This might happen if a joint is disassembled and then reassembled.) The threads on these bolts can stretch, so they may be reused only once or twice.

At this time, also install purlins, clips, and flange braces. They are bolted to the structure, but not fully tightened.

Setting rafters. Raising the rafter assemblies is done with extreme caution, even by contractors who are experienced at metal building assembly. Improper rigging can endanger workers as well as damage the structural members. Training manuals provided by the manufacturer describe proper methods of rigging and raising a load.

A crane helps erect the pre-assembled components.

A crane is used to raise the rafter assembly and guide it into place. When the rafter flanges are lined up with the column flanges, they are bolted, but not fully tightened. The crane boom is then repositioned slightly to allow slack in the rigging, and to allow the roof beam to come under its own dead load. The bolts get a final tightening, then erection cables and brace rods are installed to stabilize the bay.

The second rafter is installed in basically the same way, except that erection cables do not have to be attached. Once the second rafter is bolted in place, install the purlins to provide stability to the structure.

Purlins and Girts

Install purlins at all panel points. A panel point is the point where brace rods intersect the roof beams. Once the purlins are installed at the panel points, the remainder of the purlins are added.

Purlins can be either simple span purlins, which butt together and are installed with clips, or continuous purlins, which overlap and are bolted together.

Like purlins, girts are either simple span or continuous. Simple span girts are bolted directly to the column and terminate at the centerline of the column. Continuous girts are bolted together and to the column.

Brace rods, brace cables (if any), and flange braces complete the construction of the brace bay. Bracing rods and cables are critical to the rigidity of the structure and protect the building against swaying in the wind.

End bays. After the braced bay has been erected and stabilized, the next step is usually the erection of the end walls. The end walls can be one of two types—a lightweight post-and-beam end wall or a rigid-frame end wall. If a rigid-frame end wall is used, the building can be expanded in that direction at any time by simply extending the foundation, removing the wall panels, and adding as many bays as desired.

In the rigid-frame end wall, the intermediate columns carry only wind loads. All of the structural dead load and roof live load is carried by the rigid frame, just as in the other bays. If, on the other hand, a post-and-beam end wall is used, the post-and-beam end wall will have to be replaced if more bays are added at some time in the future.

The intermediate columns in a post-and-beam end wall carry not only the wind load, but also carry a portion of the roof live and dead loads. Lightweight post-and-beam end walls can often be completely assembled on the ground and raised as a unit, unless the span exceeds 60 feet. In this case, the wall is erected similarly to a regular bay. The columns are set in place first; then the rafter is set; and purlins to the braced bay are installed.

Enclosing the Shell

Once the framework is set and squared, you can encase the frame with whatever insulation and wall panels you like. Wall-panel material can be as simple as steel metal sheeting, or as complex as sandwich panels with rigid insulation bonded to a metal facing.

Base angles are used to anchor the wall covering sheets to the foundation. Sometimes anchor bolts are cast into the concrete, but typically base angles are secured with special masonry anchors.

Panels with exposed fasteners. If exposed-fastener wall panels are used, the insulation is usually installed along with the panels. Fiberglass insulation blankets are hung from the eaves strut and dropped to the foundation. Then the panels are installed over the blankets, with fasteners going through the insulation.

Because the panels squeeze the insulation blankets at the points where they are fastened to the girts, there is a loss of R-value at these locations. Some manufacturers have developed thermal blocks that you can install between the frame and the panel to minimize this effect.

Most panel configurations require the use of closure strips. Closure strips are metal, rubber, or foam strips that match the profile of the panel. You insert these strips in the bottom of the panel to prevent dust, rodents, and insects from entering the wall, and to act as a weather seal. If a closure strip is required, it is installed before the panel is fastened to the frame.

When installing the panels, you must line them up with the framing modules. Getting off the module even slightly will create a cumulative error that will cause problems.

Contractors must train and supervise crew members to tighten exposed panel fasteners. Too much tightening is likely to dimple the panel and may form a pocket that can collect water.

Panels with concealed fasteners. Concealed fastener panels use interlocking devices on the panels or specially designed clips. These panels are stronger and are part of the structural system of the building.

The basic installation procedures are similar. However, the insulation may be installed after the panels are installed. In this case, you have a choice of insulating materials, including fiberglass blankets or friction fit batts in the cavity.

This eliminates the problem with compressed insulation. It does, however, create an area of thermal bridging where the metal panels are attached to the metal frame with nothing to stop the heat flow.

As an alternative, the contractor can use sandwich panels, with metal and insulation made up as a unit. Sandwich panels are available in thicknesses from 2 to 5 inches depending on the desired insulation value. The panels can be of either the exposed or concealed fastener type.

Either will solve the thermal bridging problems. The disadvantage is that the openings must be planned before ordering the panels from the manufacturer. While it is possible to job-cut sandwich panels by using shears and nibblers to cut the steel, and then sawing the foam, openings are best pre-cut by the manufacturer.

Lining up sandwich panels with the frame is extremely important. Otherwise joints might not fit tightly.

What's in It for You

Because of the precision required, erecting metal buildings can be a challenge. But contractors whose crews know how to use a tape measure, level, and crane should be able to adapt to this type of construction.

A contractor with good general construction knowledge can get training and support from the manufacturer. And once the metal is up, you're home free because interior work can be done with wood. Indeed, metal buildings offer one of the easiest ways to make the shift from residential to commercial/industrial building. ■

Donna Milner is an architect and research associate at the Small Homes Council/Building Research Council in Champaign, Ill.

Steel-Stud Partitions

Steel studs are a good choice for many remodeling jobs —residential as well as commercial

by John Gaal

Most remodelers would frame up a basement family room or new bedroom with wood studs and wood plates. And why not? This is the way we've done it for years. However, I've found that in many cases steel-stud framing, normally reserved for commercial jobs, costs less and takes less time.

The 25-gauge steel studs I use come in bundles of ten and cost about $1.52 per 8-foot stud, which is very competitive with the $1.50 2x4 studs we buy here in St. Louis. Once you get familiar with the materials and the way they're installed, you'll be able to save on labor compared to building with wood.

Steel studs are commonly categorized either as non-load bearing (NLB) "drywall studs" or load bearing (LB) "structural studs." The profile of the drywall stud is narrower than the structural stud, and one flange of the drywall stud is narrower so studs can be "nested" for bundling (see Figure 1).

Steel studs come in different gauges, which are similar to the differences in lumber grades for wood studs (see "Ordering Steel Studs," page 255). The 25-gauge NLB steel studs meet the needs of most residential partitions. Steel runners, used instead of top and bottom plates, come in 10-foot lengths and should be at least .0179 inches thick.

USG

It's the drywall that gives a typical steel-stud partition its stiffness and stability.

A.M. Applebaum

Figure 1. *The structural studs on the left are 16-gauge steel; the drywall studs on the right are 25-gauge and non-load bearing.*

Installation Overview

As with wood, your objective with steel is to build a partition wall that is securely anchored to the floor and the joists above. Start by placing the bottom runner; then plumb up to install the top runner. Finish by installing the studs.

Bottom runner. If you are building the partition wall on a wood frame floor, just screw through the bottom runner into the floor joists. In the case of a concrete floor, there are several options. The fastest method is using a powder-actuated tool (PAT) such as a Ramset D-60 (ITW/Ramset/Redhead, 1300 N. Michael Dr., Wood Dale, IL 60191; 708/350-0370). Be sure to follow all manufacturers' instructions whenever operating a PAT.

If a PAT is not available, or if the concrete is soft and shatters from the blow, you can use a roto-hammer or a hammer drill and some type of lead anchor or a Tap-Con (ITW/Ramset/Redhead, see above) fastener.

As a last resort, you can use construction adhesive or mastic, but you must clean both the concrete and metal track prior to application. Use soap and water on the concrete and paint thinner on the metal. I wouldn't recommend this method for a wall with a door, though, because the swinging of the door can displace the bottom runner.

Whichever anchoring method you choose, United States Gypsum (550 N. Brand Blvd., Glendale, CA 91203; 818/956-1882), a company that markets steel studs, suggests that track be fastened 24 inches on-center and within 2 inches of an end.

Top track. Once the bottom track is in place, you can plumb up with a plumb bob or straightedge and level; then snap chalk lines for the top track.

Anyone considering working with metal studs and track must invest in a screwgun, either AC or cordless. If you are installing a room in the basement, fasten the top (ceiling) metal track to the first-floor wood joists with 1-inch Type W, bugle-head, drywall screws with Phillips heads. These screws have threads designed to penetrate wood. In remodeling, using screws sure beats pounding 12d nails through 2x4 plates into 20-year-old floor joists and eating the dirt from above.

Installing studs. The third step is to anchor the studs to the track. Phillips-head TEK screws are the most popular screws for fastening metal studs to track. One screw on each side of the stud at the top and bottom is ideal. The most commonly used TEK screw is the 8-18½-inch TEK1. The "8" refers to the screw gauge, #8; the "18" indicates 18 threads per inch; "½ inch" signifies the length of the fastener; "TEK" denotes the drill point; and "1" implies a point for penetrating sheet metal. Although the most popular screw, the 8-18½ TEK1, tends to bulge the corners of the drywall.

I prefer using a ½-inch S12 round-washer-head self-tapping screw. It leaves a much smoother surface for drywall application later. Locking C-Clamp Vise Grips (American Tool Co. Inc., 108 S. Pear St., DeWitt, NE 68341; 402/683-2315) or Adapa clamps (Adapa Inc., Box 5183, Topeka, KS 66605; 800/255-2302; 913/862-2060) come in very handy for clamping studs to the track when screwing (Figure 2). Where the finish appearance is of the utmost importance, try using pop rivets or USG's metal lock fastener tool made for this purpose.

Working Metal

Carpenters who have mainly worked residential jobs have a distinct preference for wood. Working with metal requires different tools and details. But switching over is not difficult.

To cut metal track, use aviation snips or a cutoff chopsaw. You need a real chopsaw—the kind plumbers use for cutting pipe—not the electric miter box you use for trim work. Put on a carborundum blade, and you're ready for metal cutting. Don't forget to wear eye protection.

Installing track. Track should be overlapped at corners, leaving a gap for drywall (Figure 3). Overlapping saves on fasteners and makes a stronger connection. Also leave a gap at partition Ts for drywall to slide past (Figure 4).

Track can be butted, spliced, or notched when walls are greater than 10 feet long (Figure 5). Butting the track ends is sufficient but takes twice as many fasteners. When you lap the ends with a splice or notch, you have to use only one fastener. It's mainly a matter of preference, but most carpenters now are using the lapped joints.

Length and layout. Cut each stud

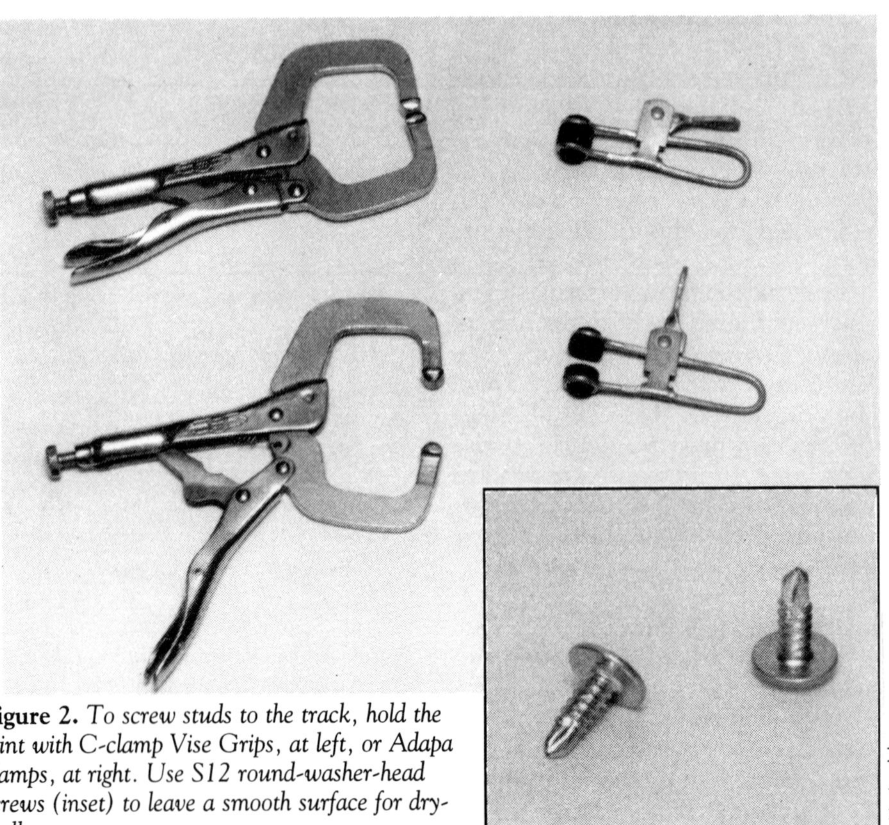

Figure 2. *To screw studs to the track, hold the joint with C-clamp Vise Grips, at left, or Adapa clamps, at right. Use S12 round-washer-head screws (inset) to leave a smooth surface for drywall.*

A.M. Applebaum

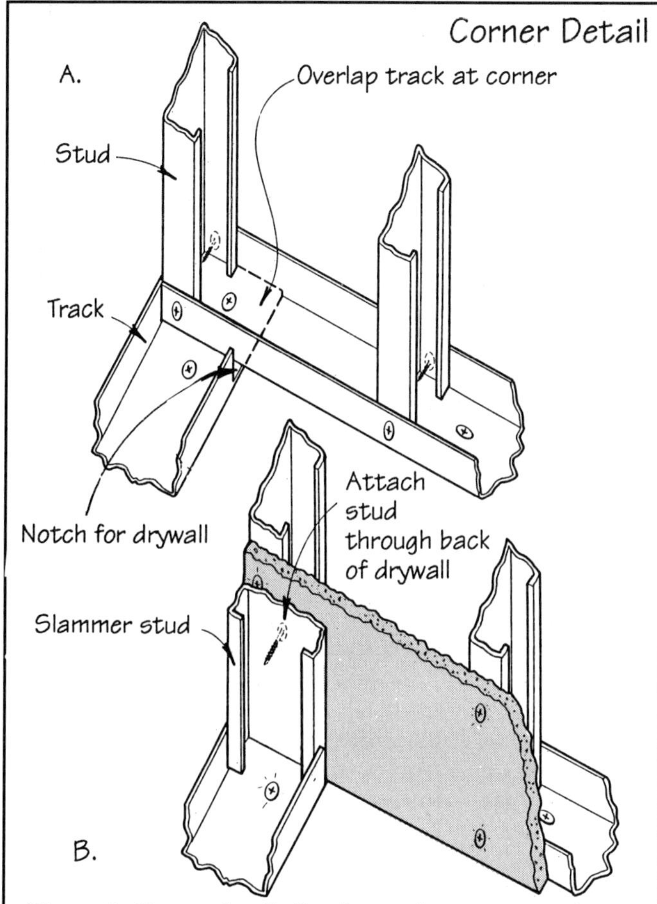

Figure 3. Corner detail: *Overlap track at corners as shown (a), notching to allow drywall to slide past the inside flange. Fasten screws through the back of the drywall into the floating "slammer" stud (b). The slammer stud isn't anchored to the track—just to the drywall.*

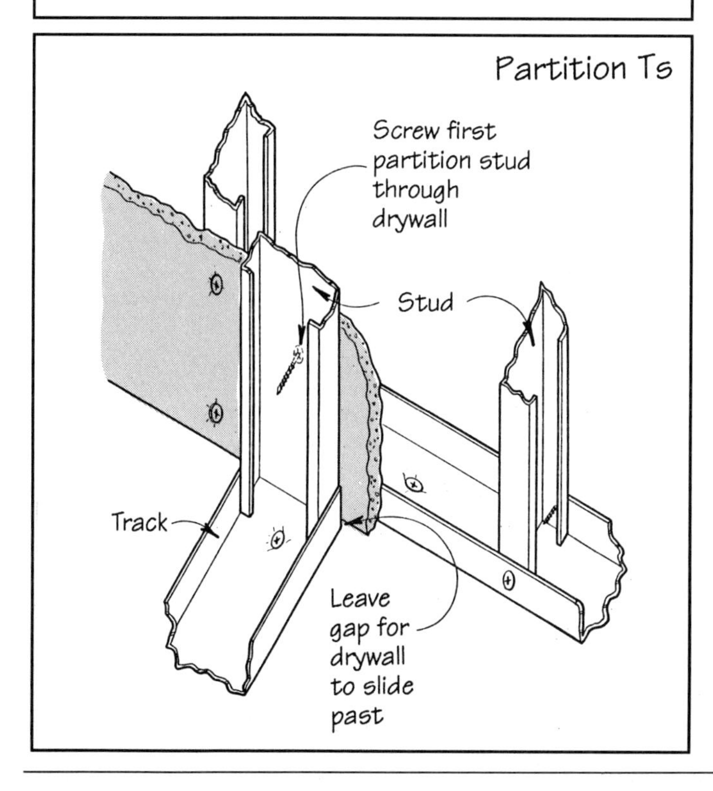

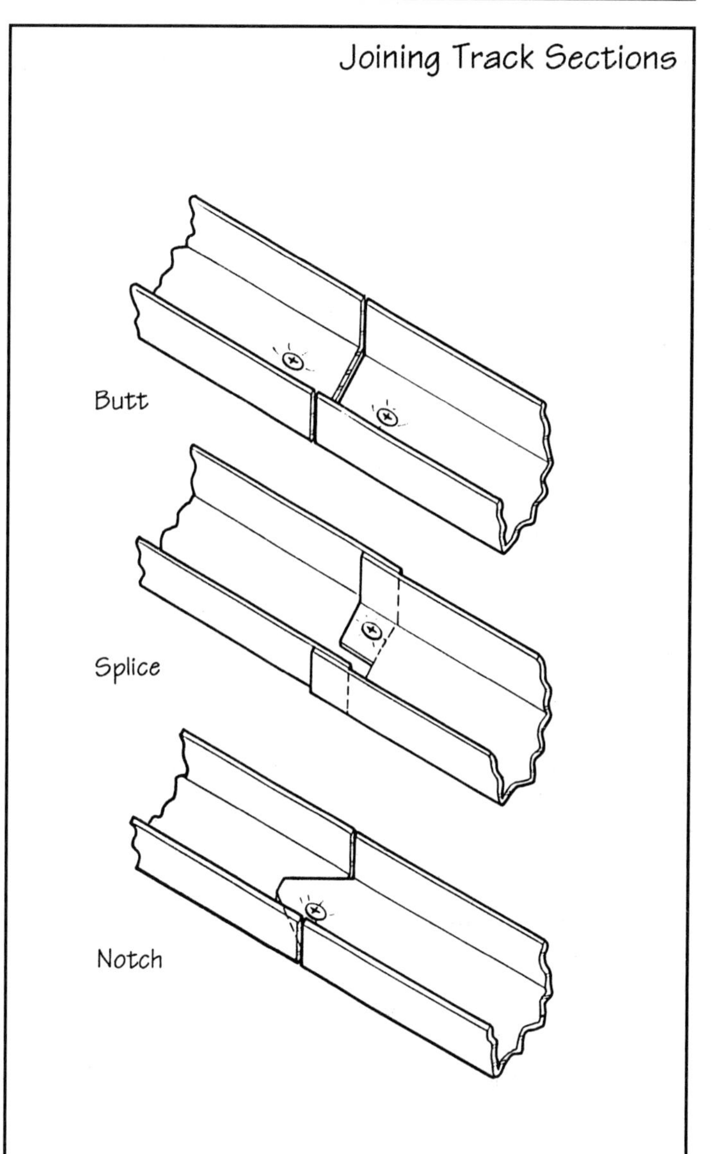

Figure 5. *Join sections of track by butting, splicing, or notching.*

Figure 4. Partition Ts: *Leave a gap in the floor and ceiling track at partition Ts for drywall to slide in. If you have access from the other side of the wall, fasten the first stud of the partition T through the back of the drywall. If not, attach the first stud at the top and bottom plates of the wall behind it, and "forkscrew" the stud to the drywall in several places. Forkscrewing is similar to toenailing: The screws are driven at opposing 45-degree angles into the drywall for better grab.*

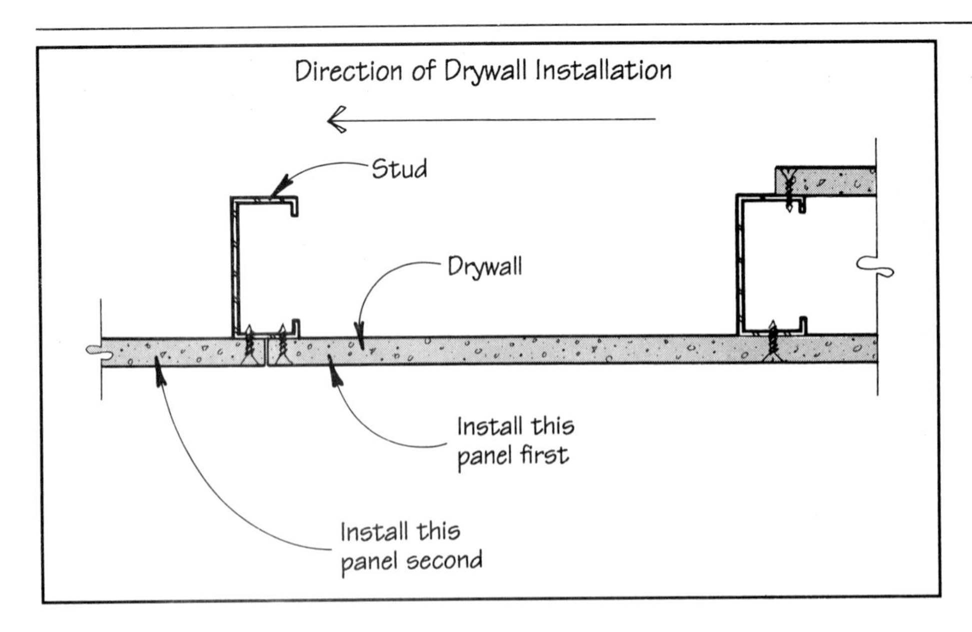

Figure 6. *Install studs so the C-sections are properly aligned for drywallers. Then install drywall in the correct direction. Otherwise, drywall will be bumpy.*

at the top with aviation snips or gang-cut them in a chopsaw. Unlike wood, you need not cut the stud to fit tight between the top and bottom track on NLB partitions. This really saves time when you're working on existing buildings. With wood studs, you would be marking each stud and cutting it to exact length.

Metal studs are laid out like their wood counterparts. A 24-inch-on-center layout is common for NLB partitions, although 12- or 16-inch centers can be used if you need closer spacing. You're better off using closer spacing if you intend to hang cabinets. If you're using 3/8- or 1/2-inch drywall, the 16-inch spacing will give a more even wall surface.

Stud throat direction. One thing is critical when you're laying out metal studs. You have to orient the studs correctly. Because the studs are C-shaped, the throats of the "on-center" studs point back towards the starting point of the layout. Drywallers will hang the gypsum board in this direction. The result will be a smooth surface for the tapers (Figure 6). The first panel that is installed anchors the stud flange and keeps it from deflecting. If you installed drywall from the other direction, the flange would deflect, and the edges of the panels would be uneven.

Door and window openings. Use metal track for header and sill material so cripple studs can interlock and follow the layout.

Studs at door and window openings must face the opening. I like to hold the metal studs in a door or window rough opening back 1 1/2 inches on each side. Likewise, I hold the metal track for the rough door header up 1 1/2 inches. This allows me to use wood 2x4s for jacks and the header. I can screw through the metal studs and track into the wood or through the wood into the metal. This gives me a nailing surface for easier installation of wood pre-hung door or window units.

Backing, bridging, and bracing. Wood and metal fastened between studs can be used as backing for sinks, grab bars, shelving cleats, etc. Bridging or bracing may be required where stud lengths become too great.

Utilities. Electric, plumbing, and telephone lines must be installed in the stud cavities before closing in both sides of the partition. One advantage of using metal studs is the pre-punched utility holes in every stud. For electrical, some residential areas will allow the use of non-metallic sheathed cable (Romex) in conjunction with plastic "grommet" insulators (Figure 7). Check your local codes. Most commercial jobs use thin-wall conduit or some type of armored cable such as B-X or Greenfield.

Metal or plastic outlet boxes can be screwed or riveted to the stud. For plumbing, plastic pipes require protection like the grommets mentioned above. Chase walls can be built for larger pipes similar to those built with wood studs.

Sound control. Prior to hanging drywall in a basement, consider installing insulation for sound control. With 4-inch, unfaced fiberglass insulation for 3 5/8-inch studs, you'll have a "friction fit" between stud cavities. This will provide a STC-44 (Sound Transmission Class) if 5/8-inch gypsum board is properly installed on both sides of the partition. The recommended rating for bedrooms in private residences is STC-37. Keep in mind that ratings for the floor/ceiling assembly must be considered too. See the references at the end of the article for more information.

Code considerations. Code officials will check on-center screw spacing for drywall into studs, and spacing along the top and bottom track. Check local codes for fastener spacing requirements.

Screw spacing is also a factor in a wall assembly's fire rating. Since screws provide up to 350 percent more holding power than nails, you can space screws further apart than nails.

Code officials see a lot of jobs where the metal studs are not fastened to the track, but are held in place by the drywall. They call this a "friction fit." Acceptance of this practice varies, so check your local code.

Drywall and Trim

After the studs are installed, begin hanging drywall on one side of the

A.M. Applebaum

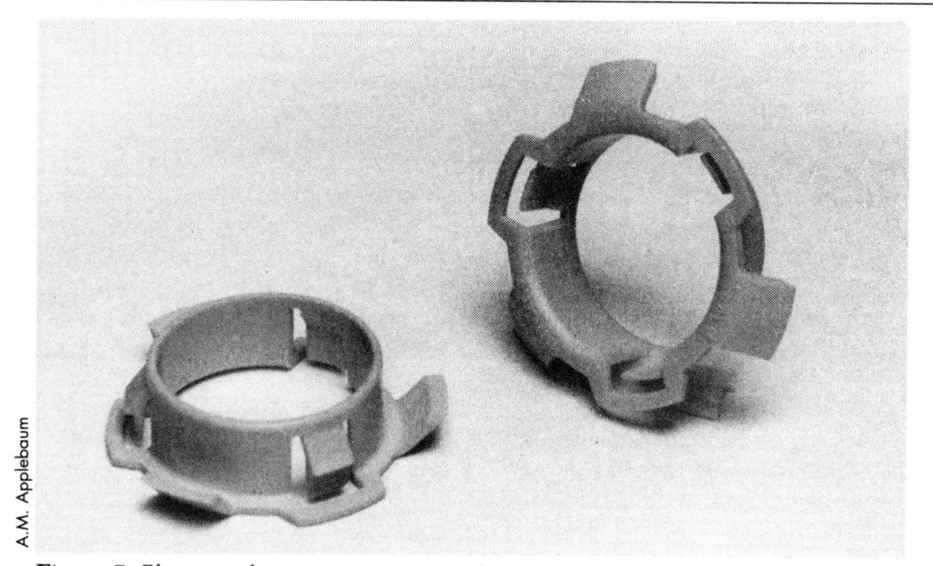

Figure 7. *Plastic insulators protect wiring and small PVC pipes running through steel studs.*

partition. Leave the other side open for any plumbing or electrical work. The drywall can be installed horizontally or vertically. The USG specs say to use 1-inch Type S, bugle-head drywall screws when applying 1/2-inch gypsumboard to metal studs. However, the 1 1/8-inch length is the most commonly sold in this area.

Wood base, chair rail, and crown mold can be attached to partitions framed with metal studs. There are several ways you can fasten trim. One method uses Type S or S12 trim-head screws. These screws resemble the drywall version mentioned above, except they have a much smaller head for countersinking. This makes it easy to fill holes with stain putty. Trim-head screws are fastened through the trim and drywall into the metal stud or track.

If you'd rather not use screws, you can use an air nailer and construction adhesive. The air nailer tacks the trim in place while the adhesive cures.

Pros and Cons

I've used metal studs on both commercial and residential jobs. I particularly like the precision of the metal studs because you don't ever have to

Ordering Steel Studs

Studs and runners are C-shaped and roll-formed from corrosion-resistant steel. Coatings are hot-dip galvanized or aluminum-zinc. You won't be able to run down to the local lumberyard to buy these studs, but most companies that manufacture them have 800 numbers. You can call and order what you need and pick them up from a local distributor. Manufacturers will even cut them to length for special orders.

Stud gauges & width. There are seven gauges available: 26, 25, 22, 20, 18, 16, and 14. The higher the gauge number, the thinner the metal.

Studs come in five widths: 1 5/8, 2 1/2, 3 5/8, 4, and 6 inch. USG has a helpful publication called USG System Folder SA-923 (1989) that lists the various stud spacings for each stud size and gauge. You should get this publication if you're going to be using steel studs on commercial jobs.

The width indicates the web dimension, which varies. However, the leg (flange) of the stud is always 1 1/4 inches. The corrosion-resistant coating is not considered part of the web and flange dimensions. Lengths range from 8 feet to 16 feet.

In my area, we commonly refer to steel studs as metal studs. USG denotes the 25-gauge 3 5/8-inch metal studs that I use in remodeling as 358ST25. They run around 19 cents per lineal foot in St. Louis.

Steel runners. Runners come in the same widths as the studs mentioned above. The length of a 25-gauge runner is limited to 10 feet. Once again, in my area, we refer to the steel runner as metal track. Metal track is available in two leg (flange) lengths; 1 inch and 1 1/4 inch. USG denotes this track style as 358CR25. The track costs about 17 cents per lineal foot.

In July 1989, Unimast bought USG's steel-stud manufacturing division. USG is the exclusive marketing agent for Unimast today.

— J.G.

Manufacturers of Steel Studs

Unimast (Formerly USG's steel stud division)
9595 West Grand Ave.
Franklin Park, IL 60131
708/451-1410

Steel Benders, Inc.
15550 West 108th Street
Lenexa, KS 66219
913/492-7274

Marino Industrial Corp.
Montrose Rd.
Westbury, N.Y. 11590
516/333-6810

send a bowed stud back to the yard.

As I see it metal studs have the following advantages:

- ease of handling
- straightness
- pre-punched utility knockouts
- fire/moisture resistance
- reduced sawdust mess

Their cost is comparable to wood's, and they can be used horizontally for ceiling framing if need be.

But there are some disadvantages too, particularly during the "learning curve" when you're getting comfortable with a different material and methods. They require some tools not common to wood framing, and their sharp edges will cut you if you're not careful. Also they're not readily available in some areas.

On your first job with metal-stud framing, plan for some extra hours. The time you take to educate yourself may pay off on some residential work and may give you the versatility to begin doing more commercial work. I believe once you have tried remodeling with metal studs, you'll find applications on upcoming projects where this system will save your company time and money. ■

John Gaal is a journeyman carpenter and teaches full-time at the Carpenter Apprenticeship School in St. Louis, Mo. Also from St. Louis, Mark Rurtz of Newger Materials and Mark Brennan of W.J. Brennan Co. assisted with portions of this article.

For More Information:

Drywall Contracting, J.T. Frane (Carlsbad, Calif.: Craftsman, 1987).

Fire Resistance Design Manual (Evanston, Ill.: Gypsum Assoc., 1984).

Gypsum Construction Handbook, (Chicago: United States Gypsum, 1982)

Sound, Noise, and Vibration Control, L.F. Yerges (Melbourne, Fla.: Krieger, 1983).

Section 12. **Energy: Design & Practice**

Owens Corning

High Performance Wall Systems

Energy-minded builders strive to save Btus while keeping construction simple

by Alex Wilson

A few years ago, during the height of energy awareness, every builder I knew used a different wall system. Builders were trying every option they could think of to minimize heat loss and produce a building shell tight enough for one of Jacques Cousteau's mini-subs.

Complexity was no concern. And as for cost, well...that was sometimes a problem. In fact, because so many extras were added in the interest of energy conservation, the builders sometimes found they couldn't charge for their own time if they wanted to come in under budget. Somehow the satisfaction of producing a zero-energy house made up for the poverty those builders often endured.

But times have changed since then, and *The Journal of Light Construction* was interested in learning how much the practices of these builders have changed. What follows is an informal survey of a handful of builders who I believe have been on the leading edge of energy-efficient home building for the past decade. Their experiments with numerous energy-conserving techniques can teach us a great deal.

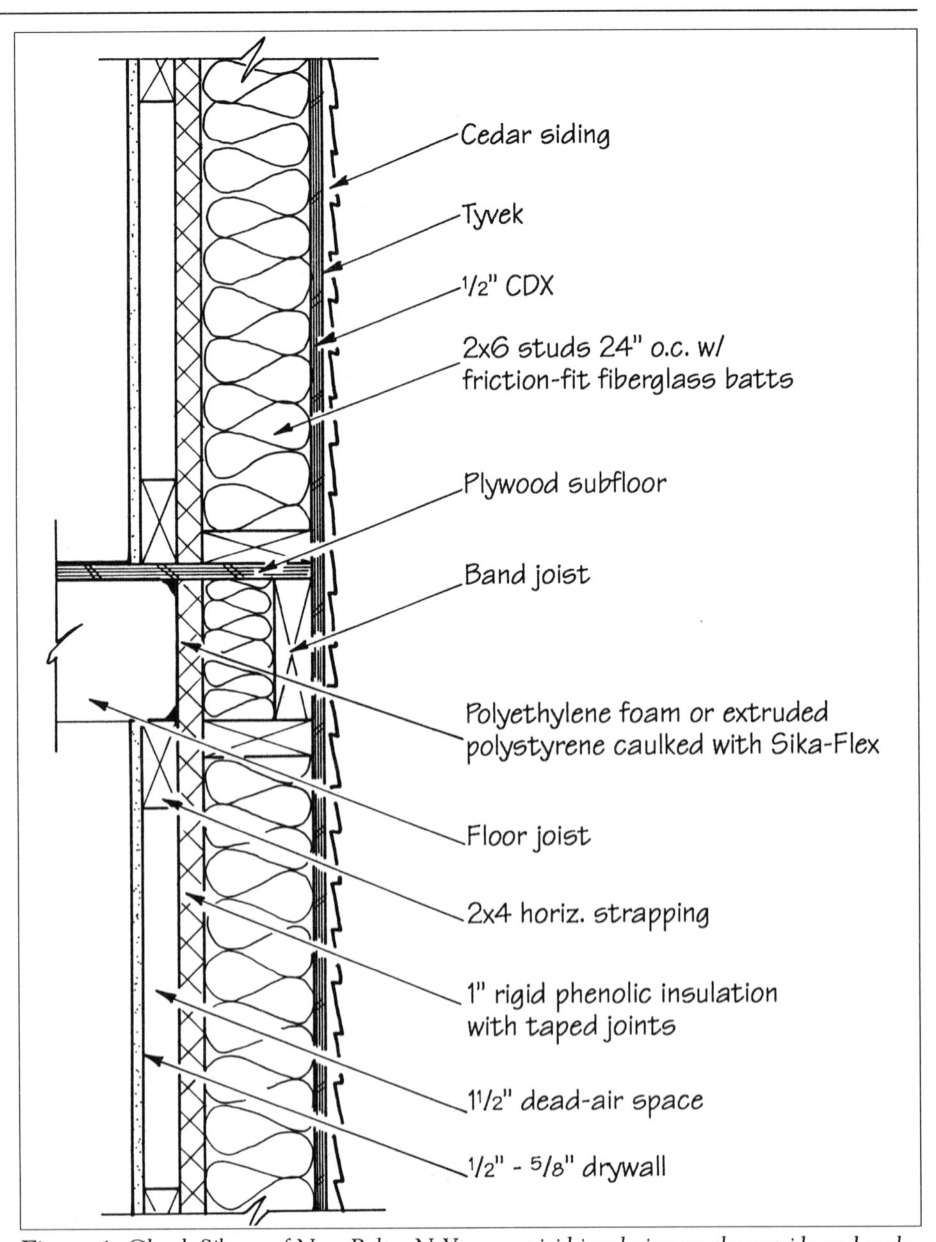

Figure 1. *Chuck Silver, of New Paltz, N.Y., uses rigid insulation on the outside and seals it with foil tape. Between the floors, flexible polyethylene foam blocks fit between joists. Wiring goes in the 2x4 strapping space. Interior strapping is run at the correct height to support outlet boxes (see photo, facing page).*

The Findings

Out of the eight builders I interviewed, there are some broad similarities in the wall systems used. Most employ some type of rigid foam sheathing over fiberglass-insulated studs. All but one builder employs 2x6 framing. Most use air barriers, such as Tyvek. And, in general, the wall systems used today are far simpler, though slightly less efficient, than the ones they used five or six years ago.

Chuck Silver, of Solaplexus, in New Paltz, N.Y., and John Rahill of Black River Design, in Montpelier, Vt., are two designer-builders who have been through the gamut of superinsulated construction details over the last ten years. They have both settled on a 2x6 framing system, 24 inches on-center, with one inch of Koppers Rx on the interior, 1/2-inch CDX plywood on the exterior, and interior strapping to provide a wiring chase and a secure nail base for drywall.

Silver seals the rigid insulation joints with foil tape to provide an uninterrupted vapor barrier, and then uses full 2x4 horizontal strapping (see Figure 1). With 2x4 strapping, he has room for shallow electrical boxes (manufactured by Bell, a Division of Square D) and thereby avoids penetrating the vapor barrier. He justifies 2x4s over 2x2s or 2x3s because the cost difference is not that significant, the wood quality tends to be a lot better with 2x4s, and it provides a wider margin of error with the drywalling.

Because of the bounciness when installing strapping over foam, Silver recommends pneumatic nailers or screw guns. By spacing the 2x4 strapping properly, horizontal sheets of drywall can be installed quickly. Protective steel plates are used wherever wiring passes through the strapping—to prevent damage to the wire or injury.

Silver uses no extra studding in exterior walls where interior partitions join. The plates are interlocked, but the last stud in the partition is left out. After insulation and strap-

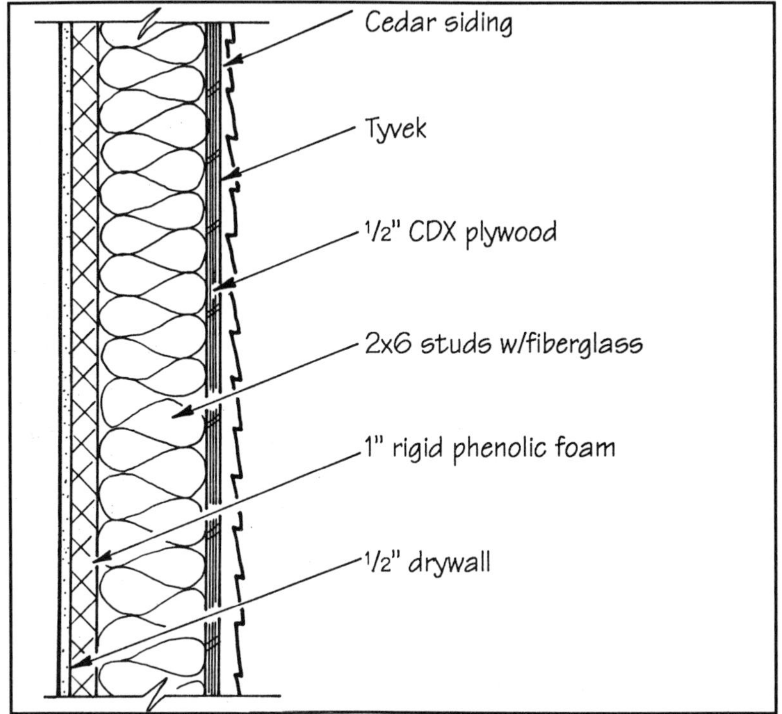

Figure 2. *Rick Schwolsky, of Grafton, Conn., keeps it simple with plywood sheathing on the outside and one-inch foam on the inside.*

ping are installed in the exterior wall, the last partition stud is simply nailed to the horizontal strapping. At wall corners, he builds a standard three-stud corner, but adds an extra stud 8 inches in from the corner in both directions, to provide a secure nailing base for the horizontal strapping.

At the rim joist, Silver employs one of two methods. One is to install fiberglass batts, capping the batts at the line of the rigid insulation with Sentinel Sill Band Sealer (Sentinel Foam Products, Airport Rd., Plant 1, Hyannis, MA 02601; 617/775-5220), and caulking the Sill Band in place with Sikaflex Multi-Caulk to make up for potential shrinkage of the Sill Band. Sikaflex is a polyurethane caulk available from Sika Products (Construction Products Division, Lyndhurst, NJ 07071; 201/933-8800). The other method is to use polystyrene blocks in place of the Sill Band sealer, also caulked in place with Sikaflex.

Silver lists a few advantages of his system as follows:

- All electrical/mechanical equipment is kept out of the insulation and not mixed with the vapor retarder.
- He gets over R-30 with a very conventional building system. Even the electricians like it (after the first one) since most of their runs are horizontal and there's less drilling.
- He gets to put the low-permeance material (the rigid foam) on the inside where it does him some good, and he avoids the need for poly by taping the seams in the Koppers—making it the vapor retarder.
- He gets some additional R-value from the foil he paid for (because of the air space).
- He gets plywood on the outside of the frame for bracing strength.
- Drywall can be glued to the strapping for a superior job. This can't be done with poly over the studs.

Rahill is not comfortable with the Koppers providing an entire vapor barrier. He prefers to add a layer of polyethylene, ideally Tu-Tuff which is a lot stronger than standard poly. Both Silver and Rahill achieve about an R-30 wall system.

Rick Schwolsky, of Grafton Builders in Grafton, Vt., employs the same system, but without the interior strapping (Figure 2). Schwolsky, who co-authored *The Builder's Guide to Solar Construction* a number of years ago, has settled on this system after using double-stud 2x4 construction (with 2x10 plates) and a 2x6 wall system with rigid insulation on the exterior. Using 2x6s, 24 inches on-center, he has encountered some waviness on the exterior wall surface. And he has switched back to 1/2-inch CDX from waferboard after swelling of the waferboard caused him problems with the window jamb extensions (they did not align with the inside wall surface).

Kate Mitchell, of Island Women Construction, Inc. on Nantucket Island, Mass., uses a very similar system, but with 2x4 studs rather than 2x6s, 16 inches on-center. She generally uses one-inch Thermax on the interior, though she sometimes goes with Dow Styrofoam on the exterior (when the rigid insulation is put on the exterior she prefers polystyrene because it breathes somewhat). Mitchell wraps the building with Tyvek.

Using interior rigid insulation with drywall right over it, Mitchell occasionally gets complaints from the drywallers, though this tends to be much less of a problem with skim-coat plaster jobs. Interestingly, veneer plaster has replaced taping and jointing almost completely on the Island.

For insulation purposes, Mitchell treats each story of the house as an individual unit. Although she used to try to wrap the band joists with a vapor barrier, she's "let go" of that. She does, however, insulate the band joist area with individual pieces of insulation fit between the joists.

Ward Smyth and Doug George both use rigid insulation on the outside of fiberglass-filled 2x6 studs. Ward Smyth of Salmon Creek Builders in Salisbury, Conn., uses 2 inches of Koppers Rx (Figure 3). He adds 3/4-inch furring strips on the Koppers as a nail base for the cedar siding, but admits that he has difficulties with the furring strips. They are attached with 4 1/2-inch annular ring nails, which is time-consuming and not all that sturdy—the furring strips "rock" a bit on the foam. At corners, Smyth installs 1/2-inch plywood for strength and uses only 1 1/2-inch Koppers to keep the outside surface even. He installs a poly vapor barrier on the interior, between the studs and drywall.

Doug George, of Conserve Associates in Dover, N.H., a firm which has won national recognition for excellence in superinsulated building design, employs a more complex version of the same wall system, but one in which moisture build-up in the stud cavities would be next to impossible.

As shown in Figure 4, George's wall system uses 2x6 studs, 24 inches on-center, with 1 1/2-inch GlasClad (Tyvek-faced rigid fiberglass insulation board from Canada) on the exterior. He attaches 1x3 furring strips over the GlasClad, nailed vertically into the studs with 2 1/2-inch galvanized roofing nails. For siding he typically uses special ship-lapped 3/4x6-inch cedar clapboards. However, George will frame at 16 inches on-center, if that's what a client wants.

For good nailing at the sill, George uses a 2x8 flush with the exterior. Similarly, he uses solid through-the-wall framing (ripped from 2x10s) around window openings. Otherwise, the GlasClad would compess and throw the window out of alignment.

George uses let-in metal bracing instead of plywood to reinforce the studs, and fills the stud cavities with 5 1/2 inches of Certainteed Insulsafe III "blown-in-batt" insulation at a controlled density to achieve R-22.55. Rather than standard poly vapor barrier, he uses 10-mil Teno-Arm poly (from Sweden) and seals all seams and edges with tape or Teno sealant. Half-inch drywall is attached right to the studs (he used to use horizontal strapping on the interior). He minimizes penetrations through the vapor barrier by using plastic receptacle boxes, sealed to the vapor barrier by cutting an undersized hole in the poly at each receptacle, and forcing the poly to stretch around the box.

To get a tight seal around the band joist area, George uses saturated foam gaskets (Denarco Sales, P.O. Box

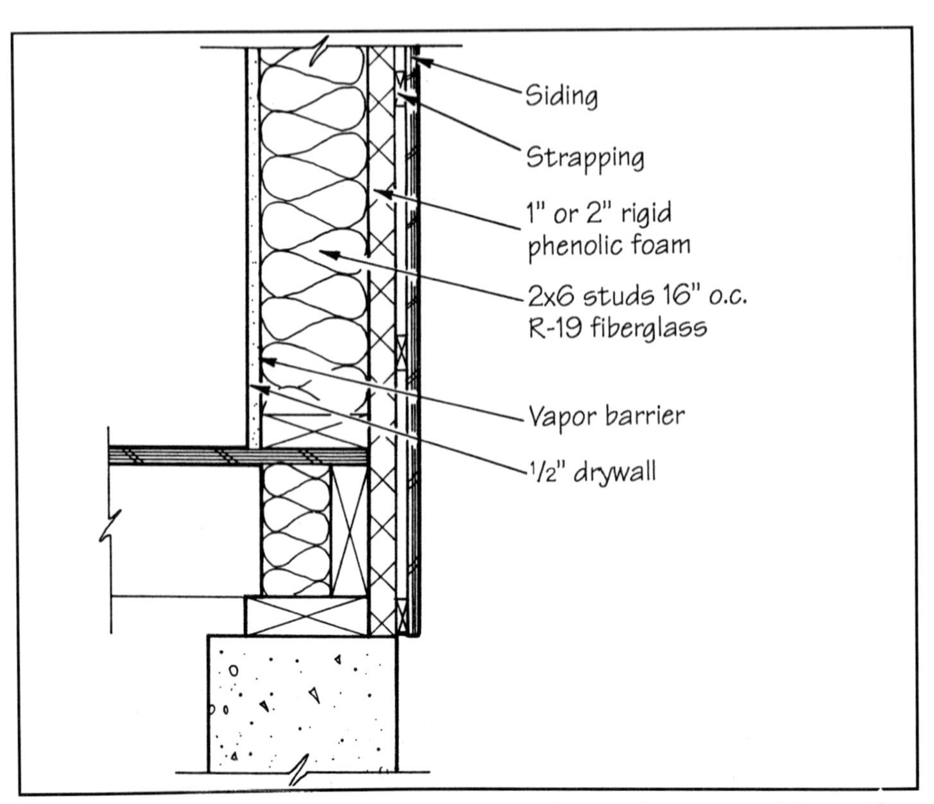

Figure 3. *Custom builder Ward Smith used to wrap the vapor barrier around the outside of the band joist, but now keeps it simple. He relies on an Aldes venting system to remove excess humidity.*

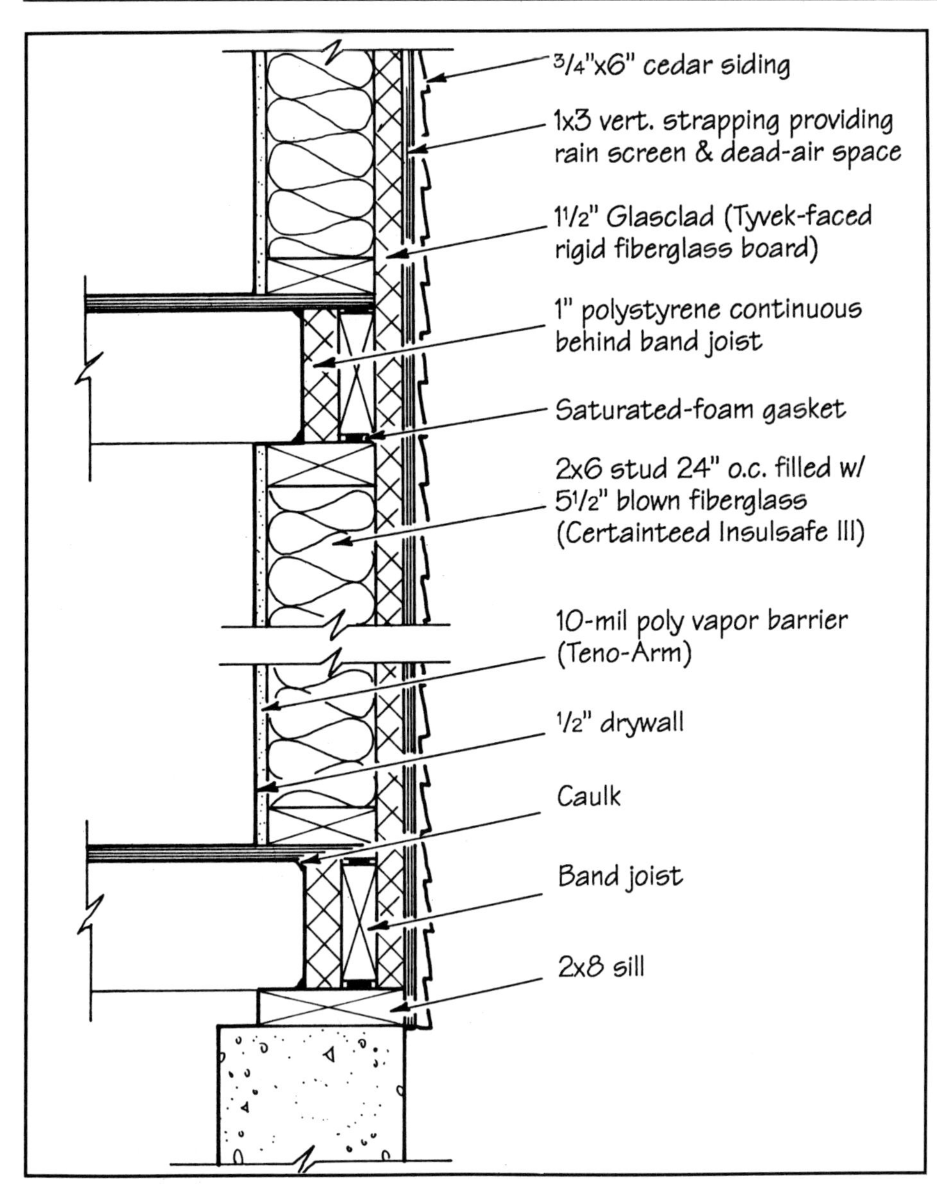

Figure 4. *Doug George, of Dover, N.H., solves the problem of between-floor transitions by running a continuous strip of foam against the inside of the band joists. Floor joists are cut short by 1 inch, but still have 3 inches of bearing.*

793, Elkhart, IN 46515; 219/294-7605) between the framing members. Against the inside face of the band joist, George places a continuous band of 1-inch-thick rigid foam and caulks it in place. The joists are cut one inch short to allow for the foam, but still have 3 inches of bearing. This eliminates the unpleasant and time-consuming job of insulating and sealing each joist space separately.

The cost for this wall system, excluding the drywall materials and labor, is about $6 per square foot. He achieves a wall insulation value of R-30 to -32. When tested with a blower door at 50 Pascals, his houses show less than one air change per hour.

Because of the high cost of GlasClad ($17/sheet), George is looking into using perforated extruded polystyrene on the exterior instead. This will require his crew to perforate the sheets, which he thinks is necessary to assure breathability. He would like to see a perforated polystyrene product brought onto the market for this application and is surprised that that no one has offered it yet. At corners, he will use 1/2-inch plywood for strength and thinner polystyrene, as Smyth does.

Jim Goodine, of Blue Heron Construction Corp. in Shaftsbury, Vt., uses the simplest wall system of any of the builders surveyed: 2x6s, 16 inches on-center. He achieves about an R-19 or R-20 with the system, about what Kate Mitchell obtains with 2x4 studs and an inch of rigid. Goodine is very conscientious about the vapor barrier on the interior and air barrier on the exterior. He tapes the Tyvek at seams and caulks it at windows.

Goodine's firm also builds timber-frame houses, which it insulates with 8 inches of fiberglass installed in wood trusses attached to the outside of the frame. He has used both "Larsen trusses" which his crew fabricated, and Trus-Joists, but would like to see economical wall trusses become available commercially.

One of Goodine's requirements for an acceptable wall system is that it does not damage the Earth's ozone layer, as most rigid-foam insulations are thought to do. He'd like to see a rigid insulation developed that insulates well, does not harm ozone, and is not toxic when burned.

The final wall system covered is also the most complicated. Bill Brodhead of Buffalo Homes in Riegelsville, Pa., frames his houses with 2x6 studs, but with 2x4 top and bottom plates, flush with the inside. He installs an inch of Koppers Rx Foam on the outside of the studs, then foams-in-place 11/2 inches of urethane insulation (see Figure 5, next page). Then he fills the rest of the stud cavity with 31/2-inch fiberglass batts. By having the top and bottom plates set in 2 inches, he gets continuous urethane coverage all the way down to the plywood deck—thus preventing any infiltration under the bottom plates. Infiltration is also blocked by using two-stud corners and no studding on exterior walls where interior partitions join (similar to Silver).

With this system, Brodhead claims he doesn't need a vapor barrier. The foamed-in-place urethane provides an excellent seal. He always installs air-to-air heat exchangers in his houses, however, to keep humidity

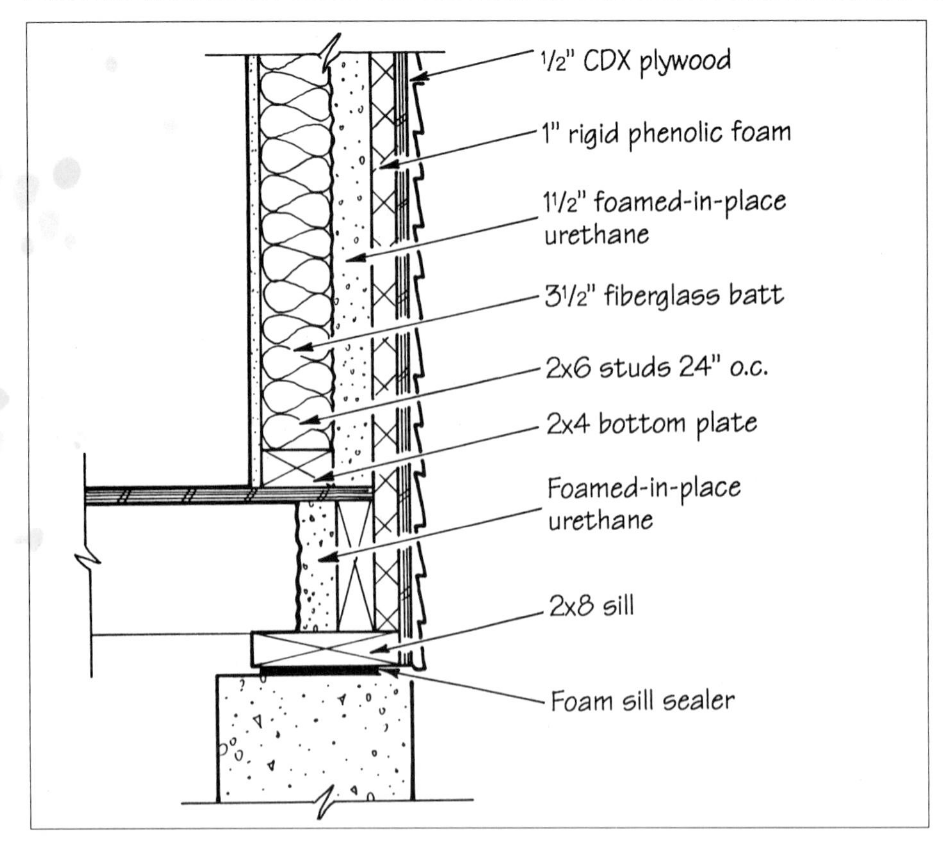

Figure 5. *Bill Brodhead of Riegelsville, Pa., uses foam-in-place polyurethane and special framing details to create a one-step air barrier and insulation. No inside vapor barrier is used.*

levels in check.

While this system sounds complicated, it is much simpler and faster than the double-stud wall system he used to use. In fact, at a customer's request, he recently built a double-wall house and he re-experienced all the hassles of double-wall framing, sealing the vapor barrier with Tremco acoustical sealant, and labor overruns. His newer system, which he has been using for about three years now, is a lot easier and comes in at a lower cost, he claims.

At R-30, his present system insulates about as well as the double wall, though it doesn't test out quite as tight in blower-door tests he has done. He gets 2 to 3 air changes per hour at 50 Pascals, as compared to 1 to 2 for the double-wall-and-poly system. He points out that the foaming must be carefully watched to ensure complete coverage.

Conclusions

As for what wall system is the best, that's up to you. I lean toward a straightforward system that uses readily available materials, and is easy to build, easy on subs, and well insulating. This seems to be the general trend among the builders I interviewed.

No matter what system you use, it pays to maintain a high level of workmanship. The best-thought-out details will do little good if they are poorly executed in the field. Because tightness counts almost as much as R-value when it comes to energy efficiency, pay particular attention to caulking, sealing air and vapor barriers, and reducing penetrations through those barriers.

The issue of ozone depletion and the CFCs (chlorofluorocarbons) used in rigid-foam insulation may have a big influence on construction details over the next several years. Because of restrictions in the use of CFCs (which are used in producing almost all insulation board stock except expanded polystyrene), rigid insulation may become more expensive, scarcer, or less effective in R-value. Keep an eye on this issue, since its outcome may affect the way we all build. ■

Alex Wilson is a technical writer based in Brattleboro, Vt., who specializes in energy and building issues.

Strapped Wall Detail

This award-winning wall design uses standard techniques to achieve top performance

by Clayton DeKorne

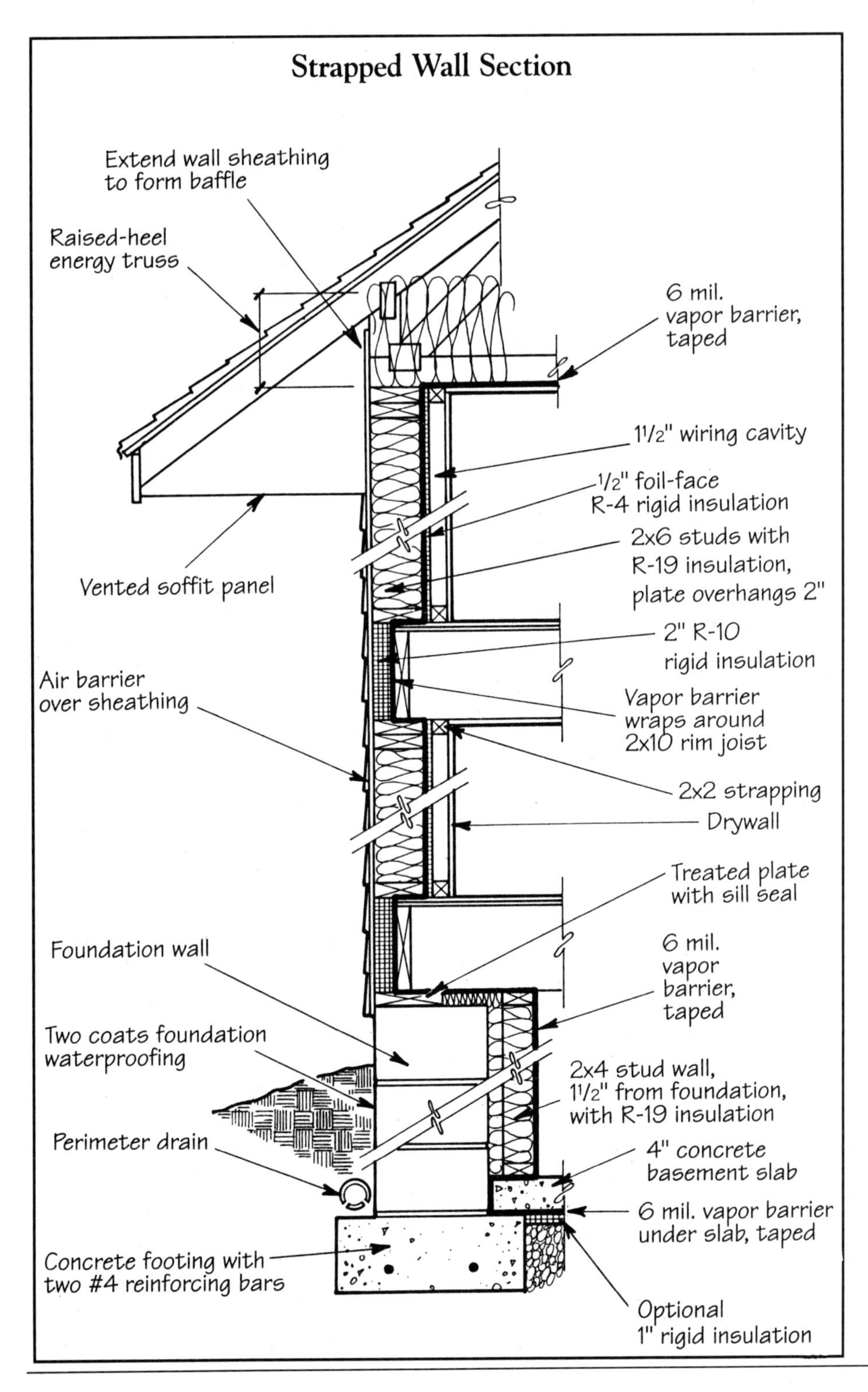

Thomas Brown, an architect from Stevens Point, Wis., won the best wall design at the 1991 Quality Building Conference held in Springfield Mass., and sponsored by the New England Sustainable Energy Association. Brown's entry for an energy-efficient strapped-wall was judged by an architect, an engineer, and a builder for "energy-efficiency, buildability, simplicity, and innovation."

Brown's design combines conventional framing methods with several energy-conserving construction practices (see illustration). Brown claims his design "is very forgiving...and allows flexibility in the completion of the project." Most of the energy-conserving measures have been added to a standard 2x6 wall, he explains, and don't require a whole new method. "In fact," he adds, "it is possible to revert to more standard practices if budget or other constraints prevail."

The basic wall consists of an insulated 2x6 stud wall covered on the inside with 1/2-inch foil-faced rigid insulation and strapped with 2x2s before the drywall is secured. With batt insulation, Brown claims the R-value of the wall materials adds up to about R-28, including an air space next to the foil-faced insulation. But Brown is quick to point out that high R-value is not the key to the wall's good performance, as on tight construction.

A tight envelope depends on following through with several sequences.

The basic wall consists of an insulated 2x6 stud wall covered with 1/2-inch foil-faced foam insulation and 2x2 strapping on the inside. Note the indented rim joist to accept 2-inch rigid foam on the outside, and the continuous vapor barrier.

Most important is careful attention to the air/vapor barrier. Here Brown urges a few departures from conventional practice:

First, the vapor barrier must form a continuous envelope around the entire house. To achieve this, Brown calls for (1) running the sub-slab vapor barrier up the foundation wall before pouring the slab; (2) draping a wide strip of 6-mil poly over the top of the foundation wall before the sill plate is laid down; and (3) running strips of poly around the rim joists and over the top plates on the upper floor. These steps leave tabs that can then be taped to the vapor barrier on the walls and ceiling.

Second, Brown calls for overhanging the 2x6 wall plates 2 inches so the rim joist can be inset to accept rigid foam insulation on the exterior.

Third, rough openings for windows are oversized by 3 inches in width and height to accommodate a 2x4 nailer to support drywall returns instead of extension jambs.

Brown has designed about 15 houses with some variation on this wall, and in many cases, he says, the builder liked it enough to adopt it on other projects. Brown attributes the success of the design to its versatility. The strapping, he says, is a high performance feature that is not absolutely necessary, but which adds an extra level of performance. Without the air space and the electrical raceway, the wall can still perform better than average if care is taken around electrical boxes and other penetrations. And even if the interior foam is eliminated, the continuous vapor barrier and rim joist detail make a difference.■

Clayton DeKorne is an associate editor at The Journal of Light Construction.

Cathedral Ceiling Solutions

Five contractors describe the sloped ceiling details that work best for them

Few components in low-energy homes produce more diverse approaches—and more disagreement—than cathedral ceilings. The reason is simple: Trying to satisfy the insulation and ventilation requirements in such a tight space is a real challenge.

It's often impossible to fit the required amount of insulation into the depth provided by a 12-inch rafter. A common solution for an R-40 ceiling is to use fiberglass batts between the rafters and then nail an inch of rigid insulation underneath the rafters. The seams of the foamboard can be taped so that it doubles as a vapor barrier. An alternative is to use deeper (but more expensive) engineered beams such as I-joists or trusses and avoid the use of foam insulation.

Another problem is providing adequate roof ventilation. Most roofs are designed to have an eaves-to-ridge flow of air. These are called cold roofs because they keep the underside of the sheathing cold in winter. In a heating climate, the flow of cold air under the sheathing discourages interior moisture buildup and ice damming. In a cooling climate, ventilation relieves heat buildup.

To complicate matters, not all builders and researchers agree that ventilation space above the insulation is necessary in heating climates. Cathedral ceilings that aren't ventilated are referred to as hot roofs, and typically rely on a well-sealed air and vapor barrier to keep moisture out of the roof. Solid foam roof systems, such as stress-skin panels, are a type of hot roof.

Thoroughly confused? Because so many builders feel caught in the crossfire of these debates, we asked five experienced, energy-conscious builders and designers from different locales to show us how they detail cathedral ceilings, and to point out why they settled on these designs.

—JLC

Packing R-38 Into 2x12 Rafters

by John Raabe
Langley, Washington

Here in the Pacific Northwest, new energy codes require all builders to meet high insulation levels and use progressive building techniques. For cathedral ceilings, this means insulation levels of R-38 to R-40.

Cathedral ceilings in this area are most often built with 2x12 rafters on 24-inch centers. Only a few years ago, insulating to R-38 with these rafters meant an additional layer of foam insulation or some extra framing. However, there are now two ways to achieve R-38 to R-40, keep the code-required 1-inch air space above the insulation, and still use standard framing.

The one I like best is using Ark-Seal's "Blown-in-Batt-System," or BIBS (Ark-Seal International Inc., 2190 S. Kalamath St., Denver, CO 80223; 800/525-8992), as shown in the drawing. This proprietary process uses chopped fiberglass in combination with a latex binder that coats the fibers and hardens to form a fluffy, semi-rigid mass. It is blown into the cavity through a hose, and is retained by mesh netting that is stapled to the framing.

This insulation does not settle, and completely fills small voids and gaps that can degrade the insulation value of fiberglass batts. The material is well documented at R-4 per inch, and a simple two-coat PVA primer on the drywall can be used as the vapor barrier. The system is fast, but it is somewhat more expensive than using fiberglass batts because it requires a special insulation contractor.

Batt manufacturers are meeting this challenge with products of their own. Manville Corporation (Box 5108, Denver, CO 80217; 303/978-4900) is beginning to market a high-density 5-inch batt. This certified R-19 batt can be doubled up in the rafter cavity to yield an R-38 roof with over an inch of vent space above. Both Owens-Corning Fiberglass (Fiberglas Tower, Toledo, OH 43659; 419/248-8000) and Certainteed Corporation (Box 860, Valley Forge, PA 19482; 215/341-7000), the other big batt manufacturers, are hard at work on high-density, 10-inch (R-38) batts for this same application.

These batts should be installed with a 6-mil poly vapor barrier underneath to provide support as well as to control moisture and air leakage.

Although I include ventilation in the cathedral ceilings I design, there is some interesting evidence that in moist climates, venting above the insulation actually increases the moisture content of the roof cavity. In some situations, it may be better to completely fill the rafter space than to ventilate the cavity. To do this successfully, you need a complete vapor barrier and no air leakage into that cavity. This means no recessed lights or other hard-to-seal openings in the cathedral ceilings.

My own suspicion is that the moisture issue will turn out to be relatively climate specific, and that roof ventilation will become more subject to review by the local inspector than it is now. For most of us, the real world dictates that it's easier to ventilate than it is to debate the point for each home and with every new building official.

John Raabe heads up Cooperative Design in Langley, Wash., and is a coauthor of Superinsulated Design and Construction.

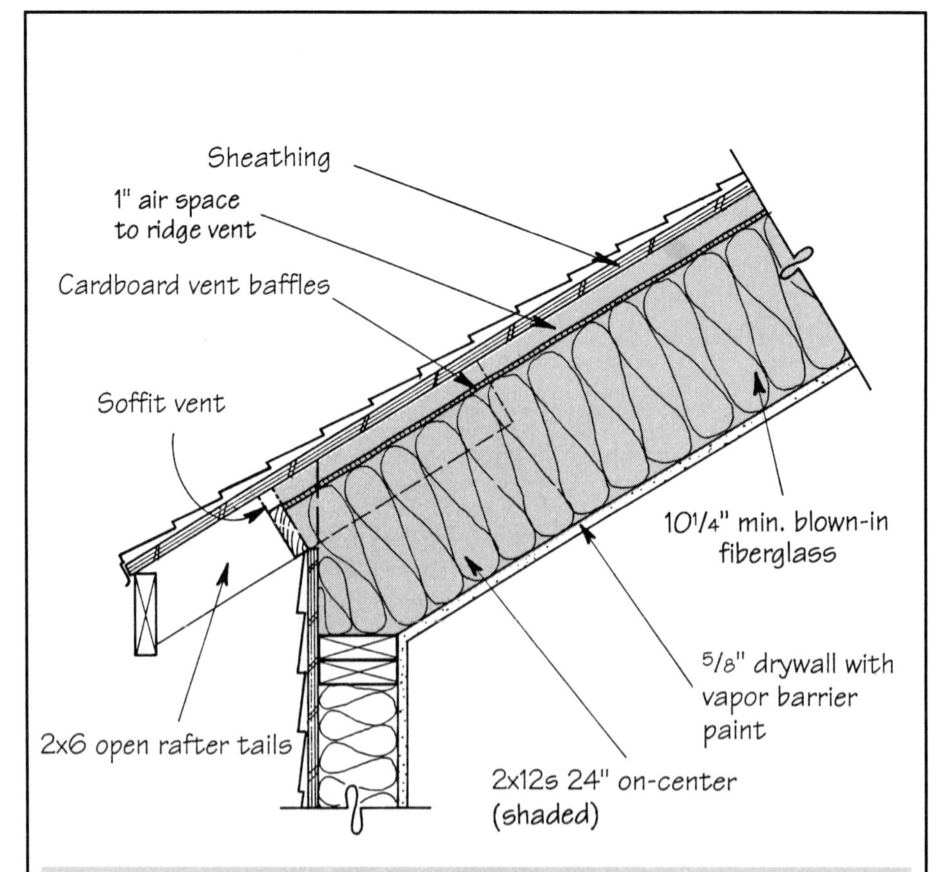

Type: 2x12 rafters with blown-in fiberglass (BIBS)
Designer: John Raabe; Langley, Wash.
R-Value: 38-40
Pros: Achieves high R-value with standard framing. Installation is fast. No settling of insulation.
Cons: Somewhat more expensive than batt insulation.

R-50 With Wood-Web I-Beams

by Chuck Silver
New Paltz, New York

I've experimented with several different ways to design an R-45 to R-50 cathedral ceiling—installing rigid insulation beneath 2x12 rafters, filling the rafters with fiberglass and using an air barrier and furring strips on top to get ventilation, and using open-web, parallel-chord trusses. None of these, however, is as easy, efficient, or cost-effective for me as using 16-inch-deep (or deeper) wood-web I-beams.

There are a lot of advantages to I-beam rafters:

- They are very light, perfectly straight, and available in any length up to 60 feet.
- They come in a large range of depths to accommodate any insulation thickness.
- They offer minimal interruption to the insulation (a ½-inch plywood web every 2 feet).
- They eliminate the need for rigid insulation on the bottom of the rafters, giving you a superior fastening surface for the drywall.
- They have terrific span capabilities. This can sometimes mean the elimination of bearing walls which can help offset the additional expense of I-beams over solid-sawn rafters.

I use 16-inch TJIs (Trus-Joist MacMillan, 9777 W. Chinden Blvd., Boise, ID 83714; 800/338-0515) on 24-inch centers for an R-50 ceiling. A structural ridge beam is typically necessary. Wood I-beams also require 2x4 vertical stiffeners between rafters where they bear on exterior walls. We use the stiffeners as nailers for plywood baffles, which provide bracing, as well as directing ventilation air from the soffit up over the fiberglass. The baffles prevent cold air from penetrating the edge of the batts and chilling the ceiling corners.

However, when I use cedar shingles, as shown in the drawing, I don't include soffit-to-ridge ventilation. My reasoning is this: The primary

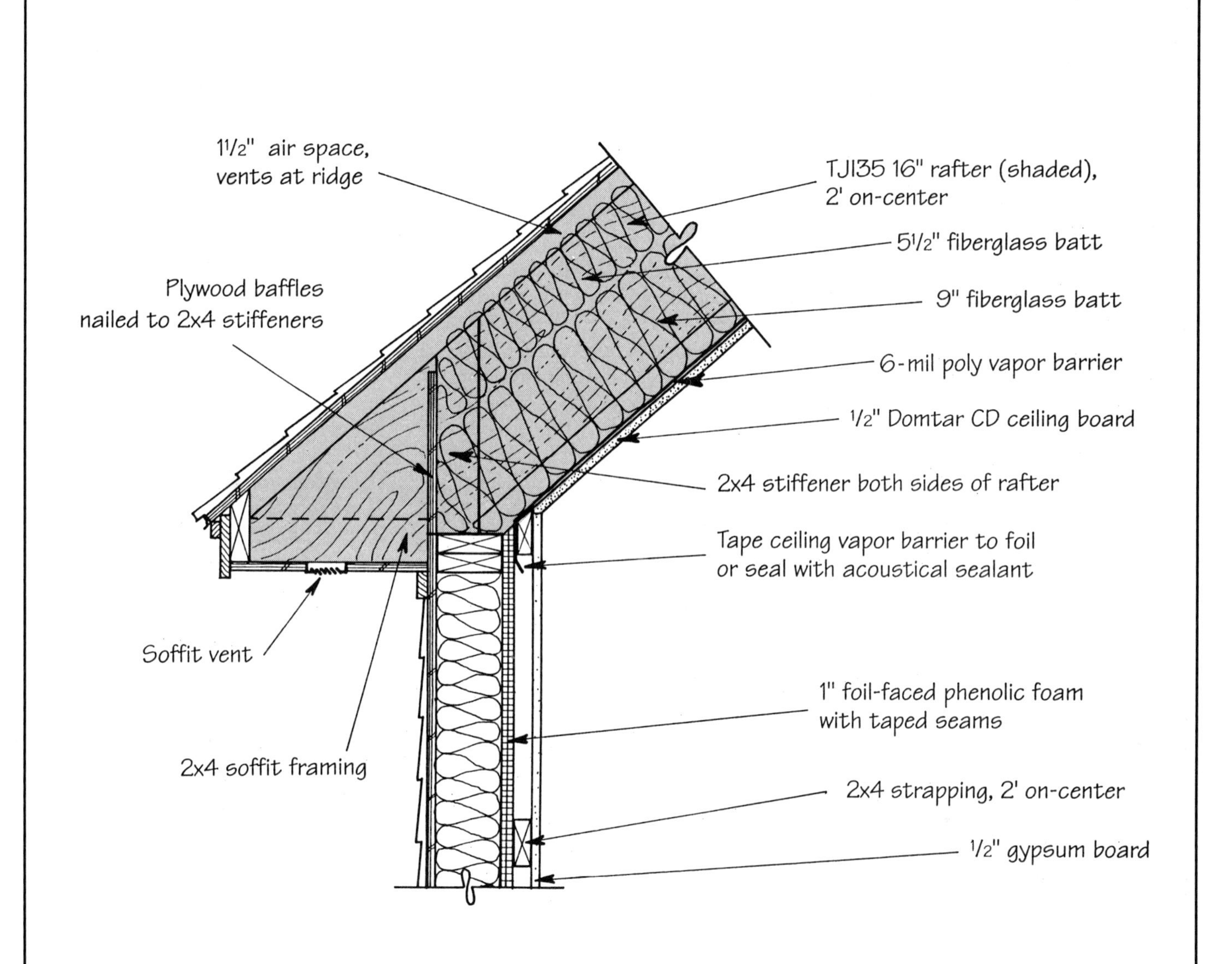

Type: 16-inch wood I-beams, fiberglass batts
Designer: Chuck Silver; New Paltz, N.Y.
R-Value: 45-50
Pros: Wood I-beams are stable, straight, and offer plenty of room for insulation and ventilation. No foam is needed.
Cons: I-beams are costly and require special detailing at bearing points.

difference between walls and roofs, besides orientation, is that roofs typically have an impermeable membrane attached to them, which makes it important to vent. This isn't true of cedar shingles on spaced sheathing, so I don't vent. I've never had any problems with this design.

With wood I-beam rafters you do have to use full-width (24-inch) insulation. I use 5½-inch and 9-inch fiberglass for a 16-inch rafter, which still leaves 1½ inches of air channel for ventilation, when required. I carefully install a continuous, 6-mil poly vapor barrier, and then finish with Domtar Gypsum's CD Ceiling Board (P.O. Box 543, Ann Arbor, MI 48106; 800/366-8274). This is a ½-inch drywall that has the rigidity of a typical ⅝-inch gypboard, and works very well on 24-inch centers.

Chuck Silver is president of Solaplexus, a design and consulting firm in New Paltz, N.Y.

Scissor Trusses Keep It Simple

by William J. Baldwin
Johnston, Rhode Island

We're residential developers with a strong interest in energy-efficiency because of what it offers to the client. The roof system is just one of many integral parts of this package, and like the whole, must be simple in design, easy to install, cost-efficient, and durable.

We typically use simple, wood scissor trusses with raised heels. Since the trusses stand "on edge" like a wall stud and are placed 24 inches on-center, there is only a small percentage of thermal short-circuiting through the truss chord members. Still, we are very careful about fitting the fiberglass batts within the truss bays, making sure all the voids are filled and the insulation is fully expanded.

We extend the exterior sheathing up the wall to the top point of the insulation to direct air brought in by the soffit vents above the insulation. You can substitute cardboard wind baffles if you like.

We also take a lot of care with the vapor barrier. It should be continuous and overlapped, and sealed at all joints and stack pipe penetrations. On top of the vapor barrier we nail strapping—utility-grade 2x3 furring strips—on 24-inch centers to the underside of the bottom chords. This $1^1/_2$-inch space allows ample room for all electrical wiring and ceiling light boxes. If clients ask for recessed (can) lights, we explain the obvious advantages of substituting track lights.

A lot of what makes this system work is that it's straightforward. This allows us to expect high quality workmanship and provide proper supervision. We do a lot of training, with both employees and subcontractors, and we include the "whys" as well as the "hows." This saves countless hours of work-site time and energy, particularly in the volume we build—over 15 houses per year.

We also feel strongly that energy-efficient features should be cost-effective. We like to be able to point to the direct relationship between increased building costs and energy savings.

Bill Baldwin is president of the Baldwin Corporation, a residential development company in Johnston, R.I.

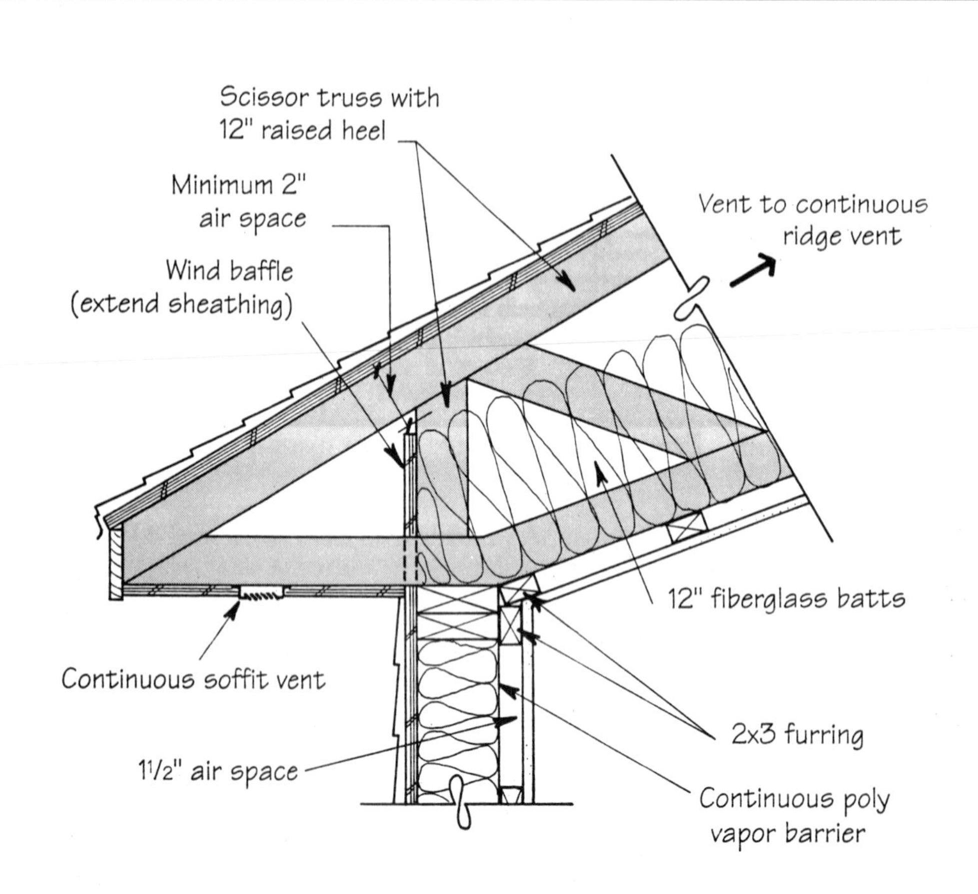

Type: Scissors trusses with raised heels, fiberglass batts.
Builder: William Baldwin; Johnston, R.I.
R-Value: 40
Pros: Simple detailing is cost-effective. Raised heels ensure full insulation at eaves.
Cons: Batts must be carefully fitted to minimize short-circuiting at truss chords.

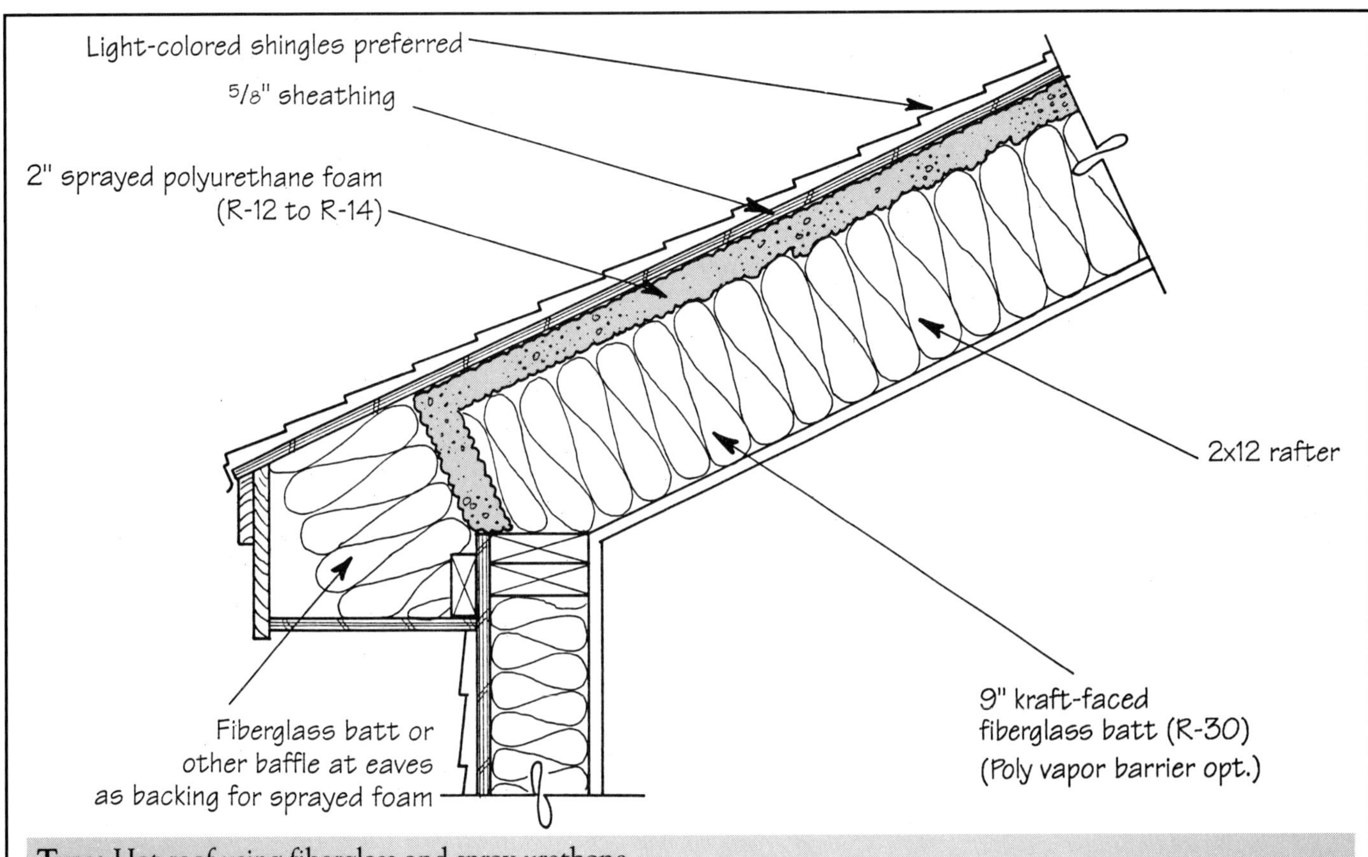

Type: Hot roof using fiberglass and spray urethane
Builder: Paul Bourke; Leverett, Mass.
R-Value: 45 plus
Pros: No venting needed, which simplifies complex roofs with hips, valleys, skylights, etc. If properly designed, condensation should not be a problem.
Cons: Not accepted by some code officials. Ice damming an issue in high-snow areas. CFC concerns.

A "Hot" Roof For a Cold Climate

by Paul Bourke
Leverett, Massachusetts

I have used this "hot roof" detail on my last eight homes, and I couldn't be more pleased with its performance and simplicity. Here are some of its features:

- Allows for higher insulation levels in the same size rafter cavity.
- Eliminates the costs of installing soffit and ridge vents, and insulation baffles.
- Doesn't require venting for those almost impossible-to-vent rafter bays created by hips, valleys, gambrel roofs, and skylights.
- Raises the temperature of the condensing surface on the underside of the roof sheathing, reducing the risk of condensation problems.

However, "hot roofs" are not widely accepted, particularly among code officials. When I run into resistance, I often add a detail drawing signed by a local architect or engineer, or hand the official a copy of "Hot Roofs and Cold Roofs," an article from the May 1986 edition of the newsletter *Energy Design Update*. I also point out that stress-skin panels, which are widely accepted in commercial and residential work around here, operate on this same principle.

Code officials and others have three primary concerns: ice damming, shingle deterioration, and moisture buildup in the cavity. I have not had problems with any of these.

The ice damming question is directly related to the level of insulation used (higher levels reduce the problem), the care taken with the foam, and the average depth of snow on the roof. My roofs are typically R-45 (2 inches of sprayed urethane foam and 9 inches of fiberglass). I have a very good foam sub, and I limit this design to roofs of at least a 4/12 pitch, which keeps the snow depth to less than 13 inches.

As for shingle life, venting an attic or ceiling will cool down the roof surface only a few degrees. The more important step is to use light colored shingles, which reflect more sunlight and stay significantly cooler.

Condensation is not a problem for a couple of reasons. First, very little moisture gets into the ceiling due to the vapor retarder on the warm side and the near-perfect air barrier (the foam) on the outside. Second, any water vapor that does leak into the cavity is unlikely to condense since the underside of the foam where condensation would occur is relatively warm.

To insulate the roof, I bring in my foam sub once I have the 5/8-inch sheathing on, and all plumbing vents, electrical, ductwork, etc., are in place. The 2 inches of foam is sprayed on the *underside* of the sheathing. The foam does two things: It seals the roof system, making it virtually airtight, and it adds R-12 to R-14 to the cavity.

Next I install 9-inch kraft-faced fiberglass batts. These need to be stapled tightly to the rafters. The paper facing meets my code for a vapor retarder on the warm side of the ceiling, but you can also use a poly barrier or vapor retarding paint on the drywall if you choose.

Increasing the thickness of foam, and decreasing the amount of fiberglass below it, will raise the R-value of the system. But with foam-in-place polyurethane coming in at 50¢ to $1 per board foot, you need to use it wisely.

The only concern I have with this system is that the blowing agent for polyurethane foam currently contains CFCs. However, a new formulation using H-CFC-141B is now available for residential use. This is only 5% as harmful to the Earth's ozone layer as CFCs. And that number can be further reduced by using it in conjunction with a 35% water-blown system, which will be available soon. A 100% water-blown system with no CFCs will be available within one to two years.

Paul Bourke, of Bourke Builders, designs and builds energy-efficient homes in Leverett, Mass.

Holding the Heat Out

by Lawrence Maxwell
Cape Canaveral, Florida

In a hot climate, heat gain through the roof of a cathedral ceiling to the living space below is a primary concern. The first step in blocking the heat is to choose the roofing material and its color wisely.

The ideal roofing in our hot southeast climate is concrete or clay tile.

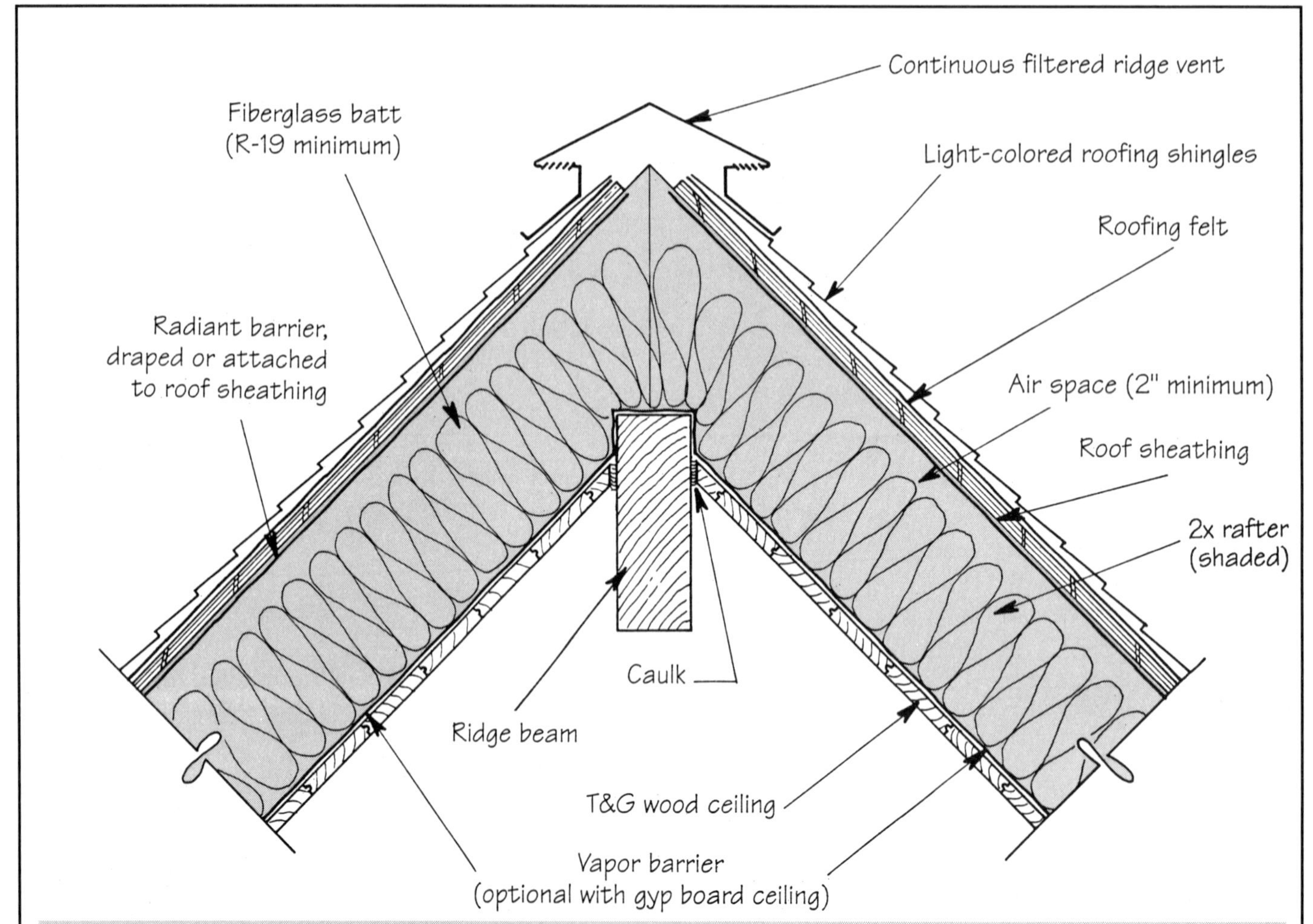

Type: Radiant barrier with fiberglass batts (for hot climate)
Designer: Lawrence Maxwell; Cape Canaveral, Fla.
R-Value: R-19 plus
Pros: Combined with concrete, clay, or white asphalt shingles, radiant barrier effectively blocks heat gain.
Cons: Draping radiant barrier over rafters can be tricky. (Option is to fasten to sheathing before installation.)

These materials work very well at reducing heat gain because of their mass and the fact that they sit on battens. The mass lets the tiles absorb and store the heat (the lag effect) and the battens allow the heat to escape through ventilation.

If the budget can't afford tiles, asphalt/fiberglass shingles are okay, but it's important to stay with very light colors; white is best. The principle is simple: The lighter the color, the more heat the surface reflects.

Cathedral ceilings in hot climates also need to be ventilated to discourage heat gain. This requires continuous soffit and ridge vents, and a *minimum* of 2 inches between the top of the insulation and the underside of the radiant barrier.

This barrier is important in hot climates, since as much as 90% of the heat transfer from the roof deck to the ceiling drywall is by *radiant energy transfer*.

There are two common methods of radiant barrier installation; both require that the "shiny side" go face down (toward the living space). The first method is draping the radiant barrier over the tops of the rafters. The second, which requires a good deal less work, is to fasten it directly to the plywood prior to sheathing the roof. A perforated radiant barrier is best for this second method.

Installing the insulation is pretty straightforward, although you should place it carefully, and fasten it to the ridge beam so it doesn't slip downhill over time and leave an uninsulated space at the peak.

If you are using drywall as a ceiling finish, a vapor barrier is optional in this climate. However, if you're nailing a tongue-and-groove wood ceiling directly to the underside of the rafters, you should first place a vapor barrier to control the infiltration and exfiltration of air through the joints between boards. Make sure the vapor barrier is continuous under the ridge beam, and use backer rod and caulk between the topmost board and the ridge beam. ■

Larry Maxwell is the senior research architect with the Florida Solar Energy Center in Cape Canaveral.

Tight Construction Simplified

If you hated poly and caulk, weren't sure about ADA, but still want to build tight houses, maybe it's time you tried "simplecs"

by Jim Maloney

Although I am now a builder and designer, I spent three years with the Oregon Department of Energy trying to encourage builders to build more energy-efficient homes. I learned that going against the grain of conventional practice is not an easy task and does not make you very popular on building sites. But the old way is not always the best way to do things. In fact, my experience in Oregon has shown me that conventional wisdom would often be better described as "prevalent nonsense."

Before I describe any building innovations, I should note some peculiarities in the way we do things here in Oregon.

- About 80 percent of the homes built in the state have crawlspace foundations.
- Half the new homes use electric heat, because electricity only costs 2.6 cents per kilowatt hour. The rest use gas at 63 cents a therm.
- The first-floor structural system over the crawlspace is usually a post-and-deck floor using random length T&G 2x8s.
- We build year 'round here, and get 50 to 60 inches of rain a year—making for a lot of damp job sites.

The regional program I worked for (the Residential Standards Demonstration Program) paid builders a financial incentive to meet a set of standards called the Model Conservation Standards. While most of the standards dealt with insulation levels, window U-values, and heating equipment, the ones that met with the most resistance and caused the most problems had to do with air leakage and moisture movement.

At that time, builders interested in airtight construction used polyethylene, which they wrapped around the entire house and sealed meticulously. Although this system was used successfully in Canada and some Scandinavian countries, a vast majority of the Oregon builders balked at the idea. The most common reasons given were: "A house needs to breathe." "It will trap water in the walls and the house will rot away." "It takes too much time." "It's too tedious." "My building inspector won't allow it." "My drywall contractor can't glue his gyp board." and my favorite, "My framing crew will quit."

It was extremely clear that we had proposed a bitter pill and needed to come up with something more palatable. While searching for solutions, we came upon what is now widely known as the airtight-drywall approach or "advanced" drywall approach or just plain ADA.

This approach, developed by Toronto engineer Joseph Lstiburek, did away with the unpopular poly and used the building's drywall itself as the air barrier. The interior paint served as the vapor retarder. Or you could use 4-mil poly as the vapor retarder, but install it without all the fancy seams, nasty caulks, and general fuss that drove many builders to distraction.

At first, the builders in the program thought that ADA was great, particularly when compared to the trials and tribulations of trying to do a good job with poly. The high praise, however, was soon tempered by a host of complaints from the job site. These often took the form, "Yeah, this is a great system, but..." Some of the common "buts" were:

- My electrician had fits getting his wires in and out of the walls.
- I had to have the drywaller here twice.
- The gasket held up the rim joist so my framers had to lift each joist when nailing it.
- The gasket kept tearing away when we tried to move the wall to the chalk line.
- There must be a better way.

To make a long story short, we managed to get a number of ADA houses built in Oregon. Along the way, we made a number of modifications aimed at simplifying the system for the framers, drywallers, and mechanical subs. We found that many of the problems could be solved if the air-barrier efforts could be delayed until the building was framed and sheathed. I've combined a number of these modifications and added some new ones to come up with a system I call "Simplecs" for "simple caulk and seal."

The key to the system is to do the air sealing at the latest possible stage in construction. It turned out that it could always be done after the framing and sheathing and usually right before hanging the drywall. In some cases, such as the leak at the baseboard below the drywall, the sealing

Air Leakage Control Using "Simplecs"

Simplecs stands for "simple caulk and seal." It's an advancement on the ADA System in that it requires no sealing during the framing process. It is also flexible; it can use caulks, adhesives or gaskets depending on the specific application and the builder's preference. As in the ADA system, most of the air-barrier connections are visible and reparable.
The key details are explained below.

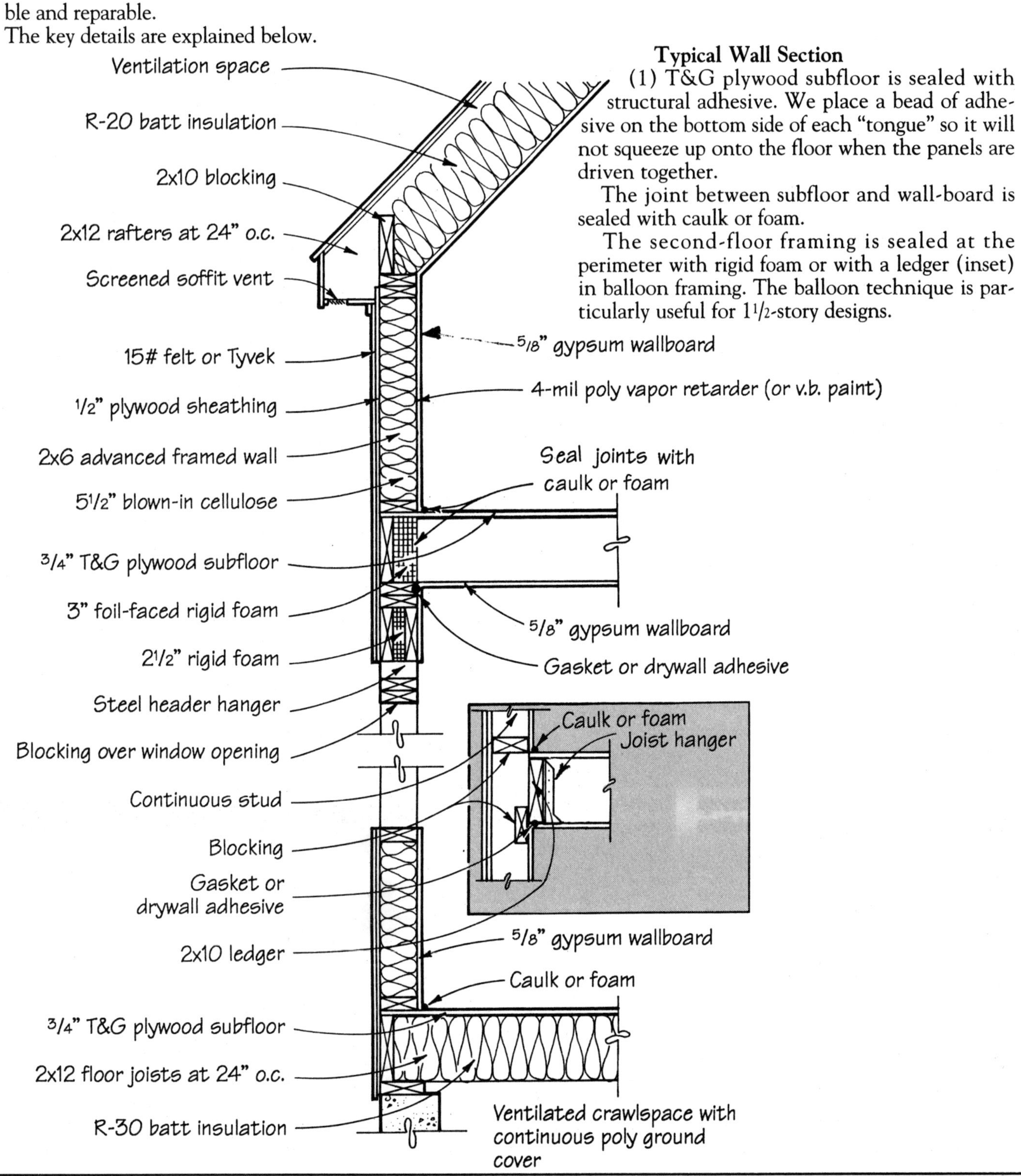

Typical Wall Section

(1) T&G plywood subfloor is sealed with structural adhesive. We place a bead of adhesive on the bottom side of each "tongue" so it will not squeeze up onto the floor when the panels are driven together.

The joint between subfloor and wall-board is sealed with caulk or foam.

The second-floor framing is sealed at the perimeter with rigid foam or with a ledger (inset) in balloon framing. The balloon technique is particularly useful for 1 1/2-story designs.

Interior Wall/Ceiling Intersection. *At the intersection of an interior partition and exterior ceiling, the top plate of the partition is gasketed, or a continuous bead of adhesive is used for the wallboard.*

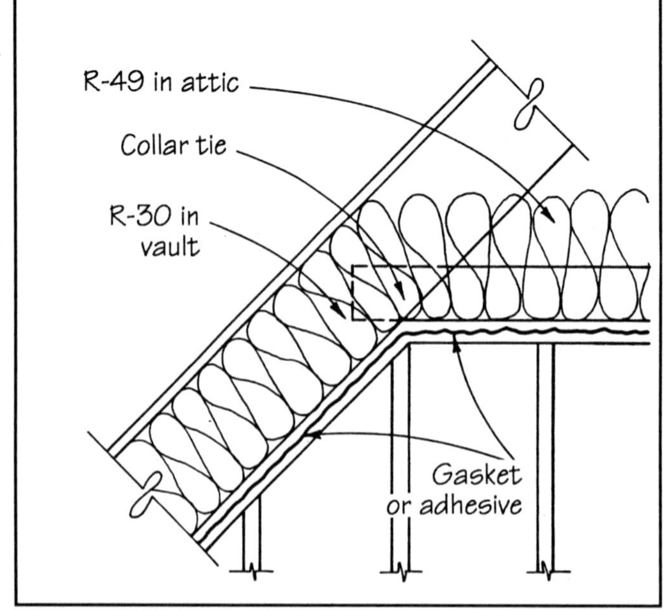

Interior/Exterior Wall Intersection. *At the intersection of an interior partition and an exterior wall, the last interior stud is gasketed, or a continuous bead of adhesive is used for the wallboard.*

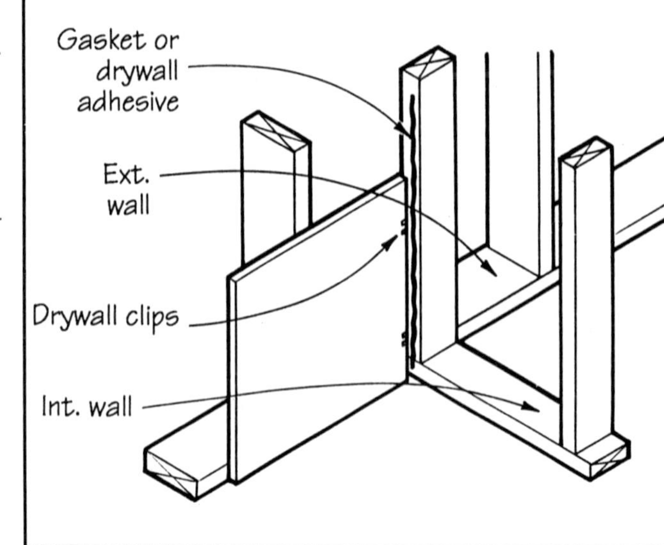

Typical Plumbing Penetration. *Plumbing penetrations of air barriers are typically sealed with long-lasting elastic sheet goods such as neoprene or EPDM. (We've had great luck with swathes cut from old truck inner tubes.) Electrical and other penetrations are sealed with foam or a compatible caulk.*

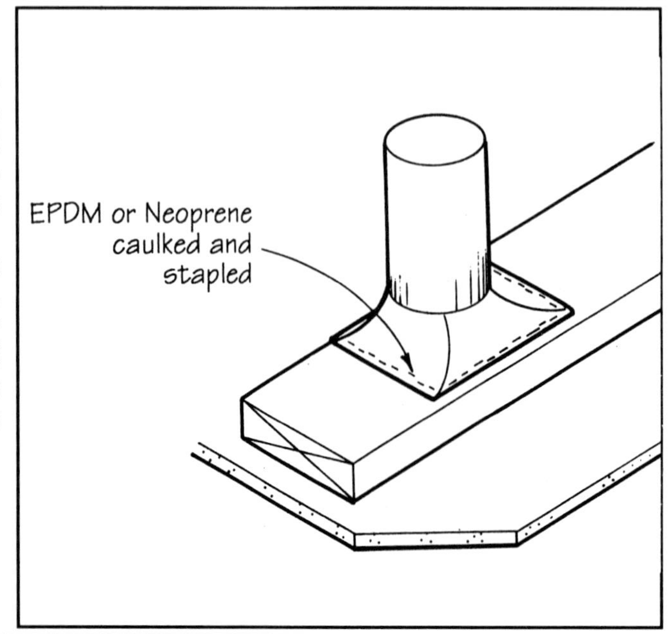

could even take place after the drywall is up. Most of the details are shown in the accompanying drawings.

One advantage of this system in our climate is that the air-barrier work is done in a more controlled, drier environment. It means that we can apply caulks, adhesives, or gaskets in a drier, warmer, perhaps even cleaner, building. It also means that a tricky sealing problem will not hold up a three-man framing crew.

To comply with local code, we still use 4-mil poly, but because we don't rely on it to stop air—only to block vapor diffusion—we can slap it up quickly with unsealed seams just before the drywall goes on. A vapor-retarding paint would work just as well.

One house we built last spring using the system tested adequately tight (3 air changes per hour at 50 Pascals), which is what we were after. The house was fairly large (2400 square feet), and quite complex—with five floor levels, curved exterior walls, vaulted ceilings, and the works.

We just started the next Simplecs house and plan to refine the system further. We're going to be evaluating how various caulks and sealants perform in the field. And since we now have our own blower door, we'll also be able to find out which materials and details stand up over time. Although few test programs address the issue of long-term durability, it's of the utmost importance to me and my clients. After all, buildings go on long after the builder leaves the job. ■

Jim Maloney is a builder, designer, and researcher of energy-efficient homes in Eugene, Ore.

Index

A

ADA. *See* **Airtight-drywall approach.**
Additions,
adding second story, 109-111
Affordable design,
using panelized walls, 202
Airtight-drywall approach, 274-276
Anchor bolts
for structural steel, 238
in seismic construction, 70
loosen in sill repair, 106
tools for layout of, 21
Arches, tools for framing, 24
Asbestos, in demolition, 98
Asphalt shingles
wear on hot roofs, 271
wear on stress-skin roofs, 194-196

B

Balloon framing,
and collar ties removed, 114-115
Band joists
insulated with rigid foam, 260-263, 265-266, 275
stress-skin details, 192
Bay windows,
framing details, 118-121
Beam pockets,
for engineered lumber, 166
Beams
cantilevers in, 37-39
undersizing of, 35
Bearing walls, misaligned, 133
Bearing
engineered lumber on columns, 167
steel on wood posts, 240-242
Bents, in timber frames, 215, 217-218
Blown-in-batt system, 267-268
Box beam, to support roof, 117
Bracing
in demolition, 98-100
of pole barns, 229-230
of roof trusses, 146-147, 149, 152
seismic, 67-72
temporary, for wood I-beams, 177
to prevent racking, 52-53
with diagonal sheathing, 58-59
Brick veneer,
where siding meets, 138-139
Building codes
for panelized homes, 207
for seismic bracing, 68
roof loads and, 113
stress-skin panels and, 190
temporary power and, 17
wind loading and, 135
Building movement,
during demolition, 100

C

Calculators
framing with, 22
roof layout with, 23
wall layout with, 51
Camber
do not force down in trusses, 140-141
in glulams, 167
Cantilevers
in joists and beams, 37-39
maximum distance allowed, 132-133
Cathedral ceilings
details for, 267-273
framing of exposed ridge, 77-78
structural ridge options, 115-117
Cementitious underlayments
self-leveling, 46-47
sources of, 46
Chain saws
framing with, 23
to cut old sills, 103
Channel marker,
to lay out plates, 22-23, 51
Chop saws, 21-22
Circular saws
blades preferred 21-22
cutting stress-skin panels with, 193
don't cut steel with, 235
used to cut curves 91
worm-drive, 21, 28
Collar ties
how they work, 113
in timber frames, 215-216
loading of, 113
removing in renovation, 112-117
Columns
bearing engineered lumber on, 167
carrying load to foundation, 133
shimming steel beams to, 243
Common rafters, layout of, 75-77
Compression wood
and truss uplift, 140
problems with, 11
Compressors
electric, 24
gas-powered, 22, 28
Costs
of pole barns, 228
of stress-skin panels, 191-192
of wood I-beam roof, 173
per square foot of framing, 30
to raise wall panels, 204
Cranes
to lift panelized walls, 204
to lift roof structure, 110-111
to lift steel beams, 242
to lift stress-skin panels, 191
to lift tower roof, 93
to set roof trusses, 148
Creep, in beams under load, 35
Cribbing, to lift steel beams, 242-243
Cripple walls,
seismic bracing of, 67, 69-70
Crown, in steel beams, 240
Curved plates
layout, 90-91
layout and cutting, 61-62
Curved walls
layout and assembly, 61-65
racking to plumb, 91

D

Decay, inspecting in sills, 102
Decks, framing cantilevered, 37-38
Deflection
in cantilevers, 37-39
problems caused by, 33
reducing in floors, 33-35
Demolition
in remodeling, 97-100
structural evaluation before, 99

Diagonal bracing,
in seismic construction, 71
Diagonal sheathing,
for wall bracing, 58-59
Dormers
framing a gable, 87-89
raising with crane, 109, 111
Drills, half-inch, 23
Drilling
of steel, 235-236
of wood joists, 131-133
Drywall clips,
to prevent ceiling cracks, 141
Drywall
nail pops, cracks, 136-137
over steel studs, 254-255,
rigid type for ceilings, 269
used as air barrier, 274-276

E

Earthquakes,
damage to buildings, 68-72
Eaves details,
with stress-skin panels, 223-224
Energy trusses, 154, 156
Energy-efficient roofs,
cathedral ceilings, 267-273
Energy-efficient walls
favorite details, 259-264
simpler technique, 274-276
using stress-skin panels, 191
with foam on interior, 265-266
Engineered lumber
connections and splices, 166-167
beams and headers, 165-170
metal connectors for, 179-185
need to align bearing walls, 133
sizing beams, 168
EPS foam, in stress-skin panels, 192
Estimating
framing costs, 30
of demolition, 98-99
truss costs, 155
Expandable design,
using panelized walls, 202
Extension cords
12-gauge preferred, 19, 21
twist-lock, 21
use of spider box, 19

F

Fiberglass insulation
high density, 267-268
in cathedral ceilings, 267-273
in walls, 259-264, 274-276
Fire safety,
of stress-skin panels, 194-196
Fireblocking
fast installation, 53
curved walls, 63
Floor joists
how far to cantilever, 37-38
recessed detail, 132
sizing for deflection, 34-35
undersized at top plate, 134
where to notch, drill, 131-132
Floor trusses,
hangers for, 180, 182-183
Floors
avoiding bounce in, 33-35
cantilevers in, 37-39
Foam insulation. *See* **Rigid foam insulation, Urethane.**
Foam-core panels. *See* **Stress-skin panels.**
Framing square, to lay out jacks, 84
Framing tips, 30, 56
Framing tools, 20-25, 28
Furring. *See* **Strapping.**

G

Gable dormer, layout, 87-89
Gable-end walls, fabricated in shop, 204
Gable roof, layout, 75-77
Gaskets, used to seal framing, 274-276
Generators, job-site use, 20-21
Glulam
as ridge beam, 116
connectors for, 180, 182
for beams and headers, 165-169
to support ridge, 116
Ground-fault,
temporary power, 17, 28, 30
Gusset, plywood for sill splice, 106
Gusset plates
for truss repair, 125-127
inspection of, 160

H

Hammers
dogyu, 24
which preferred, 24
Hardboard,
used as underlayment, 41-43
Headers
at floor openings, 131
making curved, 64-65
of ganged trusses, 229
Hip jacks, layout, 80-81
Hip rafters, layout, 78-80
Hip roofs
for bay window, 120-121
layout and cutting, 78-81, 84-85,
Hip trusses, 156, 158
Hold-downs, for seismic protection 70-71
Hot roofs, 271-272
Hurricane anchors, 135
Hydraulic jacks, for jacking buildings, 100, 104-105

I

Ice dams, on hot roofs, 271
Insulation
high-density batts, 267-268
in cantilevered floors, 38-39
in cathedral ceilings, 267-273
of metal buildings, 250
of walls, 259-264

J

Jack rafters
cutting, 83-86
layout, 89
Jacking buildings, 100, 102-108
Jacking techniques, 104-106
Jacks, equipment choices, 104-105
Joist hangers
for engineered lumber, 179-185
solution to settling, 137
to hang trusses from beam, 229
undersized, 134
with stress-skin panels, 192
Joists. *See* **Floor joists.**
Juvenile wood
problems with, 11
truss uplift and, 140

K

Kneewalls
support roof with, 114-115
support timber roof with, 216
Knots, effect on trusses, 159-161

L

Laminated-veneer lumber
as ridge beam, 116
connectors for, 180, 182-183
for beams and headers, 169-170
Layout
of common rafters, 75-77
of curved plates, 90-91
of hip jacks, 80-81
of hip rafters, 78-80
of hip roofs, 84
of roof valleys, 85
of roof with calculators, 23
of tower rafters, 91-93
of valley jacks, 81-82
of valley rafters, 87-89
of wall plates, 51-52
with calculators, 51
Layout stick, use of, 51-52
Let-in bracing
fast technique for, 52-53

for seismic reinforcing, 69
strength of, 58
vs. plywood sheathing, 204
Levels, choices for framing, 23
Lumber shrinkage. *See* **Wood shrinkage**.
Lumber
characteristics of, 9-13
choices for sill repair, 107
heartwood, 9
moisture content of, 10, 12
plainsawn, 10
quartersawn, 10
sapwood, 9
LVL. *See* **Laminated-veneer lumber.**

M

Metal buildings
assembly, 245-250
insulation of, 250
manufacturers' support, 248
post-and-beam, 246
rigid frame, 245-247
Metal connectors
at rafter ends, 135
for engineered lumber, 179-185
for glulam, 180, 182
for LVL, 180, 182-183
for seismic bracing, 70-72
for sill repair, 107
suppliers of, 185
Moment connections,
in metal buildings, 245-247
Mortise and tenon,
strength of joint, 217-218

N

Nail guns. *See* **Pneumatic nailers.**
Nail pops, causes in drywall, 136
Notching, of floor joists, 131-132

O

Oriented-strand board, used as underlayment, 44
OSB. *See* **Oriented-strand board.**

P

Panelization
advantages of, 203-204, 206
erection on site, 202-204
of walls, 201-204, 205-209
Panelized homes
suppliers of, 199
vs. stick-built, 198-200
Parallam, for beams and headers, 170
Particleboard,
used as underlayment, 43-44
Pick-up work, tools for, 24
Pier foundation, for pole barns, 228
Plates, layout of curved, 61-62
Plumb and line
of walls, 54-55
tools for, 23
Plumbing, in stress-skin panels, 193
Plywood
as curved plates, headers, 63-65
as shearwall, 69-72
as underlayment, 44-45
for wall bracing, 58-59
gusset for sill splice, 106
in roof diaphragm, 116
installing on curved roof, 93
problems with
southern-yellow pine, 137
sheathing curved walls, 63-64
spacing on subfloor, 137
to reinforce truss, 127
Pneumatic nailers
compressors for, 22, 24
finish work with, 24
framing with, 22-23
panelization with, 208-209
Pole barns,
construction techniques, 227-231
Polyurethane foam. *See* **Urethane foam.**
Posts
caps for engineered lumber, 167
maximum spacing under floors, 35
Powder-actuated fasteners,
used on steel, 236
Purlins
in metal buildings, 249
in pole barns, 229
in timber frames, 217-219
on a structural ridge, 116

R

Racking
bracing to prevent, 57-60
preventing in timber frames, 217
Radial-arm saws, 22, 28
Radiant barriers, in roofs, 272-273
Rafter cutting,
production techniques, 74-82
Rafter layout
production techniques, 74-82
using stair gauges, 83-84
with speed square, 23
See also **Roof framing, Roof layout.**
Rafters
bracing with struts, 133
for bay window, 120-121
for tower roof, 91-93
gang-cutting, 76-77
layout of commons, 75-77
undersized at top plate, 135
Railroad jacks, 104-106
Reaction wood, problems with, 11
Ridge, structural design of, 116
Rigid foam insulation
and wall racking, 58-59
on exterior, 262-263
on interior, 260-261, 269
to seal band joints, 275
Rim joist. *See* **Band joist.**
Roof framing
cripple jacks, 86
cutting jack rafters, 83-86
for towers, 90-93
hip layout, 84
seismic reinforcing of, 71
valley jacks, 85
with wood I-beams, 172-177
Roof joints,
visible with foam panels, 196
Roof layout
calculators for, 23
of hips, 84
of valley rafters, 87-89
of valleys, 85
production techniques, 74-82
Roof raising,
lifting with crane, 110-111
Roof shingles, *See* **Asphalt shingles**.
Roof trusses
adding a skylight in, 125-127
bracing, importance of, 160-161
bracing, resources, 161
collapse of, 159-161
design of, 11
failure from wind, 149
field repair of, 125
for pole barns, 228-229
glossary, 157
hangers for, 180, 183-185
inspection of, 160
installation, 29, 145-151
manual installation, 29
metal anchors for, 180, 183-185
modifications on-site, 122-127
ordering, 145, 148, 155
production techniques, 150-152
proper bracing, 146-147, 149, 152
proper lifting, 146
raised heel, 270
setting with crane, 148
specialty types, 153-158
terms defined, 158
theory behind, 123
vs. wood I-beams, 173
Routers
to cut arches, 24
to cut splines in foam panels, 224-225

S

Safety, during demolition, 100
Scaffolding, fast and light, 28
Scissors trusses
insulation of, 270
types available, 154
Screw jacks,
for jacking buildings,
100, 104-105
Sealing
air leaks in subfloor, 275
joints in stress skin panels,
193, 221, 225
Seismic bracing
codes, 68
retrofitting, 67-72
Settling, over center beam, 137
Shear, in cantilevers, 37-39
Shear panels
for wall bracing, 58-59
temporary, in demolition, 100
Shearwall
in demolition, 100
installation, 69-71
Sheathing
curved walls, 63-64
with diagonal boards, 58-59
Sheetrock. *See* **Drywall.**
Shimming
of steel beams and columns, 243
to re-level floor, 137
Shoring, during demolition, 99
Shrinkage. *See* **Wood shrinkage.**
Siding, installing with walls flat,
203
Sills
inspecting for damage, 102
repairing and jacking, 102-107
shimming to level, 28
Sound control, with steel studs, 254
Speed square, for rafter layout, 23
Splines, to join stress skin panels,
190, 221, 224-225
Squangle, used in rafter layout, 77-79
Stainless-steel nails, in sill repair, 107
Steel beams. *See* **Structural steel.**
Steel connectors
sources of supply, 185.
See also **Metal connectors.**
Steel plates, used in sill repair, 107
Steel strapping,
to brace tower roof, 93
Steel studs
as ridge beams, 116
installation of, 251-256
ordering, 255
resources for, 256
sound control using, 254
tools for, 252
Strapping
on cathedral ceilings, 270
on exterior over foam, 262
on interior over foam,
260-261, 265-266
Stress-skin panels
durability of, 194-196
energy performance of, 191
insect damage and, 196
installing on timber
frame, 220-225
manufacturers and suppliers, 195
splined joints, 190
strength of, 191
structural uses, 189-196
vs. stick building, 194
Structural evaluation,
before demolition, 99
Structural ridge, to support roof, 116
Structural steel
bolted connections, 236-238
buying used, 234
cutting, 235-241
drilling, 235-236
finishing, 238-240
lifting, 240, 242-243
to prevent settling, 137
to support ridge, 116
types and sizes, 238-240
used in residential, 234-244
welding, 236
Stucco, shear strength of, 69
Styrofoam. *See* **Rigid
foam insulation.**
Subcontractors
in demolition, 98-99
in steelwork, 244
Subfloor
buckling from expansion, 137
sealing air leaks, 275
vs. underlayment, 41

T

T-strapping, for wall bracing, 59
Temporary bracing
for wood I-beams, 177
in demolition, 98-100
Temporary power
borrowing from neighbor, 18
codes and permits, 17
ground-fault
protection, 17, 28,30
renting a pole, 15-16
using a power pole, 15-18
Tight construction
simple techniques, 274-276
with foam and strapping, 265-266
Timber frames,
components of, 217-218
layout of foam panels on, 220, 222
sill repair, 102-106
Timber framing
mixed with conventional
framing, 213-218
preventing racking in, 217
use of bents in, 215
TJIs. *See* **Wood I-beams.**
Truss plates
how they work, 123
inspection of, 160
installing on site, 125
resources for, 161
See also **Gusset plates.**
Truss rise. *See* **Truss uplift.**
Truss uplift,causes and cures,
137-138, 140-141
Trusses.
See **Floor trusses, Roof trusses.**
Two-story walls,
and drywall cracks, 138

U

Underlayments
cementitious, 46-47
for resilient flooring, 41-47
nailing schedule, 42
Urethane foam
in stress-skin panels, 192, 196
sprayed into cathedral
ceilings, 271-272
sprayed into walls, 263-264

V

Valley jacks,
layout and cutting, 81-82, 85-86
Valley rafter, layout, 81, 87-89
Valley trusses, 156
Valleys, layout, 81-82
Vapor barriers
in cantilevered floors, 38-39
in cathedral ceilings, 267-273
in wall framing, 262-263, 265-266
Ventilation, of cathedral ceilings,
267-273

W

Wall jacks, 28-29, 204
Walls
automated assembly, 208-209
bracing against racking 57-60
building curved, 61-65
energy-efficient, 259-264, 265-
266, 274-276
fast framing of, 50-55
framing with 2x6s,
260-263, 265-266
panelized, 205-209
panelized vs. stick, 198-200

panelized, 205-209
seismic bracing for, 67-72
with foam on exterior, 262-263
with foam on interior, 265-266
Welding, of structural steel,
236, 241-242
Wind bracing
in pole barns, 229-230
of roofs to walls, 135
Wind loads,
and roof connections, 135
Window openings
in curved walls, 64-65
in stress-skin panels, 222, 193
Wiring, in stress-skin panels, 193
Wood I-beams
code acceptance as rafters, 174
cutting, 173
for roof framing, 172-177
hangers for, 180, 182-183
in cathedral ceilings, 268-269
joining at ridge, 174-175
suppliers of, 177
Wood shrinkage
and relative humidity, 10, 12
distortion caused by, 11-13
effects on frame and finish,
136-139
of selected species, 13